Lecture Notes in Computational Science and Engineering

135

Editors:

Timothy J. Barth
Michael Griebel
David E. Keyes
Risto M. Nieminen
Dirk Roose
Tamar Schlick

More information about this series at http://www.springer.com/series/3527

Gabriel R. Barrenechea • John Mackenzie

Editors

Boundary and Interior Layers, Computational and Asymptotic Methods BAIL 2018

 Springer

Editors

Gabriel R. Barrenechea
Department of Mathematics and Statistics
University of Strathclyde
Glasgow, UK

John Mackenzie
Department of Mathematics and Statistics
University of Strathclyde
G1 1XH, UK

ISSN 1439-7358   ISSN 2197-7100   (electronic)
Lecture Notes in Computational Science and Engineering
ISBN 978-3-030-41802-1   ISBN 978-3-030-41800-7   (eBook)
https://doi.org/10.1007/978-3-030-41800-7

Mathematics Subject Classification: Primary 65N12, 65N30, 76R99, 76D10, 65M15, 65M12, 65M60; Secondary 65L06; 35M10, 35B35, 35B27, 65M12, 65M60

This Springer imprint is published by the registered company Springer Nature Switzerland AG.
The registered company address is: Gewerbestrasse 11, 6330 Cham, Switzerland

Preface

These proceedings contain contributions reflecting a selection of the lectures presented at the conference BAIL 2018: Boundary and Interior Layers—Computational and Asymptotic Methods, which was held from 18th to 22nd of June 2018 at the University of Strathclyde, Glasgow, UK. Apart from two hiatuses, the BAIL conferences have been held every 2 years and BAIL 2018 was the 16th in this series. The first three BAIL conferences were organized by Professor John Miller in Dublin (1980, 1982, and 1984). The next twelve were held in Novosibirsk (1986), Shanghai (1988), Colorado (1992), Beijing (1994), Perth (2002), Toulouse (2004), Göttingen (2006), Limerick (2008), Zaragoza (2010), Pohang (2012), Prague (2014), and Beijing (2016). BAIL 2020 will take place in Buenos Aires.

The conference BAIL 2018 attracted 50 participants from 16 countries across four continents. They presented 41 lectures, five of which were invited. As usual, both mathematicians (pure and applied) and engineers participated in this BAIL conference, which led to a very fruitful exchange of ideas. The participation of several PhD students was another highlight of the conference, showing that the topics of the meeting remain attractive for young scientists. The lectures at the conference comprised both rigorous mathematical results on problems with layers and their numerical solution and sophisticated computational techniques for advanced applications.

The contributions in these proceedings are devoted to theoretical and/or numerical analysis of problems involving boundary and interior layers and to strategies for solving these problems efficiently. These problems usually arise in incompressible fluid mechanics but the examples studied here also include convection–diffusion, reaction–diffusion systems, fluid–structure interaction, and homogenization problems. In addition, the numerical methods studied, other than the standard finite element method (or a standard stabilized version of it), include alternatives spanning from discontinuous Galerkin methods, a posteriori error bounds, fictitious domain problems, monotonicity-preserving strategies, and tailored methods derived using layer-adapted meshes. In addition, one contribution on the behavior of a thin flow is also presented, mixing both the numerical studies with more qualitative asymptotic developments. Almost all contributions contain numerical results illustrating the respective theoretical considerations. In view of the wide variety of topics treated in the contributions, these proceedings provide a very good overview of current research into the theory and numerical solution of problems involving boundary and interior layers.

All contributions gathered in this volume were subject to a rigorous refereeing process. We would like to thank the authors of the papers for their cooperation and the unnamed referees for their valuable suggestions that contributed to the high quality of the papers.

In addition, we also wish to thank the organizers of the mini-symposia at BAIL 2018, all the attendees for their active participation in the conference, and especially the Glasgow Mathematical Journal Trust and the Edinburgh Mathematical Society for their very valuable financial support. Finally, we would like to thank our colleague Phil Knight for all his help in several organizational aspects and John Devlin for his help with the day-to-day aspects of running the conference.

Glasgow, UK   Gabriel R. Barrenechea
October 2019   John MacKenzie

Contents

Dual Weighted Residual Based Error Control for Nonstationary Convection-Dominated Equations: Potential or Ballast?

Marius Paul Bruchhäuser, Kristina Schwegler, and Markus Bause

Abstract Even though substantial progress has been made in the numerical approximation of convection-dominated problems, its major challenges remain in the scope of current research by John et al. (Comput Vis Sci 19(5–6):1–17, 2018). In particular, parameter robust a posteriori error estimates for quantities of physical interest and adaptive mesh refinement strategies with proved convergence are still missing. Here, we study numerically the potential of the Dual Weighted Residual (DWR) approach applied to stabilized finite element methods to further enhance the quality of approximations. The impact of a strict application of the DWR methodology is particularly focused rather than the reduction of computational costs for solving the dual problem by interpolation or localization.

1 Introduction

In their recent review paper [18] the authors nicely survey the current state of research in the numerical approximation of convection-dominated equations and incompressible flow. These problems with prominent applications in many branches of technology have strongly attracted researchers' interest not just since the pioneering works of the 1980s (cf., e.g., [11, 15]). The introduction of various families of linear and nonlinear residual-based stabilization techniques and of algebraic stabilization techniques, usually referred to as algebraic flux correction schemes, are regarded as milestones in the development of discretization techniques that are able to reduce spurious nonphysical oscillations close to sharp layers

M. P. Bruchhäuser (✉) · K. Schwegler · M. Bause
Helmut Schmidt University, Hamburg, Germany
e-mail: bruchhaeuser@hsu-hh.de; bause@hsu-hh.de

© Springer Nature Switzerland AG 2020
G. R. Barrenechea, J. Mackenzie (eds.), *Boundary and Interior Layers, Computational and Asymptotic Methods BAIL 2018*, Lecture Notes in Computational Science and Engineering 135,
https://doi.org/10.1007/978-3-030-41800-7_1

of convection-dominated problems; cf. [17] for a comparative study of those techniques. For a general review of those methods as well as a list of references we refer to, e.g., [18, 24]. In [18], the authors further identify numerous problems that are still unresolved in this field of research. In particular, the non-availability of parameter-robust a posteriori error estimates for quantities of physical interest and in general situations is stressed. Moreover, the authors point out that adaptive mesh refinement strategies that are based on such a posteriori error estimates and guarantee convergence in appropriate norms are desirable and indispensable for further improvement.

One possible technique for those adaptive strategies is the DWR method [4, 8, 9], where the error is estimated in an arbitrary user-chosen goal quantity of physical interest. The DWR approach relies on a space-time variational formulation of the discrete problem and uses duality techniques to find rigorous a posteriori error estimates, obtained through the approximation of an additional dual problem. Early studies for adaptive mesh refinement applied to various stationary stabilized equations date back to the end of the last century; cf. [9, Sect. 3.3 and 8] for a brief overview and further literature. For the nonstationary Navier-Stokes equations the DWR approach was applied together with local projection stabilization (LPS) in [10]. In addition, several approaches using an a posteriori error estimation for a quantity of interest have been proposed; cf. for example [2, 14, 20–23, 25]. In [1] an adaptive algorithm in time is presented for convection-dominated problems, where the time step control uses a post-processed solution. Furthermore, in [13] a space-time adaptive method is proposed that aims at an L^2 equal-distribution of the error in time.

In this work, we study numerically the potential of combining the DWR approach with SUPG [11, 15] stabilized finite element methods for the efficient and reliable approximation of nonstationary convection-dominated problems. Here, a *first dualize and then stabilize* principle is applied; cf. Sect. 3. For the approximation of the dual problem higher-order schemes are used, due to the recently received results for stationary convection-dominated problems in [12], where a comparative study to an approximation by higher-order interpolation was done. The presented numerical results illustrate the performance properties and robustness of the proposed algorithm with respect to a vanishing perturbation or diffusion parameter. Thereby, the *potential* of a DWR approach is demonstrated and it is shown that the DWR based adaptivity is not *ballast* for the approximation of convection-dominated problems.

This work is organized as follows. In Sect. 2 we present our model problem as well as the stabilized space-time discretization. A first dualize and then stabilize DWR approach as well as a localized error representation is presented in Sect. 3. Finally, in Sect. 4 the results of our numerical experiments are presented.

2 Model Problem and Stabilized Space-Time Discretizations

In this work we consider the following convection-diffusion-reaction equation

$$\partial_t u - \nabla \cdot (\varepsilon \nabla u) + \mathbf{b} \cdot \nabla u + \alpha u = f \quad \text{in } \Omega \times I,$$
$$u = g_D \text{ on } \partial\Omega \times I, \tag{1}$$
$$u(0) = u_0 \text{ in } \Omega,$$

where $\Omega \subset \mathbf{R}^d$, with $d = 2$ or $d = 3$ is a polygonal or polyhedral bounded domain with Lipschitz boundary $\partial\Omega$ and $I = (0, T)$, $T > 0$, is a bounded domain in time. To ensure the well-posedness of problem (1) we assume that $0 < \varepsilon \leq 1$ is a constant diffusion coefficient, $\mathbf{b} \in \mathbf{H}^1(\Omega) \cap \mathbf{L}^\infty(\Omega)$ is the flow field or convection tensor, $\alpha \in L^\infty(\Omega)$ is the reaction coefficient, $u_0 \in H_0^1(\Omega)$ is a given initial condition, $f \in L^2(I; L^2(\Omega))$ is a given source of the unknown scalar quantity u and $g \in L^2(I; H^{\frac{1}{2}}(\partial\Omega))$ is a given function specifying the Dirichlet boundary condition. Furthermore, we assume that the conditions $\nabla \cdot \mathbf{b}(\mathbf{x}) = 0$ and $\alpha(\mathbf{x}) \geq 0$ are fulfilled for all $\mathbf{x} \in \Omega$. Henceforth, for the sake of simplicity, we deal with homogeneous Dirichlet boundary conditions only. In our numerical examples in Sect. 4, we also consider more general boundary conditions; cf. also Remark 3.

It is well known that problem (1) along with the above conditions admits a unique weak solution $u \in V := \left\{ v \in L^2(I; H_0^1(\Omega)) \middle| \partial_t v \in L^2(I; H^{-1}(\Omega)) \right\}$ that satisfies the following variational formulation; cf., e.g. [24].

Find $u \in V$, satisfying $u(0) = u_0$, such that

$$A(u)(\varphi) = F(\varphi) \quad \forall \varphi \in V, \tag{2}$$

where the bilinear form $A : V \times V \to \mathbf{R}$ and the linear form $F : V \to \mathbf{R}$ are

$$A(u)(\varphi) := \sum_{n=1}^{N} \int_{I_n} \left\{ (\partial_t u, \varphi) + a(u)(\varphi) \right\} \mathrm{d}t + \sum_{n=2}^{N} ([u]_{n-1}, \varphi_{n-1}^+), \tag{3}$$

$$F(\varphi) := \int_I (f, \varphi) \, \mathrm{d}t. \tag{4}$$

Here, $a(u)(\varphi) := (\varepsilon \nabla u, \nabla \varphi) + (\mathbf{b} \cdot \nabla u, \varphi) + (\alpha u, \varphi)$ and $(\cdot, \cdot)$ denotes the standard inner product of $L^2(\Omega)$. For the discretization in time we divide the time interval I into not necessarily equidistant, left-open subintervals $I_n := (t_{n-1}, t_n]$, with $n = 1, \ldots, N$, where $0 = t_0 < t_1 < \ldots < t_N = T$ with step size $\tau_n = t_n - t_{n-1}$ and $\tau = \max_n \tau_n$. Next, we introduce the time-discrete function spaces.

$$V_\tau^{cG(r)} := \left\{ v \in C(\bar{I}; H_0^1(\Omega)) \middle| v|_{I_n} \in \mathbf{P}_r(\bar{I}_n; H_0^1(\Omega)) \right\}. \tag{5}$$

$$V_\tau^{dG(r)} := \left\{ v \in L^2(I; H_0^1(\Omega)) \middle| v|_{I_n} \in \mathbf{P}_r(I_n; H_0^1(\Omega)), v_\tau(0) \in L^2(\Omega) \right\}, \tag{6}$$

where $\mathbf{P}_r(\bar{I}_n; H_0^1(\Omega))$ denotes the space of all polynomials in time up to degree $r \geq 0$ on I_n with values in $H_0^1(\Omega)$. For some function $v_\tau \in V_\tau^{dG(r)}$ we define the limits $v_\tau^\pm$ from above and below of v_τ as well as their jump at t_n by

$$v_{\tau,n}^\pm := \lim_{t \to t_n \pm 0} v_\tau(t), \qquad [v_\tau]_n := v_{\tau,n}^+ - v_{\tau,n}^-.$$

Using the discontinuous Galerkin method for the time discretization of the so called primal problem (2) leads to the following time-discrete variational approximation.

Find $u_\tau \in V_\tau^{dG(r)}$ such that

$$A(u_\tau)(\varphi_\tau) + (u_{\tau,0}^+, \varphi_{\tau,0}^+) = F(\varphi_\tau) + (u_0, \varphi_{\tau,0}^+) \quad \forall \varphi_\tau \in V_\tau^{dG(r)}, \tag{7}$$

with $A(\cdot)(\cdot)$ and $F(\cdot)$ being defined by (3) and (4), respectively.

We note that the initial condition is incorporated into the variational problem. Next, we describe the Galerkin finite element approximation in space of the semidiscrete problem (7). We use Lagrange type finite element spaces of continuous functions that are piecewise polynomials. For the discretization in space, we consider a decomposition $\mathcal{T}_h$ of the domain Ω into disjoint elements K, such that $\bar{\Omega} = \cup_{K \in \mathcal{T}_h} \bar{K}$. Here, we choose the elements $K \in \mathcal{T}_h$ to be quadrilaterals for $d = 2$ and hexahedrals for $d = 3$. We denote by h_K the diameter of the element K. The global space discretization parameter h is given by $h := \max_{K \in \mathcal{T}_h} h_K$. Our mesh adaptation process yields locally refined cells, which is enabled by using hanging nodes. We point out that the global conformity of the finite element approach is preserved since the unknowns at such hanging nodes are eliminated by interpolation between the neighboring 'regular' nodes; cf. [4]. On $\mathcal{T}_h$ we define the discrete finite element space by $V_h^{p,n} := \{v \in V \cap C(\bar{\Omega}) \,|\, v|_K \in Q_h^p(K), \forall K \in \mathcal{T}_h, \}$, $n = 1, \ldots, N$, where $Q_h^p(K)$ is the space of polynomials that are of degree less than or equal to p with respect to each variable $x_1, \ldots, x_d$. By replacing $H_0^1(\Omega)$ in the definition of the semidiscrete function spaces $V_\tau^{cG(r)}$ and $V_\tau^{dG(r)}$ in (5) and (6), respectively, by $V_h^{p,n}$, we obtain the fully discrete function spaces

$$V_{\tau h}^{cG(r),p} := \left\{ v_{\tau h} \in V_\tau^{cG(r)} \,\middle|\, v_{|I_n} \in \mathbf{P}_r(\bar{I}_n; V_h^{p,n}) \right\} \tag{8}$$

$$V_{\tau h}^{dG(r),p} := \left\{ v_{\tau h} \in V_\tau^{dG(r)} \,\middle|\, v_{|I_n} \in \mathbf{P}_r(I_n; V_h^{p,n}), v_{\tau h}(0) \in V_h^p \right\} \tag{9}$$

with $V_{\tau h}^{cG(r),p} \subset V_\tau^{cG(r)}$ and $V_{\tau h}^{dG(r),p} \subset V_\tau^{dG(r)}$. We note that the spatial finite element space $V_h^{p,n}$ is allowed to be different on all intervals I_n which is natural in the context of a discontinuous Galerkin approximation of the time variable and allows dynamic mesh changes in time. The fully discrete discontinuous in time scheme then reads as follows.

Find $u_{\tau h} \in V_{\tau h}^{dG(r),p}$ such that

$$A(u_{\tau h})(\varphi_{\tau h}) + (u_{\tau h,0}^+, \varphi_{\tau h,0}^+) = F(\varphi_{\tau h}) + (u_0, \varphi_{\tau h,0}^+) \quad \forall \varphi_{\tau h} \in V_{\tau h}^{dG(r),p}, \quad (10)$$

with $A(\cdot)(\cdot)$ and $F(\cdot)$ being defined in (3) and (4), respectively.

In this work we focus on convection-dominated problems with small difussion parameter ε. Then the finite element approximation needs to be stabilized in order to reduce spurious and non-physical oscillations of the discrete solution arising close to sharp fronts or layers. Here, we apply the streamline upwind Petrov-Galerkin method (for short SUPG); cf. [5, 11, 15, 17, 24]. The stabilized variant of the fully discrete scheme (10) then reads as follows.

Find $u_{\tau h} \in V_{\tau h}^{dG(r),p}$ such that

$$A_S(u_{\tau h})(\varphi_{\tau h}) + (u_{\tau h,0}^+, \varphi_{\tau h,0}^+) = F(\varphi_{\tau h}) + (u_0, \varphi_{\tau h,0}^+) \quad \forall \varphi_{\tau h} \in V_{\tau h}^{dG(r),p}, \quad (11)$$

with $A_S(u)(\varphi) := A(u)(\varphi) + S(u)(\varphi)$ and stabilization terms

$$S(u_{\tau h})(\varphi_{\tau h}) := \sum_{n=1}^{N} \int_{I_n} \sum_{K \in \mathscr{T}_h} \delta_K \left(R(u_{\tau h}), \mathbf{b} \cdot \nabla \varphi_{\tau h} \right)_K \, dt$$

$$+ \sum_{n=2}^{N} \sum_{K \in \mathscr{T}_h} \delta_K \left([u_{\tau h}]_{n-1}, \mathbf{b} \cdot \nabla \varphi_{\tau h,n-1}^+ \right)_K + \sum_{K \in \mathscr{T}_h} \delta_K \left(u_{\tau h,0}^+ - u_0, \mathbf{b} \cdot \nabla \varphi_{\tau h,0}^+ \right)_K,$$

$$R(u_{\tau h}) := \partial_t u_{\tau h} + \mathbf{b} \cdot \nabla u_{\tau h} - \nabla (\varepsilon \nabla u_{\tau h}) + \alpha u_{\tau h} - f.$$

Remark 1 The proper choice of the stabilization parameter δ_K is an important issue in the application of the SUPG approach; cf., e.g., [16] and the discussion therein. For the situation of steady-state convection and reaction, an optimal error estimate for $\delta_K = O(h)$ was derived in [16].

3   A DWR Approach and A Posteriori Error Estimation

The DWR method aims at the control of an error in an arbitrary user-chosen target functional J of physical interest. To get an error representation with respect to this target functional, an additional dual problem has to be solved. Before we focus on the error representation, we introduce the dual problem of (2) whose derivation is based on the Euler-Lagrangian method of constrained optimization. For a detailed derivation we refer to [4, Chapter 6,9]. We note that here a so-called first dualize and then stabilize principle is used, where the stabilization is applied to the discrete dual problem after its derivation via the Euler-Lagrangian method of constrained optimization; cf. [12, Remark 2].

For some given functional $J : V \to \mathbf{R}$ we consider solving

$$J(u) = \min\{J(v), \ v \in V, \ \text{where } A(v)(\varphi) = F(\varphi) \ \forall \varphi \in V\}.$$

We assume that the functional J is Fréchet differentiable. i.e. $J'(y) \in V'$ for $y \in V$. For the derivation of the error representation we define the corresponding Lagrangian functional $\mathscr{L} : V \times V \to \mathbf{R}$ by

$$\mathscr{L}(u, z) := J(u) + F(z) - A(u)(z) - (u(0) - u_0, z(0)), \tag{12}$$

where we refer to $z \in V$ as the dual variable (or Lagrangian multiplier); cf. [4]. We determine a stationary point $\{u, z\} \in V \times V$ of $\mathscr{L}(\cdot, \cdot)$ by the condition $\mathscr{L}'(u, z)(\psi, \varphi) = 0$, or equivalently by the system of equations

$$A'(u)(\psi, z) = J'(u)(\psi) \quad \forall \psi \in V,$$

$$A(u)(\varphi) = F(\varphi) \qquad \forall \varphi \in V.$$

The second of these equations, the z-component of the stationary condition, is just the given primal problem (2), whereas the u-component of the stationary condition, is called the dual or adjoint equation with $A'(u)(\psi, z) = \int_I \{(\partial_t \psi, z) + a(\psi)(z)\} dt$ and $J'(u)(\psi) = \int_I \{(j(u), \psi)\} dt$ for some function $j(\cdot) \in L^2(I; L^2(\Omega))$. Applying integration by parts in time to the first term of A' and taking the condition $\nabla \cdot \mathbf{b(x)} = 0$ into account (cf. Sect. 2) yields the representation $A^*(z)(\psi) := A'(u)(\psi, z) = \int_I \{-(\partial_t z, \psi) + (\varepsilon \nabla z, \nabla \psi) - (\mathbf{b} \cdot \nabla z, \psi) + (\alpha z, \psi)\} dt$. Finally, we find by using the proposed stabilized Galerkin discretization scheme (11) the following stabilized discrete dual problem.

Find $z_{\tau h} \in V_{\tau h}^{dG(r),p}$ *such that*

$$A_S^*(z_{\tau h})(\psi_{\tau h}) + (z_{\tau h,T}^-, \psi_{\tau h,T}^-) = J'(u_{\tau h})(\psi_{\tau h}) \quad \forall \psi_{\tau h} \in V_{\tau h}^{dG(r),p}. \tag{13}$$

In (13), we put $A_S^*(z_{\tau h})(\psi_{\tau h}) := A^*(z_{\tau h})(\psi_{\tau h}) + S^*(z_{\tau h})(\psi_{\tau h})$ with

$$A^*(z_{\tau h})(\psi_{\tau h}) := \sum_{n=1}^{N} \int_{I_n} \{ -(\partial_t z_{\tau h}, \psi_{\tau h}) - (\mathbf{b} \cdot \nabla z_{\tau h}, \psi_{\tau h})$$
$$+ (\varepsilon \nabla z_{\tau h}, \nabla \psi_{\tau h}) + (\alpha z_{\tau h}, \psi_{\tau h}) \} dt$$
$$- \sum_{n=2}^{N} ([z_{\tau h}]_{n-1}, \psi_{\tau h,n-1}^-),$$

$$S^*(z_{\tau h})(\psi_{\tau h}) := \sum_{n=1}^{N} \int_{I_n} \sum_{K \in \mathscr{T}_h} \delta_K^* (R^*(z_{\tau h}), -\mathbf{b} \cdot \nabla \psi_{\tau h})_K \, dt$$

$$- \sum_{n=2}^{N} \sum_{K \in \mathscr{T}_h} \delta_K^* ([z_{\tau h}]_{n-1}, -\mathbf{b} \cdot \nabla \psi_{\tau h,n-1}^-)_K + \sum_{K \in \mathscr{T}_h} \delta_K^* (z_{\tau h,N}^-, -\mathbf{b} \cdot \nabla \psi_{\tau h,N}^-)_K,$$

$$R^*(z_{\tau h}) := -\partial_t z_{\tau h} - \mathbf{b} \cdot \nabla z_{\tau h} - \nabla (\varepsilon \nabla z_{\tau h}) + \alpha z_{\tau h} - j(u_{\tau h}).$$

To derive a representation of the error $J(e) = J(u) - J(u_{\tau h})$ we need some abstract results. In order to keep this work self-contained we pare down to the key arguments of the DWR approach applied to the stabilized model problem. We follow the lines of [4, Chapter 6 and 9] and [7], where all of the proofs can be found. To start with, we need to extend the definition of the Lagrangian functional to arguments of $(V + V_{\tau h}^{dG(r),p}) \times V$. In the following we let $\mathscr{L} : (V + V_{\tau h}^{dG(r),p}) \times V$ be defined by

$$\mathscr{L}(u, z) := J(u) + F(z) - A(u)(z) - \sum_{n=2}^{N} \left([u]_{n-1}, z_{n-1}^+\right) - \left(u(0) - u_0, z(0)\right).$$

(14)

Then it follows that

$$\mathscr{L}_u(u, z)(\psi) + \mathscr{L}_z(u, z)(\varphi) = 0 \quad \forall \{\psi, \varphi\} \in V \times V.$$

(15)

The discrete solution $\{u_{\tau h}, z_{\tau h}\} \in V_{\tau h}^{dG(r),p} \times V_{\tau h}^{dG(r),p}$ then satisfies

$$\mathscr{L}_u(u_{\tau h}, z_{\tau h})(\psi_{\tau h}) + \mathscr{L}_z(u_{\tau h}, z_{\tau h})(\varphi_{\tau h}) = S(u_{kh})(\varphi_{kh}) + S^*(z_{kh})(\psi_{kh})$$

(16)

for all $\{\psi_{\tau h}, \varphi_{\tau h}\} \in V_{\tau h}^{dG(r),p} \times V_{\tau h}^{dG(r),p}$. For the defect of the discrete solution in the stationary condition (16) we use the notation

$$\tilde{S}(x_{\tau h})(y_{\tau h}) := S(u_{\tau h})(\varphi_{\tau h}) + S^*(z_{\tau h})(\psi_{\tau h}),$$

with $x_{\tau h} := \{u_{\tau h}, z_{\tau h}\} \in V_{\tau h}^{dG(r),p} \times V_{\tau h}^{dG(r),p}$ and $y_{\tau h} := \{\psi_{\tau h}, \varphi_{\tau h}\} \in V_{\tau h}^{dG(r),p} \times V_{\tau h}^{dG(r),p}$. To derive a representation of the error $J(u) - J(u_{\tau h})$ we need the following abstract theorem that develops the error in terms of the Lagrangian functional.

Theorem 1 *Let X be a function space and $\mathscr{L} : X \to \mathbf{R}$ be a three times differentiable functional on X. Suppose that $x_c \in X_c$ with some ("continuous") function space $X_c \subset X$ is a stationary point of $\mathscr{L}$. Suppose that $x_d \in X_d$ with some ("discrete") function space $X_d \subset X$, with not necessarily $X_d \subset X_c$, is a Galerkin approximation to x_c being defined by the equation*

$$\mathscr{L}'(x_d)(y_d) = \tilde{S}(x_d)(y_d) \quad \forall y_d \in X_d.$$

In addition, suppose that the auxiliary condition $\mathscr{L}'(x_c)(x_d) = 0$ is satisfied. Then there holds the error representation

$$\mathscr{L}(x_c) - \mathscr{L}(x_d) = \frac{1}{2} \mathscr{L}'(x_d)(x_c - y_d) + \frac{1}{2} \tilde{S}(x_d)(y_d - x_d) + \mathscr{R} \quad \forall y_d \in X_d,$$

where the remainder $\mathscr{R}$ is defined by $\mathscr{R} = \frac{1}{2} \int_0^1 \mathscr{L}'''(x_d + se)(e, e, e) \cdot s \cdot (s - 1) \, ds$, with the notation $e := x_c - x_d$.

For the subsequent theorem we introduce the primal and dual residuals by

$$\rho(u_{\tau h})(\varphi) := F(\varphi) - A(u_{\tau h})(\varphi) - (u_{\tau h,0}^+ - u_0, \varphi(0)) \qquad \forall \varphi \in V, \quad (17)$$

$$\rho^*(z_{\tau h})(\psi) := J'(u_{\tau h})(\psi) - A^*(z_{\tau h})(\psi) - (z_{\tau h,N}^-, \psi(T)) \quad \forall \psi \in V. \quad (18)$$

Theorem 2 *Suppose that $\{u, z\} \in V \times V$ is a stationary point of the Lagrangian functional $\mathscr{L}$ defined in (14) such that (15) is satisfied. Let $\{u_{\tau h}, z_{\tau h}\} \in V_{\tau h}^{dG(r),p} \times V_{\tau h}^{dG(r),p}$ denote its Galerkin approximation being defined by (11) and (13), respectively, such that (16) is satisfied. Then there holds the error representation that*

$$J(u) - J(u_{\tau h}) = \frac{1}{2}\rho(u_{\tau h})(z - \varphi_{\tau h}) + \frac{1}{2}\rho^*(z_{\tau h})(u - \psi_h) + \mathscr{R}_{\tilde{s}} + \mathscr{R}_J \quad (19)$$

for arbitrary functions $\{\varphi_{\tau h}, \psi_{\tau h}\} \in V_{\tau h}^{dG(r),p} \times V_{\tau h}^{dG(r),p}$, where the remainder terms are $\mathscr{R}_{\tilde{s}} := \frac{1}{2}S(u_{\tau h})(\varphi_{\tau h} + z_{\tau h}) + \frac{1}{2}S^*(z_{\tau h})(\psi_{\tau h} - u_{\tau h})$ and $\mathscr{R}_J := \frac{1}{2}\int_0^1 J'''(u_{\tau h} + s \cdot e)(e, e, e) \cdot s \cdot (s - 1) \, ds$, with $e = u - u_{\tau h}$.

In the error respresentation (19) the continuous solution u is required for the evaluation of the dual residual. The following theorem shows the equivalence of the primal and dual residual up to a quadratic remainder. This observation will be used below to find our final error respresentation in terms of the goal quantity J and a suitable linearization for its computational evaluation or approximation, respectively.

Theorem 3 *Under the assumptions of Theorem 2, and with the definitions (17) and (18) of the primal and dual residual, respectively, there holds that*

$$\rho^*(z_{\tau h})(u - \psi_{\tau h}) = \rho(u_{\tau h})(z - \varphi_{\tau h}) + S(u_{\tau h})(\varphi_{\tau h} - z_{\tau h}) + S^*(z_{\tau h})(u_{\tau h} - \psi_{\tau h}) + \Delta\rho_J,$$

for all $\{\psi_{\tau h}, \varphi_{\tau h}\} \in V_{\tau h}^{dG(r),p} \times V_{\tau h}^{dG(r),p}$, where the remainder term is given by $\Delta\rho_J := -\int_0^1 J''(u_{\tau h} + s \cdot e)(e, e) \, ds$ with $e := u - u_{\tau h}$.

In a final step we combine the results of the previous two theorems to get a localized approximation of the error that is then used for the design of the adaptive algorithm. We note that the final result (20) is a slight modification of Theorem 5.2 for the nonstationary Navier-Stokes equations stabilized by LPS in [10]. The difference comes through using a first dualize and then stabilize approach as well as SUPG stabilization here.

Theorem 4 (Localized Error Representation) *Let the assumptions of Theorem 2 be satisfied. Neglecting the higher-order error terms, then there holds as a linear* *approximation the cell-wise error representation that*

$$
\begin{aligned}
J(u) - J(u_{\tau h}) \doteq \sum_{n=1}^{N} \int_{I_n} \sum_{K \in \mathscr{T}_h} \Big\{ & \big(R(u_{\tau h}), z - \varphi_{\tau h}\big)_K - \delta_K \big(R(u_{\tau h}), \mathbf{b} \cdot \nabla\varphi_{\tau h}\big)_K \\
& - \big(E(u_{\tau h}), z - \varphi_{\tau h}\big)_{\partial K} \Big\} dt \\
& - \sum_{K \in \mathscr{T}_h} \big(u_{\tau h,0}^+ - u_0, z(t_0) - \varphi_{\tau h,0}^+\big)_K \\
& - \sum_{n=2}^{N} \sum_{K \in \mathscr{T}_h} \big([u_{\tau h}]_{n-1}, z(t_{n-1}) - \varphi_{\tau h,n-1}^+\big)_K \\
& + \sum_{K \in \mathscr{T}_h} \delta_K \big(u_{\tau h,0}^+ - u_0, \mathbf{b} \cdot \nabla\varphi_{\tau h,0}^+\big)_K \\
& + \sum_{n=2}^{N} \sum_{K \in \mathscr{T}_h} \delta_K \big([u_{\tau h}]_{n-1}, \mathbf{b} \cdot \nabla\varphi_{\tau h,n-1}^+\big)_K .
\end{aligned}
$$

(20)

The cell- and edge-wise residuals are defined by

$$
R(u_{\tau h})|_K := f - \partial_t u_{\tau h} + \nabla \cdot (\varepsilon \nabla u_{\tau h}) - \mathbf{b} \cdot \nabla u_{\tau h} - \alpha u_{\tau h} ,
$$

(21)

$$
E(u_{\tau h})|_\Gamma := \begin{cases} \frac{1}{2}\mathbf{n} \cdot [\varepsilon \nabla u_{\tau h}] & \text{if } \Gamma \subset \partial K \backslash \partial\Omega , \\ 0 & \text{if } \Gamma \subset \partial\Omega , \end{cases}
$$

(22)

where $[\nabla u_{\tau h}] := \nabla u_{\tau h}|_{\Gamma \cap K} - \nabla u_{vh}|_{\Gamma \cap K'}$ *defines the jump of* $\nabla u_{\tau h}$ *over the inner edges* Γ *with normal unit vector* $\mathbf{n}$ *pointing from* K *to* K'.

Remark 2 We have neglected the remainder terms $\mathscr{R}_J$ and $\Delta\rho_J$ in Theorem 4 as well as in the numerical computation of the space-time error indicators defined at the beginning of Sect. 4. This is due to the higher-order character of these terms compared to the remaining terms in (20); cf. [4, Rem. 6.5]. Nevertheless, an actual upper bound on the error may be violated when these additional terms are significant, at least for low numbers of degrees of freedom, as it was pointed out in [2] and [20] by taking these terms into account.

Remark 3 (Nonhomogeneous Dirichlet Boundary Conditions) In the case of non-homogeneous Dirichlet boundary conditions the following additional term has to be added to the error representation (20)

$$
\sum_{n=1}^{N} \int_{I_n} -\big((g_D - \tilde{g}_{D,\tau h}), \varepsilon\nabla z \cdot \mathbf{n}\big)_{\partial\Omega} dt ,
$$

where the discrete function $\tilde{g}_{D,\tau h}$ is an appropriate finite element approximation of the extension $\tilde{g}_D$ in the sense that the trace of $\tilde{g}_D$ equals g_D on $\partial\Omega$; cf. [4, 12].

4 Numerical Studies

In this section we illustrate and investigate the performance properties of the proposed approach of combining the DWR method with SUPG stabilized finite element approximations of nonstationary convection-dominated problems. Therefore some general indications are needed. The error representation (20), written as

$$J(u) - J(u_{\tau h}) \doteq \eta := \sum_{n=1}^{N} \sum_{K \in \mathscr{T}_h} \eta_K^n , \qquad (23)$$

depends on the discrete primal solution $u_{\tau h}$ as well as on the exact dual solution z. For solving the primal problem (1) we use the discontinuous in time scheme (11) to get a discrete solution $u_{\tau h} \in V_{\tau h}^{dG(r),p}$. For the application of (23) in computations, the unknown dual solution z has to be approximated, which results in an approximate error indicator $\tilde{\eta}$. This approximation cannot be done in the same finite element space as used for the primal problem, since this would result in an useless vanishing error representation $\tilde{\eta} = 0$, due to Galerkin orthogonality. In contrast to the approximation by higher-order interpolation which is widespread used in the literature, cf. [4, 7, 10], we use an approximation by higher-order finite elements here. This is done due to the results in [12], where a comparative study between these two approaches is presented for steady convection-dominated problems. In this study the superiority of using higher-order finite elements was shown for an increasing convection dominance. Thus, we use for the discretization of the dual problem a finite element space that consists of polynomials in space and time that are at least of one polynomial degree higher than its primal counterpart, more precisely we compute a discrete dual solution $z_{\tau h} \in V_{\tau h}^{cG(r+1),p+1}$. For the now following example we briefly present our algorithm. For further details we refer to [19].

Remark 4 Space-time adaptive methods splitting the error indicators in space and time are investigated, for instance, in [10] or [13], where the latter refinement strategy is motivated by the principle of equally distributing the a posteriori error indicators in time instead of using a DWR approach.

For the implementation of the adaptive algorithm we use our DTM++ frontend software [19] that is based on the open source finite element library deal.II; cf. [3]. For measuring the accuracy of the error estimator, we will study the effectivity index

$$I_{\text{eff}} = \left| \frac{\tilde{\eta}^\ell}{J(u) - J(u_{\tau h}^\ell)} \right| \qquad (24)$$

as the ratio of the estimated error $\tilde{\eta}$ of (23) over the exact error. Desirably, the index I_{eff} should be close to one.

Adaptive solution algorithm (Refining in space and time)

Initialization: Set $\ell = 1$ and generate the initial space-time slab $Q_n^{\ell=1} = \Omega_h^\ell \times I_n^\ell$.

1. Compute the **primal** and **dual** solution $u_{\tau h}^\ell \in V_{\tau h}^{dG(r),p}$ and $z_{\tau h}^\ell \in V_{\tau h}^{cG(r+1),p+1}$
2. Evaluate the **a posteriori space-time error indicator** $\tilde{\eta}^\ell = \sum_{n=1}^N \tilde{\eta}^{n,\ell}$, with $\tilde{\eta}^{n,\ell} = \sum_{K \in \mathcal{T}_h} \tilde{\eta}_K^{n,\ell}$.
   Mark a space-time slab n for refinement in time for which the corresponding value of $\tilde{\eta}^{n,\ell}$ belongs to the top fraction $0 \le \theta_\tau \le 1$ of largest values, then, on each slab Q_n^ℓ mark a mesh cell $K \in \mathcal{T}_h$ for $d-$dimensional isotropic refinement in space for which the corresponding value $|\tilde{\eta}_K^{n,\ell}|$ belongs to the top fraction $\theta_{h,1}$ or $\theta_{h,2}$, for a slab that is not or is marked for time refinement, of largest values, with $0 \le \theta_{h,1} \le \theta_{h,2} \le 1$.
3. Check the **goal** (for instance $\tilde{\eta}^\ell < \mathrm{tol}$, $\|u - u_{\tau h}^\ell\| < \mathrm{tol}$, or maximum of $N_{\mathrm{DoF}}^{\mathrm{tot}}$ is reached): If the goal is reached, then the adaptive solution algorithm is terminated.
4. Else, **adapt** the space-time slab Q simultaneously in space and time in the following manner: Execute the spatial refinement first and afterwards execute the temporal refinement. Increase ℓ to $\ell + 1$ and return to Step 1; cf. [19, Sect. 1.2] including the corresponding open-source software.

Example 1 (Rotating Hill with Changing Orientation) In this example we analyze the performance properties of our algorithm for a global target quantity. We study problem (1) with the prescribed solution

$$u(t, x, y) := \frac{v_1(t) \cdot s \cdot \arctan(v_2(t))}{1 + a_0\left(x - \frac{1}{2} - \frac{1}{4}\cos(2\pi t)\right)^2 + a_0\left(y - \frac{1}{2} - \frac{1}{4}\sin(2\pi t)\right)^2}, \quad (25)$$

where $\Omega \times I := (0,1)^2 \times (0,1]$ and $v_1(\hat{t}) := -1, v_2(\hat{t}) := 5\pi \cdot (4\hat{t} - 1)$, for $\hat{t} \in [0, 0.5)$ and $v_1(\hat{t}) := 1, v_2(\hat{t}) := 5\pi \cdot (4(\hat{t} - 0.5) - 1)$, for $\hat{t} \in [0.5, 1)$, $\hat{t} = t - k, k \in \mathbb{N}_0$, and, scalars $a_0 = 50, s = -\frac{1}{3}$. We choose the flow field $\mathbf{b} = (2,3)^\top$ and the reaction coefficient $\alpha = 1.0$. The solution (25) is characterized by a counterclockwise rotating hill and designed in such a way that the orientation of the hill changes its sign from positive to negative at $t = 0.25$ and again from negative to positive at $t = 0.75$ so that the final position equals the initial one. For the solution (25) the right-hand side function f is calculated from the partial differential equation. Boundary conditions are given by the exact solution. Our target quantity is chosen to control the global L^2-error in space and time, given by

$$J(u) = \frac{1}{\|e\|_{(0,T) \times \Omega}} \int_I (u, e) \mathrm{d}t, \quad \text{with } \| \cdot \|_{(0,T) \times \Omega} = \left(\int_I (\cdot, \cdot) \, \mathrm{d}t \right)^{\frac{1}{2}}. \quad (26)$$

In our first test we investigate problem (1) for $\varepsilon = 1$ and without any stabilization to verify our algorithm and demonstrate its properties. In Fig. 1b we monitor the development of the total discretization error $J(e)$, the space-time error estimator $\tilde{\eta}$ as well as the effectivity index I_{eff} during the adaptive refinement process. Here

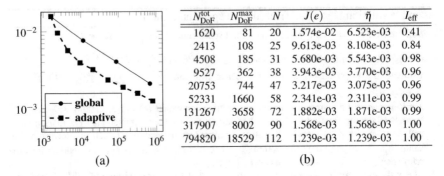

$N_{\mathrm{DoF}}^{\mathrm{tot}}$	$N_{\mathrm{DoF}}^{\mathrm{max}}$	N	$J(e)$	$\tilde{\eta}$	I_{eff}
1620	81	20	1.574e-02	6.523e-03	0.41
2413	108	25	9.613e-03	8.108e-03	0.84
4508	185	31	5.680e-03	5.543e-03	0.98
9527	362	38	3.943e-03	3.770e-03	0.96
20753	744	47	3.217e-03	3.075e-03	0.96
52331	1660	58	2.341e-03	2.311e-03	0.99
131267	3658	72	1.882e-03	1.871e-03	0.99
317907	8002	90	1.568e-03	1.568e-03	1.00
794820	18529	112	1.239e-03	1.239e-03	1.00

(a)   (b)

Fig. 1 Comparison of global space-time L^2-errors and table for adaptive refinement. (**a**) Global L^2-error in space and time over $N_{\mathrm{DoF}}^{\mathrm{tot}}$ for uniform and DWR adaptive mesh refinement. (**b**) Effectivity indices I_{eff} for the goal quantity (26) for adaptive refinement in space and time

and in the following, $N_{\mathrm{DoF}}^{\mathrm{tot}}$ denotes the total number of degrees of freedom in space and time for one DWR loop while $N_{\mathrm{DoF}}^{\mathrm{max}}$ denotes the maximum number of degrees of freedom of a spatial mesh used within one DWR loop. Furthermore, N denotes the total number of space-time slabs Q_n used for one DWR loop; cf. the adaptive solution algorithm at the beginning of this chapter. Figure 1a compares the received convergence behavior with a uniform mesh refinement strategy. The DWR based adaptive mesh adaptation is clearly superior to the uniform refinement in terms of accuray over the total number of degrees of freedom in space and time.

In Fig. 2 we illustrate the distribution of the time subinterval lengths of Q_n^ℓ over I as well as the adaptive spatial meshes after the last DWR loop at selected time points. Since a global target function is used here, one would expect an almost equal distribution of the temporal step size on I for a solution that acts smooth in time. Here, the prescribed solution (25) is chosen to change its orientation at $t = 0.25$ as well as $t = 0.75$, so that the temporal step size should be smaller close to these time points. This behavior is confirmed by Fig. 2a. Considering the underlying spatial meshes, we note that the total number of the spatial cells is nearly equal to each other comparing positions Fig. 2c with 2d,b with 2e, respectively, but the distribution of the spatial cells differs depending on the related position of the hill. Thus, for the chosen target functional (26) the spatial mesh runs as expected with the rotation of the hill in a synchronous way. In addition, we note that the mesh refinement is slightly weaker at the final time point compared to the initial position. This is due to the error propagation of the underlying problem which is captured by the dual weights in the error estimate. This effect is in good agreement to the results obtained for the heat equation in [4, p. 122].

Example 2 (Hump with Circularly Layer) In a second example we investigate a convection-dominated case for a sequence of decreasing diffusion coefficients by means of a local tarquet functional for a diffusive parameter dependent exact

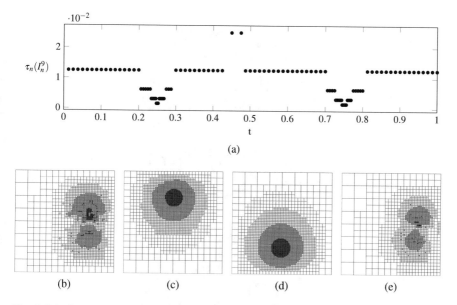

Fig. 2 Distribution of the temporal step size τ_n of the space-time slab Q_n for DWR-loop $\ell = 9$ (a) and related adaptive spatial meshes at time points $t_n = 0$ (b), $t_n = 0.25$ (c), $t_n = 0.75$ (d) and $t_n = 1$ (e)

solution given by (cf. [1, 17])

$$u(t, x, y) := 16 \sin(\pi t) x_1 (1 - x_1) x_2 (1 - x_2) \tag{27}$$

$$\cdot \left(\frac{1}{2} + \frac{\arctan \left(2\varepsilon^{-\frac{1}{2}} (0.25^2 - (x_1 - 0.5)^2 - (x_2 - 0.5)^2) \right)}{\pi} \right),$$

where $\Omega \times I := (0, 1)^2 \times (0, 0.5]$. We choose $\mathbf{b} = (2, 3)^\top$ and $\alpha = 1.0$. The solution is characterized by a hump changing its height in the course of the time. The target quantity is chosen to control the L^2-error at the final time point $T = 0.5$, given by

$$J(u) = \frac{(u_N^-, e_N^-)_\Omega}{\|e_N^-\|_\Omega}. \tag{28}$$

In Table 1 we present selected effectivity indices of the proposed DWR approach applied to the stabilized approximation scheme (11) for different diffusion coefficients. For $\varepsilon = 10^{-3}$ as well as for the more challenging case of $\varepsilon = 10^{-6}$ the effectivity indices are close to one.

Table 1 Effectivity indices for the goal quantity (28) for different values of ε

$\varepsilon = 10^{-3}$						$\varepsilon = 10^{-6}$					
$N_{\mathrm{DoF}}^{\mathrm{tot}}$	$N_{\mathrm{DoF}}^{\mathrm{max}}$	N	$J(e)$	$\tilde{\eta}$	I_{eff}	$N_{\mathrm{DoF}}^{\mathrm{tot}}$	$N_{\mathrm{DoF}}^{\mathrm{max}}$	N	$J(e)$	$\tilde{\eta}$	I_{eff}
1620	81	20	6.128e−02	6.671e−02	1.08	11,912	463	30	8.158e−02	4.003e−01	4.90
3583	168	23	3.169e−02	2.616e−02	0.82	20,701	727	36	5.660e−02	2.080e−01	3.67
8531	384	26	1.604e−02	1.363e−02	0.85	61,290	2194	51	2.861e−02	4.556e−02	1.59
19,147	782	29	7.816e−03	7.184e−03	0.92	101,975	4045	61	2.095e−02	2.218e−02	1.05
42,391	1487	33	3.672e−03	3.055e−03	0.83	277,115	7146	87	1.112e−02	1.407e−02	1.26
96,000	3056	37	2.738e−03	2.389e−03	0.87	729,740	16,014	124	5.649e−03	5.626e−03	0.99
204,933	5828	42	2.291e−03	2.200e−03	0.96	1,927,200	24,678	177	2.867e−03	2.815e−03	0.98
468,540	11,661	48	2.191e−03	2.270e−03	1.03	3,163,125	33,507	212	2.115e−03	1.936e−03	0.92

Finally, we present the distribution of the temporal time step size in Fig. 3 as well as selected solution profiles and the related spatial meshes in Fig. 4, all exemplarily for $\varepsilon = 10^{-6}$.

We observe that the first time steps are chosen relatively large whereas the time step sizes close to the final time point $T = 0.5$ become much smaller. At the beginning and even after half of the time the solution is still strongly perturbed in the backward part of the hump's layer. The mesh is coarse in that part of the domain (cf. Fig. 4a, b). For $T = 0.5$ an almost perfect solution profile is obtained and the finite element mesh cells are concentrated close to the layer of the hump (cf. Fig. 4c). We note that the spurious oscillations are reduced significantly.

Such a behaviour is admissible since the chosen target functional (28) aims to control the solution profile at the final time point $T = 0.5$ only.

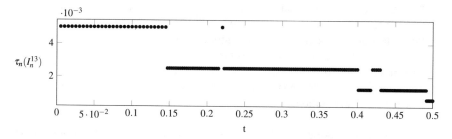

Fig. 3 Distribution of the temporal step size τ_n of the space-time slab Q_n^{13}

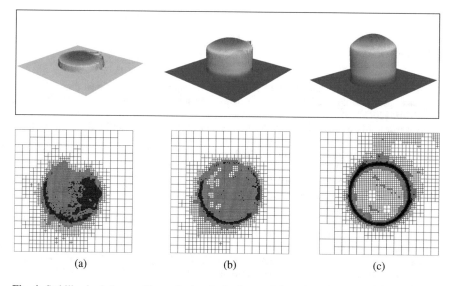

(a)   (b)   (c)

Fig. 4 Stabilized solution profiles and related adaptive spatial meshes after 13 DWR loops at time points $t_n = 0.07$ (**a**), $t_n = 0.25$ (**b**) and $t_n = 0.5$ (**c**) for $\varepsilon = 10^{-6}$

5 Conclusions and Outlook

In this work we presented an adaptive solution algorithm for SUPG stabilized finite element approximations of time-dependent convection-diffusion-reaction equations. The underlying approach is based on the Dual Weighted Residual method for goal-oriented error control. A *first dualize and then stabilize* philosophy was applied for combining the space-time adaptation process in the course of the DWR approach with the stabilization of the finite element techniques. We used a higher-order finite element approximation in space and time in order to compute the dual solution. In numerical experiments we could prove that spurious oscillations that typically arise in numerical approximations of convection-dominated problems could be reduced significantly. Effectivity indices close to one were obtained for different values of the diffusion coefficient. It was shown that the DWR based adaptivity is no *ballast* on the way to solve convection-dominated problems. Conversely, it offers *potential* for further improvements in handling those problems.

We note that recent results in post-processing variational time discretization schemes (cf., e.g., [1, 6]) allow the computation of improved solutions admitting an additional order of convergence for the discretization in time by negligible computational costs, and thus offer further potential for reducing the costs of computing the dual solution.

Acknowledgements The authors wish to thank the anonymous reviewers for their help to improve the presentation of this paper. Furthermore, we acknowledge Uwe Köcher for his help in the design and implementation of the underlying software DTM++/dwr-diffusion; cf.[19].

References

1. Ahmed, N., John, V.: Adaptive time step control for higher order variational time discretizations applied to convection-diffusion equations. Comput. Methods Appl. Mech. Eng. **285**, 83–101 (2015)
2. Ainsworth, M., Rankin, R.: Guaranteed computable bounds on quantities of interest in finite element computations. Int. J. Numer. Methods Eng. **89**(13), 1605–1634 (2012)
3. Alzetta, G., Arndt, D., Bangerth, W., Boddu, V., Brands, B., Davydov, D., Gassmoeller, R., Heister, T., Heltai, L., Kormann, K., Kronbichler, M., Maier, M., Pelteret, J.-P., Turcksin, B., Wells, D.: The deal.II library, Version 9.0. J. Numer. Math. **26**(4), 173–183 (2018)
4. Bangerth, W., Rannacher, R.: Adaptive Finite Element Methods for Differential Equations. Birkhäuser, Basel (2003)
5. Bause, M., Schwegler, K.: Analysis of stabilized higher order finite element approximation of nonstationary and nonlinear convection-diffusion-reaction equations. Comput. Methods Appl. Mech. Eng. **209–212**, 184–196 (2012)
6. Bause, M., Köcher, U., Radu, F.A., Schieweck, F.: Post-processed Galerkin approximation of improved order for wave equations. Math. Comput. **89**(322), 595–627 (2018). arXiv:1803.03005
7. Becker, R.: An optimal-control approach to a posteriori error estimation for finite element discretizations of the Navier–Stokes equations. East-West J. Numer. Math. **8**, 257–274 (2000)

8. Becker, R., Rannacher, R.: Weighted a posteriori error control in FE methods. In: Bock, H.G., et al. (eds.) Proceedings of the 2nd European Conference on Numerical Mathematics and Advanced Applications (ENUMATH 97), pp. 621–637. World Scientific, Singapore (1998)

9. Becker, R., Rannacher, R.: An optimal control approach to a posteriori error estimation in finite element methods. In: Iserles, A. (ed.) Acta Numerica, vol. 10, pp. 1–102. Cambridge University, Cambridge (2001)

10. Besier, M., Rannacher, R.: Goal-oriented space-time adaptivity in the finite element Galerkin method for the computation of nonstationary incompressible flow. Int. J. Numer. Methods Fluids **70**(9), 1139–1166 (2012)

11. Brooks, A.N., Hughes, T.J.R.: Streamline upwind/Petrov-Galerkin formulations for convection dominated flows with particular emphasis on the incompressible Navier-Stokes equations. Comput. Methods Appl. Mech. Eng. **32**(1–3), 199–259 (1982)

12. Bruchhäuser, M.P., Schwegler, K., Bause, M.: Numerical study of goal- oriented error control for stabilized finite element methods. In Apel, T., et al. (eds.) Advanced Finite Element Methods with Applications: Selected Papers from the 30th Chemnitz Finite Element Symposium 2017. Lecture Notes in Computational Science and Engineering, vol. 128, p. 85 (2019). https://doi.org/10.1007/978-3-030-14244-5_5

13. Gaspoz, F., Kreuzer, C., Siebert, K., Ziegler, D.: A convergent time-space adaptive dg (s) finite element method for parabolic problems motivated by equal error distribution (2016). arXiv preprint:1610.06814

14. Giles, M.B., Pierce, N.A.: Adjoint error correction for integral outputs. In Barth, T.J., Deconinck, H. (eds.) Error estimation and adaptive discretization methods in computational fluid dynamics. Lecture Notes in Computational Science and Engineering, vol. 25, pp. 47–95. Springer, Berlin (2003)

15. Hughes, T.J.R., Brooks, A.N.: A multidimensional upwind scheme with no crosswind diffusion. In: Hughes, T.J.R. (eds.) Finite Element Methods for Convection Dominated Flows, AMD. American Society of Mechanical Engineers (ASME), vol. 34, pp. 19–35 (1979)

16. John, V., Novo, J.: Error analysis of the SUPG finite element discretization of evolutionary convection-diffusion-reaction equations. SIAM J. Numer. Anal. **49**(3), 1149–1176 (2011)

17. John, V., Schmeyer, E.: Finite element methods for time-dependent convection-diffusion-reaction equations with small diffusion. Comput. Methods Appl. Mech. Eng. **198**, 173–181 (2009)

18. John, V., Knobloch, P., Novo, J.: Finite elements for scalar convection-dominated equations and incompressible flow problems: a never ending story? Comput. Vis. Sci. **19**(5–6), 1–17 (2018). https://doi.org/10.1007/s00791-018-0290-5

19. Köcher, U, Bruchhäuser, M.P., Bause, M.: Efficient and scalable data structures and algorithms for goal-oriented adaptivity of space–time FEM codes. SoftwareX **10**, 100239 (2019). https://doi.org/10.1016/j.softx.2019.100239

20. Nochetto, R.H., Veeser, A., Verani, M.: A safeguarded dual weighted residual method. IMA J. Numer. Anal. **29**(1), 126–140 (2009)

21. Parés, N., Bonet, J., Huerta, A., Peraire, J.: The computation of bounds for linear-functional outputs of weak solutions to the two-dimensional elasticity equations. Comput. Methods Appl. Mech. Eng. **195**(4–6), 406–429 (2006)

22. Peraire, J., Patera, A.T.: Bounds for linear-functional outputs of coercive partial differential equations: local indicators and adaptive refinement. In Ladeveze, P., Oden, J.T. (eds.) Advances in Adaptive Computational Methods in Mechanics (Cachan, 1997). Studies in Applied Mechanics, vol. 47, pp. 199–216. Elsevier, Amsterdam (1998)

23. Prudhomme, S., Oden, T.J.: On goal-oriented error estimation for elliptic problems: application to the control of pointwise errors. Comput. Methods Appl. Mech. Eng. **176**(1–4), 313–331 (1999)

24. Roos, H.-G., Stynes, M., Tobiska, L.: Robust Numerical Methods for Singularly Perturbed Differential Equations. Springer, Berlin (2008)

25. Sauer-Budge, A.M., Bonet, J., Huerta, A., Peraire, J.: Computing bounds for linear functionals of exact weak solutions to Poissons equation. SIAM J. Numer. Anal. **42**(4), 1610–1630 (2004)

Automatic Variationally Stable Analysis for FE Computations: An Introduction

Victor M. Calo, Albert Romkes, and Eirik Valseth

Abstract We introduce an automatic variationally stable analysis (AVS) for finite element (FE) computations of scalar-valued convection-diffusion equations with non-constant and highly oscillatory coefficients. In the spirit of least squares FE methods (Bochev and Gunzburger, Least-Squares Finite Element Methods, vol 166, Springer Science & Business Media, Berlin, 2009), the AVS-FE method recasts the governing second order partial differential equation (PDE) into a system of first-order PDEs. However, in the subsequent derivation of the equivalent weak formulation, a Petrov-Galerkin technique is applied by using different regularities for the trial and test function spaces. We use standard FE approximation spaces for the trial spaces, which are C^0, and broken Hilbert spaces for the test functions. Thus, we seek to compute pointwise continuous solutions for both the primal variable and its flux (as in least squares FE methods), while the test functions are piecewise discontinuous. To ensure the numerical stability of the subsequent FE discretizations, we apply the philosophy of the discontinuous Petrov-Galerkin (DPG) method by Demkowicz and Gopalakrishnan (Comput Methods Appl Mech Eng 199(23):1558–1572, 2010; Discontinuous Petrov-Galerkin (DPG) method, Tech. rep., The Institute for Computational Engineering and Sciences, The University of Texas at Austin, 2015; SIAM J Numer Anal 49(5):1788–1809, 2011; Numer Methods Partial Differ Equ 27(1):70–105, 2011; Appl Numer Math 62(4):396–427,2012; Carstensen et al., SIAM J Numer Anal 52(3):1335–1353, 2014), by invoking test functions that lead to unconditionally stable numerical systems (if the kernel of the underlying differential operator is trivial). In the AVS-FE method, the discontinuous test functions are ascertained per the DPG approach from local, decoupled, and well-

V. M. Calo
Applied Geology Department, Curtin University, Perth, WA, Australia
e-mail: victor.calo@curtin.edu.au

A. Romkes (✉) · E. Valseth
Department of Mechanical Engineering, South Dakota School of Mines & Technology, Rapid City, SD, USA
e-mail: Albert.Romkes@sdsmt.edu; Eirik.Valseth@mines.sdsmt.edu

© Springer Nature Switzerland AG 2020
G. R. Barrenechea, J. Mackenzie (eds.), *Boundary and Interior Layers, Computational and Asymptotic Methods BAIL 2018*, Lecture Notes in Computational Science and Engineering 135,
https://doi.org/10.1007/978-3-030-41800-7_2

posed variational problems, which lead to best approximation properties in terms of the energy norm. We present various 2D numerical verifications, including convection-diffusion problems with highly oscillatory coefficients and extremely high Peclet numbers, up to $O(10^9)$. These show the unconditional stability without the need for any upwind schemes nor any other artificial numerical stabilization. The results are not highly diffused for convection-dominated problems nor show any strong oscillations, but adequately capture and indicate the presence of boundary layers, even for very coarse meshes and low polynomial degrees of approximation, p. Remarkably, we can compute the test functions by using the same p level as the trial functions without significantly impacting the numerical accuracy or asymptotic convergence of the numerical results. In addition, the AVS method delivers high numerical accuracy for the computed flux. Importantly, the AVS methodology delivers optimal asymptotic error convergence rates of order $p+1$ and p are obtained in the L^2 and H^1 norms for the primal variable. Our experience indicates that for convection-dominated problems we often observe a convergence rate of $p + 1$ for the L^2 norm of the flux variable.

1   Introduction

Singularly perturbed problems are ubiquitous in many engineering applications. We seek to develop a framework to tackle this large class of problems in a constructive manner. We start with a common model problem, that is, the convection-diffusion problem which is relevant to many engineering applications where transport mechanisms play a significant role, e.g., subsurface flow through porous media, dynamics of viscous flow, convective transfer of heat, drug delivery, turbulence modeling, etc. In this paper, we focus on the stationary version of the scalar-valued convection-diffusion equation and therefore limit our consideration to solutions which only depend on spatial variables and not the temporal variable. To date, the numerical analysis of even the stationary problem poses significant challenges due to the presence of the convection term, which dominates the diffusion processes. Classical FE methodologies, such as the Bubnov-Galerkin FE method [16, 45, 57, 61], mixed FE methods [14, 60], and Petrov-Galerkin method [58], struggle in their numerical analysis due to the numerical instability introduced by the convection term. The corresponding discrete systems of equations can be ill-posed (i.e., a discrete solution does not exist) or lead to either spurious solutions or solutions with severe oscillations. These generally do not tend to attenuate with continued mesh refinements and/or enrichments until the boundary layers are resolved, which in many applications is prohibitively expensive. The least squares FE methods (LSFEMs) [13], the k-version of the FE method by Surana et al. [1, 2, 64, 65], and the DPG method by Demkowicz and Gopalakrishnan [20, 29–33] resolve the numerical instability issues by choosing test/weight functions that lead to unconditionally stable systems of equations governing the FE discretizations. However, in the case of LSFEM and the k-version FE method, the numerical solutions, while stable, can be overly diffusive, particularly for coarse mesh partitions, and therefore fail to indicate the presence and/or location of any sharp boundary layers or other local solution features. As a result, the corresponding adaptive mesh strategies can be ineffective in the presence of strong convection and require overly refined mesh partitions with large numbers of degrees of freedom to resolve boundary layers or other local phenomena. Contrarily, the DPG method does not suffer from overly diffused solutions but also requires edge fluxes and traces (referred to as numerical fluxes and traces). The number of degrees of freedom, once you statically condense the degrees of freedom internal to each element, is similar to the count of the method we propose herein. In addition, although the DPG method provides unconditionally stable FE discretizations, the stabilization is problem-dependent. To ensure the numerical stability and asymptotic convergence of the FE process, the numerical fluxes and traces have to be numerically stabilized through multiplication by mesh dependent terms. This stabilization is akin to upwind-schemes used in other FE methodologies and depends highly on the form/nature of the diffusion and convection coefficients. It is therefore problem-dependent.

Another technique which enlarges the approximation, introduced in [18], extends the use of the generalized multiscale finite elements to stabilize the advection-diffusion model problem. Alternatively, stabilized finite element methods do not add extra degrees of freedom to the global system, but require problem specific modifications of the stabilization parameter. The original stabilization technique is the streamlined-upwind Petrov-Galerkin (SUPG) stabilization, introduced by Brooks and Hughes [15] for the Navier-Stokes system. Using the analytical framework proposed by Hughes [44], we can interpret many stabilization methods as residual-based modifications of the discrete weak forms where a locally scaled differential operator acts on the test function to weight the residual of each trial function. The multiscale interpretation of the stabilization process was illuminating and opened many application opportunities [47, 51, 52], but did not simplify the design process of the stabilization technique. Effectively, this design process is arduous, and problem specific. Among the many successful stabilized methods we cite several that were applied to the transport and Navier-Stokes equations [12, 22, 23, 37–39, 42, 46, 48–50, 54, 55, 62, 63].

In this manuscript, we introduce the automatic variationally stable (AVS) analysis for FE computations of the convection-diffusion equation in which the diffusion and convection coefficients can be highly oscillatory. The method is essentially a hybrid of the LSFEM, Petrov-Galerkin, and the DPG methods by employing the strength and benefits of each approach separately, leading to a FE process that is unconditionally stable and produces numerical solutions that are not overly diffusive, even for coarse FE mesh partitions and low polynomial degrees of approximation. There is no need for the determination of any mesh- and problem-dependent stabilization parameters to warrant unconditionally stable numerical schemes nor overly refined/enriched initial FE mesh partitions to ascertain the presence and location of any boundary layers or local phenomena.

Firstly, we follow mixed FE approaches, by introducing the fluxes as auxiliary variables and thereby recast the second order, scalar-valued convection, diffusion problem into a first order vector-valued PDE. We subsequently apply the Petrov-Galerkin philosophy in the derivation of the equivalent integral formulation (i.e. the weak form) of the established vector-valued PDE by allowing a different regularity for the trial and test spaces. The FE discretization of the weak form is then applied such that the base variable and the fluxes are classical global C^0 functions. However, we apply broken (i.e., discontinuous) Hilbert spaces for the test functions in an effort to allow a maximum flexibility in choosing test functions that lead to unconditionally stable FE processes. To do so, we invoke the philosophy of the DPG method in the FE discretization of the weak form by constructing a test function for every C^0 trial function, which is a solution to decoupled element-wise local variational problems; called 'test problems.' Conforming to the DPG philosophy, the test problems employ bilinear forms which define a local inner product on each element. In the AVS-FE method, we apply a local H^1 inner product as the bilinear form in the test problems. As in DPG, the resulting test functions lead to unconditionally stable systems of equations governing the FE approximation of the problem. In addition, the specific choice for local H^1 inner products in the test problems, appears to result in FE approximations that are not overly diffusive; even for convection-dominated problems with Peclet numbers of order 10^9. Remarkably, the numerical solutions we obtain for the flux variables with the AVS-FE method are highly accurate.

Our choice for C^0 trial functions is motivated by the fact that it enables us to enforce the continuity of all variables strongly and in a straightforward manner. This is of particular benefit for the analysis of the fluxes in the presence of highly oscillatory diffusion coefficients. Moreover, it negates the need to introduce numerical (edge) fluxes and traces as auxiliary variables and thereby reduces the computational cost and removes the need for any problem-dependent numerical stabilization of such variables. A key benefit of this functional choice is that legacy software for pre- and post-processing the data for the simulations can be directly used to prepare and analyze the data required and produced by AVS-FE. Importantly, our simulations rely on continuous discretizations which facilitate solution interpretations and analyses from an engineering point of view. That is, from the user point of view, they are standard finite element solutions where all variables are continuous, simplifying the adoption of the technique by the end-user community.

As in the DPG method, we establish a best approximation property in terms of the energy norm that is induced by the bilinear form of the integral formulation of the AVS-FE method and obtain optimal asymptotic convergence rates in L^2 and H^1 for the base variable and in L^2 for the flux variables.

In the following, we present the derivation of the AVS-FE weak formulation for the convection-diffusion problem in Sect. 2.1 and its subsequent FE discretization in Sect. 2.2, and various two-dimensional verifications in Sect. 3. Concluding remarks and future efforts are discussed in Sect. 4.

2   Derivation of Integral Statement and FE Discretization

Let $\Omega \subset \mathbb{R}^2$ be an open bounded domain (see Fig. 1) with Lipschitz boundary $\partial\Omega$ and outward unit normal vector $\mathbf{n}$. The boundary $\partial\Omega$ consists of open subsections $\Gamma_D, \Gamma_N \subset \partial\Omega$, such that $\Gamma_D \cap \Gamma_N = \emptyset$ and $\partial\Omega = \overline{\Gamma_D \cup \Gamma_N}$. For our model problem, we consider the following convection-diffusion equation in Ω with homogeneous Dirichlet boundary conditions applied on Γ_D and (possibly) non-homogeneous Neumann boundary conditions on Γ_N:

$$
\begin{aligned}
\text{Find } u \text{ such that:} \\
-\nabla \cdot (\mathbf{D}\nabla u) + \mathbf{b} \cdot \nabla u = f, & \quad \text{in } \Omega, \\
u = 0, & \quad \text{on } \Gamma_D, \\
\mathbf{D}\nabla u \cdot \mathbf{n} = g, & \quad \text{on } \Gamma_N,
\end{aligned}
\tag{1}
$$

where $\mathbf{D}$ denotes the second order diffusion tensor, with symmetric, bounded, and positive definite coefficients $D_{ij} \in L^\infty(\Omega)$; $\mathbf{b} \in [L^2(\Omega)]^2$ the convection coefficient; $f \in L^2(\Omega)$ the source function; and $g \in H^{-1/2}(\Gamma_N)$ the Neumann boundary data. We consider the scenario in which the diffusion coefficients D_{ij} can be highly heterogeneous and therefore can change many orders in magnitude over small length scales throughout Ω (e.g., in Fig. 1, the differently colored subdomains represent areas with different values of the diffusions coefficients).

In this work, we seek to derive a DPG weak formulation of (1) by using a regular partition $\mathscr{P}_h$ of Ω into open subdomains, or elements, K_m (see Fig. 1), with diameters h_m, such that :

$$
\Omega = \text{int}(\bigcup_{K_m \in \mathscr{P}_h} \overline{K_m}).
$$

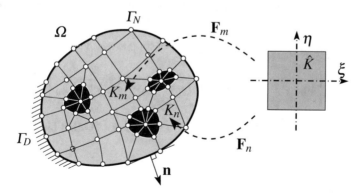

Fig. 1 The model problem

Any such partition $\mathscr{P}_h$ of Ω is applied such that any discontinuities in the diffusion coefficient D_{ij} or convection coefficient $\mathbf{b}$ are restricted to the boundaries ∂K_m of the elements $K_m \in \mathscr{P}_h$ (see Fig. 1). That is, we assume our mesh fully resolves these spatial features, while it may not resolve the induced internal layers.

We apply a mixed FE methodology and introduce the flux $\mathbf{q} = \{q_x, q_y\}^T = \mathbf{D}\nabla u$ as an auxiliary variable, then, accordingly, $\mathbf{q} \in H(\text{div}, \Omega)$ and (1) can be recast equivalently as a first-order system of PDEs, where the regularity of u can be relaxed to be in $H^1(\Omega)$:

$$
\boxed{
\begin{aligned}
&\text{Find } (u, \mathbf{q}) \in H^1(\Omega) \times H(\text{div}, \Omega) \text{ such that:}\\
&\qquad \mathbf{q} - \mathbf{D}\nabla u = 0, \qquad \text{in } \Omega,\\
&\quad -\nabla \cdot \mathbf{q} + \mathbf{b} \cdot \nabla u = f, \qquad \text{in } \Omega,\\
&\qquad\qquad\qquad u = 0, \qquad \text{on } \Gamma_D,\\
&\qquad\qquad \mathbf{q} \cdot \mathbf{n} = g, \qquad \text{on } \Gamma_N.
\end{aligned}
}
\tag{2}
$$

2.1 Derivation of Integral Formulation

To start the derivation of the DPG formulation of (2), we enforce the PDE weakly on each element $K_m \in \mathscr{P}_h$, i.e., we seek the restrictions u_m and $\mathbf{q}_m$ of u and $\mathbf{q}$ to each K_m, such that:

$$
\int_{K_m} \left\{ [\mathbf{q}_m - \mathbf{D}\nabla u_m] \cdot \mathbf{w}_m + [-\nabla \cdot \mathbf{q}_m + \mathbf{b} \cdot \nabla u_m] \, v_m \right\} \, d\mathbf{x} = \int_{K_m} f \, v_m \, d\mathbf{x},
$$
$$
\forall (v_m, \mathbf{w}_m) \in L^2(K_m) \times [L^2(K_m)]^2.
\tag{3}
$$

By repeating this process for all $K_m \in \mathscr{P}_h$ and summing the resulting local integral formulations, we get:

Find $(u, \mathbf{q}) \in H^1(\Omega) \times H(\text{div}, \Omega)$:

$$
\sum_{K_m \in \mathscr{P}_h} \int_{K_m} \left\{ [\mathbf{q}_m - \mathbf{D}\nabla u_m] \cdot \mathbf{w}_m + [-\nabla \cdot \mathbf{q}_m + \mathbf{b} \cdot \nabla u_m] \, v_m \right\} \, d\mathbf{x}
\tag{4}
$$
$$
= \sum_{K_m \in \mathscr{P}_h} \int_{K_m} f \, v_m \, d\mathbf{x}, \qquad \forall (v, \mathbf{w}) \in L^2(\Omega) \times [L^2(\Omega)]^2
$$

Next, we apply Green's identity to the $(\nabla \cdot \mathbf{q}_m)\,v_m$ terms, which demands that we increase the regularity of each v to be in H^1 locally for every $K_m \in \mathscr{P}_h$, i.e.,

Find $(u, \mathbf{q}) \in H^1(\Omega) \times H(\mathrm{div}, \Omega)$:

$$
\sum_{K_m \in \mathscr{P}_h} \left\{ \int_{K_m} \left[(\mathbf{q}_m - \mathbf{D}\nabla u_m) \cdot \mathbf{w}_m + \mathbf{q}_m \cdot \nabla v_m + (\mathbf{b} \cdot \nabla u_m)\, v_m \right] dx \right.
$$
$$
\left. - \oint_{\partial K_m} \gamma_{\mathbf{n}}^m(\mathbf{q}_m)\, \gamma_0^m(v_m)\, ds \right\} = \sum_{K_m \in \mathscr{P}_h} \int_{K_m} f\, v_m\, dx, \tag{5}
$$
$$
\forall (v, \mathbf{w}) \in H^1(\mathscr{P}_h) \times [L^2(\Omega)]^2
$$

where the broken H^1 Hilbert space on $\mathscr{P}_h$ is defined as follows:

$$
H^1(\mathscr{P}_h) \overset{\text{def}}{=} \left\{ v \in L^2(\Omega) : \quad v_m \in H^1(K_m),\ \forall K_m \in \mathscr{P}_h \right\}, \tag{6}
$$

and $\gamma_0^m : H^1(K_m) \longrightarrow H^{1/2}(\partial K_m)$ and $\gamma_{\mathbf{n}}^m : H(\mathrm{div}, K_m) \longrightarrow H^{-1/2}(\partial K_m)$ denote the trace and normal trace operators (e.g., see [40]) on K_m; and $\mathbf{n}_m$ is the outward unit normal vector to the element boundary ∂K_m of K_m. Strictly speaking, the edge integral on ∂K_m in (5) is to be interpreted as the duality pairing in $H^{-1/2}(\partial K_m) \times H^{1/2}(\partial K_m)$ of $\gamma_{\mathbf{n}}^m(\mathbf{q}_m)$ and $\gamma_0^m(v_m)$, but we apply the engineering notation here by using an integral representation.

Now, by decomposing each edge term in (5) into a sum of several terms, i.e., one term concerning the portion of the edge ∂K_m that intersects with neighboring elements and possibly one or two additional terms concerning the portion of ∂K_m that intersects with Γ_D or Γ_N, we can rewrite (5) as follows:

Find $(u, \mathbf{q}) \in H^1(\Omega) \times H(\mathrm{div}, \Omega)$:

$$
\sum_{K_m \in \mathscr{P}_h} \left\{ \int_{K_m} \left[(\mathbf{q}_m - \mathbf{D}\nabla u_m) \cdot \mathbf{w}_m + \mathbf{q}_m \cdot \nabla v_m + (\mathbf{b} \cdot \nabla u_m)\, v_m \right] dx \right.
$$
$$
- \int_{\partial K_m \setminus \overline{\Gamma_D \cup \Gamma_N}} \gamma_{\mathbf{n}}^m(\mathbf{q}_m)\, \gamma_0^m(v_m)\, ds - \int_{\partial K_m \cap \Gamma_D} \gamma_{\mathbf{n}}^m(\mathbf{q}_m)\, \gamma_0^m(v_m)\, ds
$$
$$
\left. - \int_{\partial K_m \cap \Gamma_N} \gamma_{\mathbf{n}}^m(\mathbf{q}_m)\, \gamma_0^m(v_m)\, ds \right\} = \sum_{K_m \in \mathscr{P}_h} \int_{K_m} f\, v_m\, dx,
$$
$$
\forall (v, \mathbf{w}) \in H^1(\mathscr{P}_h) \times [L^2(\Omega)]^2
$$

By subsequently enforcing the Neumann boundary condition on the normal trace of $\mathbf{q}$ as well as constraining the traces of the test function v_m on the Dirichlet boundary (since we apply the Dirichlet condition on u strongly), we arrive at the final variational statement:

Find $(u, \mathbf{q}) \in U(\Omega)$:

$$
\sum_{K_m \in \mathscr{P}_h} \left\{ \int_{K_m} \left[(\mathbf{q}_m - \mathbf{D}\nabla u_m) \cdot \mathbf{w}_m + \mathbf{q}_m \cdot \nabla v_m + (\mathbf{b} \cdot \nabla u_m) v_m \right] d\mathbf{x} \right.
$$
$$
\left. - \int_{\partial K_m \setminus \overline{\Gamma_D \cup \Gamma_N}} \gamma_{\mathbf{n}}^m(\mathbf{q}_m) \, \gamma_0^m(v_m) \, ds \right\} = \sum_{K_m \in \mathscr{P}_h} \left\{ \int_{K_m} f \, v_m \, d\mathbf{x} + \int_{\partial K_m \cap \Gamma_N} g \, \gamma_0^m(v_m) \, ds \right\},
$$
$$
\forall (v, \mathbf{w}) \in V(\mathscr{P}_h)
$$
$$
(7)
$$

where the trial and test function spaces, $U(\Omega)$ and $V(\mathscr{P}_h)$, are defined as follows:

$$
U(\Omega) \overset{\text{def}}{=} \left\{ (u, \mathbf{q}) \in H^1(\Omega) \times H(\text{div}, \Omega) : \; \gamma_0^m(u_m)_{|\partial K_m \cap \Gamma_D} = 0, \; \forall K_m \in \mathscr{P}_h \right\},
$$
$$
V(\mathscr{P}_h) \overset{\text{def}}{=} \left\{ (v, \mathbf{w}) \in H^1(\mathscr{P}_h) \times [L^2(\Omega)]^2 : \; \gamma_0^m(v_m)_{|\partial K_m \cap \Gamma_D} = 0, \; \forall K_m \in \mathscr{P}_h \right\},
$$
$$
(8)
$$

with norms $\|\cdot\|_{U(\Omega)} : U(\Omega) \longrightarrow [0, \infty)$ and $\|\cdot\|_{V(\mathscr{P}_h)} : V(\mathscr{P}_h) \longrightarrow [0, \infty)$ defined as:

$$
\|(u, \mathbf{q})\|_{U(\Omega)} \overset{\text{def}}{=} \sqrt{\int_\Omega \left[\nabla u \cdot \nabla u + u^2 + (\nabla \cdot \mathbf{q})^2 + \mathbf{q} \cdot \mathbf{q} \right] d\mathbf{x}}.
$$
$$
(9)
$$
$$
\|(v, \mathbf{w})\|_{V(\mathscr{P}_h)} \overset{\text{def}}{=} \sqrt{\sum_{K_m \in \mathscr{P}_h} \int_{K_m} \left[h_m^2 \nabla v_m \cdot \nabla v_m + v_m^2 + \mathbf{w}_m \cdot \mathbf{w}_m \right] d\mathbf{x}}.
$$

By introducing the bilinear form, $B : U(\Omega) \times V(\mathscr{P}_h) \longrightarrow \mathbb{R}$, and linear functional, $F : V(\mathscr{P}_h) \longrightarrow \mathbb{R}$, i.e.,

$$
B((u, \mathbf{q}); (v.\mathbf{w})) \overset{\text{def}}{=} \sum_{K_m \in \mathscr{P}_h} \left\{ \int_{K_m} \left[(\mathbf{q}_m - \mathbf{D}\nabla u_m) \cdot \mathbf{w}_m + \mathbf{q}_m \cdot \nabla v_m + (\mathbf{b} \cdot \nabla u_m) v_m \right] d\mathbf{x} \right.
$$
$$
\left. - \int_{\partial K_m \setminus \overline{\Gamma_D \cup \Gamma_N}} \gamma_{\mathbf{n}}^m(\mathbf{q}_m) \, \gamma_0^m(v_m) \, ds \right\},
$$
$$
F((v, \mathbf{w})) \overset{\text{def}}{=} \sum_{K_m \in \mathscr{P}_h} \left\{ \int_{K_m} f \, v_m \, d\mathbf{x} + \int_{\partial K_m \cap \Gamma_N} g \, \gamma_0^m(v_m) \, ds \right\},
$$
$$
(10)
$$

we can rewrite the weak formulation (7) in compact form as follows:

Find $(u, \mathbf{q}) \in U(\Omega)$ such that:
$$
B((u, \mathbf{q}); (v, \mathbf{w})) = F((v, \mathbf{w})), \quad \forall (v, \mathbf{w}) \in V(\mathscr{P}_h).
$$
$$
(11)
$$

Lemma 2.1 *Let $f \in (H^1(\mathscr{P}_h))'$ and $g \in H^{-1/2}(\Gamma_N)$. Then there exists a unique solution $(u, \mathbf{q}) \in U(\Omega)$ of the weak formulation (11).*

We refer to [19] for a proof of this lemma.   □

Now, (11) essentially represents a DPG formulation [20, 29–33, 56], as the spaces $U(\Omega)$ and $V(\mathscr{P}_h)$ have different regularities. However, it differs significantly by using (weakly) globally continuous trial spaces. Currently existing DPG methods require weak enforcement of continuity conditions across inter-element edges by introducing numerical traces and fluxes as auxiliary variables. Thus, by employing trial spaces in which continuity of the primal variable and the normal fluxes is inherent (weakly), we attempt to keep the formulation, from the point of view of the user, as close as possible to a standard FE discretization. Lastly, the discrete description of the solution behaves like standard finite element discretizations, which will accelerate the adoption of this discretization technique by practitioners and paves the way to extend it to solutions with higher order global continuity, such as the ones produced by isogeometric analysis [3–11, 17, 21, 24–28, 34–36, 41, 43, 53, 59], to show just a few of the relevant applications of this powerful simulation technique.

2.2 AVS-FE Discretization

We now seek numerical approximations $(u^h, \mathbf{q}^h)$ of solutions $(u, \mathbf{q})$ of the weak form (11) by using classical *globally continuous*, $C^0(\Omega)$, trial functions for $(u^h, \mathbf{q}^h)$. However, the discontinuous topology is maintained for the space of test functions, as this allows the maximum flexibility in constructing test functions that lead to unconditionally numerically stable discrete systems and provide best approximation properties in terms of the energy norm, $\|\cdot\|_B : U(\Omega) \longrightarrow [0, \infty]$, of the error, i.e.:

$$\|(u, \mathbf{q})\|_B \stackrel{\text{def}}{=} \sup_{(v,\mathbf{w})\in V(\mathscr{P}_h)\setminus\{(0,\mathbf{0})\}} \frac{|B((u, \mathbf{q}); (v, \mathbf{w}))|}{\|(v, \mathbf{w})\|_{V(\mathscr{P}_h)}}. \tag{12}$$

The discrete fluxes we use, are more regular than is required by the minimal topology we described in the previous section. We apply discrete fluxes that belong to $H^1(\Omega)$ rather than $H(\text{div}, \Omega)$. Our experience indicates that the numerical solutions we obtain when we use approximations in $H(\text{div}, \Omega)$ yield similar accuracy, as long as the domain does not exhibit any re-entrant corners and/or cracks. Convergence is observed in the latter case, but the onset of asymptotic convergence is then generally observed at a higher number of mesh refinements. Raviart-Thomas discretizations most likely resolve this and will be investigated in an upcoming manuscript. Currently, using discrete fluxes in $C^0(\Omega)$ is certainly less challenging to implement $H^1(\Omega)$ partitions on standard meshes. Possibly more importantly, this will allow the use of AVS formulations in commercial simulation software by redefining the user-defined elemental routines.

Let us now proceed by deriving the FE discretization of (11) by first introducing the family of invertible maps, $\{\mathbf{F}_m : \hat{K} \subset \mathbb{R}^2 \longrightarrow \Omega\}$, such that every $K_m \in \mathscr{P}_h$ is the image of a master element $\hat{K}$ through one of the mappings $\mathbf{F}_m$ (see Fig. 1). The (conforming) space of trial functions, $U^h(\Omega) \subset U(\Omega)$, is then defined as:

$$
U^h(\Omega) \stackrel{\text{def}}{=} \Big\{ (\varphi^h, \boldsymbol{\theta}^h) \in C^0(\Omega) \times [C^0(\Omega)]^2 : (\varphi^h_{|K_m}, \boldsymbol{\theta}^h_{|K_m}) = (\hat{\varphi}, \hat{\boldsymbol{\theta}}) \circ \mathbf{F}_m, \\
\hat{\varphi} \in P^{p_m}(\hat{K}) \wedge \hat{\boldsymbol{\theta}} \in [P^{p_m}(\hat{K})]^2, \quad \forall K_m \in \mathscr{P}_h \Big\},
$$

(13)

where p_m denotes the local polynomial degree of approximation on K_m. We are essentially following the classical FE method here and therefore accordingly represent the FE approximations, u^h and $\mathbf{q}^h = \{q_x^h, q_y^h\}^T$, as linear combinations of trial functions $(e^i(\mathbf{x}), (E_x^j(\mathbf{x}), E_y^k(\mathbf{x}))) \in U^h(\Omega)$ and corresponding degrees of freedom, $\{u_i^h \in \mathbb{R}, i = 1, 2, \ldots, N\}$, $\{q_x^{h,j} \in \mathbb{R}, j = 1, 2, \ldots, N\}$ and $\{q_y^{h,k} \in \mathbb{R}, k = 1, 2, \ldots, N\}$; i.e.,

$$
u^h(\mathbf{x}) = \sum_{i=1}^N u_i^h e^i(\mathbf{x}), \quad q_x^h(\mathbf{x}) = \sum_{j=1}^N q_x^{h,j} E_x^j(\mathbf{x}), \quad q_y^h(\mathbf{x}) = \sum_{k=1}^N q_y^{h,k} E_y^k(\mathbf{x}).
$$

(14)

As mentioned previously, contrary to the trial functions (which are global C^0 functions), the test functions are to be piecewise discontinuous and constructed by invoking the DPG strategy [20, 29–33, 56]. Each of the $3N$ trial functions $e^i(\mathbf{x})$, $E_x^j(\mathbf{x})$, and $E_y^k(\mathbf{x})$, is paired with a vector-valued test function. Thus, $e^i(\mathbf{x})$ is paired with $(\tilde{e}^i, \tilde{\mathbf{E}}^i) \in V(\mathscr{P}_h)$, $E_x^j(\mathbf{x})$ with $(\tilde{e}_x^j, \tilde{\mathbf{E}}_x^j) \in V(\mathscr{P}_h)$, and $e_y^k(\mathbf{x})$ with $(\tilde{e}_y^k, \tilde{\mathbf{E}}_y^k) \in V(\mathscr{P}_h)$. Following the DPG philosophy, these pairings are established through the following variational problems:

$$
\begin{aligned}
\Big((r, \mathbf{z}); (\tilde{e}^i, \tilde{\mathbf{E}}^i) \Big)_{V(\mathscr{P}_h)} &= B((e^i, \mathbf{0}); (r, \mathbf{z})), & \forall (r, \mathbf{z}) \in V(\mathscr{P}_h), \quad i = 1, \ldots, N, \\
\Big((r, \mathbf{z}); (\tilde{e}_x^j, \tilde{\mathbf{E}}_x^j) \Big)_{V(\mathscr{P}_h)} &= B((0, (E_x^j, 0)); (r, \mathbf{z})), & \forall (r, \mathbf{z}) \in V(\mathscr{P}_h), \quad j = 1, \ldots, N, \\
\Big((r, \mathbf{z}); (\tilde{e}_y^k, \tilde{\mathbf{E}}_y^k) \Big)_{V(\mathscr{P}_h)} &= B((0, (0, E_y^k)); (r, \mathbf{z})), & \forall (r, \mathbf{z}) \in V(\mathscr{P}_h), \quad k = 1, \ldots, N,
\end{aligned}
$$

(15)

where $(\,\cdot\,;\,\cdot\,)_{V(\mathscr{P}_h)} : V(\mathscr{P}_h) \times V(\mathscr{P}_h) \longrightarrow \mathbb{R}$, is the inner product:

$$((r, \mathbf{z}); (v, \mathbf{w}))_{V(\mathscr{P}_h)} \stackrel{\text{def}}{=} \sum_{K_m \in \mathscr{P}_h} \int_{K_m} \left[h_m^2 \nabla r_m \cdot \nabla v_m + r_m \, v_m + \mathbf{z}_m \cdot \mathbf{w}_m \right] \mathrm{d}x,$$

(16)

which induces the norm $\|\cdot\|_{V(\mathscr{P}_h)}$, as defined in (9). The solution of these Riesz representation problems in the test space norm produces the set of test functions that we use in our variational framework.

Remark 2.1 The variational statements (15) are infinite dimensional problems which we approximate numerically. To do so, we compute piecewise discontinuous polynomial approximations $(\tilde{e}_h^i, \tilde{\mathbf{E}}_h^i)$, $(\tilde{e}_{x_h}^j, \tilde{\mathbf{E}}_{x_h}^j)$, and $(\tilde{e}_{y_h}^k, \tilde{\mathbf{E}}_{y_h}^k)$ of $(\tilde{e}^i, \tilde{\mathbf{E}}^i)$, $(\tilde{e}_x^j, \tilde{\mathbf{E}}_x^j)$, and $(\tilde{e}_y^k, \tilde{\mathbf{E}}_y^k)$, respectively, by applying local polynomial degrees of approximation of order $p_m + \Delta p$.

Remark 2.2 By applying functions $(r, \mathbf{z}) \in V(\mathscr{P}_h)$ in the variational statements of (15) that vanish outside a given element K_m, the local restriction of the test functions to K_m, can easily be computed by solving the following local restrictions of (15):

$$\left((r, \mathbf{z}); (\tilde{e}_h^i, \tilde{\mathbf{E}}_h^i) \right)_{V(K_m)} = B_{|K_m}((e^i, \mathbf{0}); (r, \mathbf{z})), \qquad \forall (r, \mathbf{z}) \in V(K_m),$$

$$\left((r, \mathbf{z}); (\tilde{e}_{x_h}^j, \tilde{\mathbf{E}}_{x_h}^j) \right)_{V(K_m)} = B_{|K_m}((0, (E_x^j, 0)); (r, \mathbf{z})), \quad \forall (r, \mathbf{z}) \in V(K_m),$$

$$\left((r, \mathbf{z}); (\tilde{e}_{y_h}^k, \tilde{\mathbf{E}}_{y_h}^k) \right)_{V(K_m)} = B_{|K_m}((0, (0, E_y^k)); (r, \mathbf{z})), \quad \forall (r, \mathbf{z}) \in V(K_m),$$

(17)

where $B_{|K_m}(\cdot;\,\cdot)$ denotes the restriction of $B(\cdot;\,\cdot)$ (see (10)) to the element K_m and:

$$V(K_m) \stackrel{\text{def}}{=} \left\{ (v, \mathbf{w}) \in H^1(K_m) \times [L^2(K_m)]^2 : \gamma_0^m(v_m)_{|\partial K_m \cap \Gamma_D} = 0 \right\},$$

$$(\,\cdot\,;\,\cdot\,)_{V(K_m)} : V(K_m) \times V(K_m) \longrightarrow \mathbb{R},$$

(18)

$$((r, \mathbf{z}); (v, \mathbf{w}))_{V(K_m)} \stackrel{\text{def}}{=} \int_{K_m} \left[h_m^2 \nabla r \cdot \nabla v + r \, v + \mathbf{z} \cdot \mathbf{w} \right] \mathrm{d}x.$$

If we look at the action of the local restriction of the bilinear form $B(\cdot;\,\cdot)$ onto functions $(\varphi, \boldsymbol{\theta})$, that have the same regularity as our FE trial functions (i.e., they belong to $C^0(\Omega) \times [C^0(\Omega)]^2$), and test functions $(r, \mathbf{z}) \in V(\mathscr{P}_h)$ that vanish outside

K_m, we get from (10):

$$B_{|K_m}((\varphi, \theta); (r.\mathbf{z})) = \int_{K_m} \left[(\theta_m - \mathbf{D}\nabla\varphi_m) \cdot \mathbf{z}_m + \theta_m \cdot \nabla r_m + (\mathbf{b} \cdot \nabla\varphi_m) r_m \right] d\mathbf{x}$$

$$- \int_{\partial K_m \setminus \overline{\Gamma_D \cup \Gamma_N}} \gamma_{\mathbf{n}}^m (\theta_m) \gamma_0^m (r_m) \, ds$$

(19)

Thus, in the computations of the local variational statements of (17), the action of $B_{|K_m}(\cdot; \cdot)$ can be applied as shown in (19).

Remark 2.3 Since the action of the bilinear form in the RHS of (17) is entirely local to the element K_m, as given in (19), a trial function only induces a nonzero test function in elements where it has support. Hence, an additional consequence of (19) is that the support of every test function is identical to the support of the corresponding trial function.

At last, the FE discretization of (7), governing the AVS-FE approximation $(u^h, \mathbf{q}^h) \in U^h(\Omega)$ of $(u, \mathbf{q})$ can now be introduced as follows:

Find $(u^h, \mathbf{q}^h) \in U^h(\Omega)$ such that:

$$B((u^h; \mathbf{q}^h); (v^*, \mathbf{w}^*)) = F((v^*, \mathbf{w}^*)), \quad \forall (v^*, \mathbf{w}^*) \in V^*(\mathscr{P}_h),$$

(20)

---

where the finite dimensional subspace of test functions $V^*(\mathscr{P}_h) \subset V(\mathscr{P}_h)$ is spanned by the numerical approximations of the test functions $\{(\tilde{e}_h^i, \tilde{\mathbf{E}}_h^i)\}_{i=1}^N$, $\{(\tilde{e}_{x_h}^j, \tilde{\mathbf{E}}_{x_h}^j)\}_{j=1}^N$, and $\{(\tilde{e}_{y_h}^k, \tilde{\mathbf{E}}_{y_h}^k)\}_{k=1}^N$, as computed from the Riesz representation problems (15) and (17) by using local polynomial degrees of approximation $p_m + \Delta p$.

Since we essentially apply the DPG methodology [20, 29–33] in the construction of the space of test functions $V^*(\mathscr{P}_h)$ via the Riesz representation statements (15), an important consequence is that the FE discretization (20) also inherits the unconditional numerical stability property of the DPG method. Thus, there is no need for any, generally arduous, determination of problem and mesh dependent stabilization terms to stabilize the numerical scheme, as done in stabilized FE methods such as SUPG, GLS, and VMS. The discrete problem (20) is automatically and unconditionally stable for any choice of the mesh parameters h_m and p_m.

Lemma 2.2 *The FE discretization (20) is locally conservative.*

We refer to [19] for a detailed proof of this lemma   □

3 Exemplary Numerical Results

To conduct numerical studies of our new method, we consider the following simplified form of our model scalar-valued convection diffusion problem (1) on the unit square domain $\Omega = (0, 1) \times (0, 1) \subset \mathbb{R}^2$ with homogeneous Dirichlet boundary conditions:

$$-D\Delta u + \mathbf{b} \cdot \nabla u = f, \quad \text{in } \Omega,$$
$$u = 0, \quad \text{on } \partial\Omega, \tag{21}$$

where the coefficient $D \in L^\infty(\Omega)$ is a scalar-valued isotropic diffusion coefficient. In the following subsections, we first verify the asymptotic convergence behavior of the newly introduced AVS-FE method. In Sect. 3.1 we analyze a case in which convection is still rather moderate. However, since our main purpose is to investigate the intrinsic stability property of the method, we focus our attention on convection dominated problems in the subsections that follow. In Sect. 3.2, we first look at a classical scenario in which all coefficients are homogeneous, i.e., constant, throughout Ω. Next, we consider a scenario of importance to engineering applications. In Sect. 3.3, the diffusion coefficient is heterogeneous and therefore varies throughout the domain. Lastly, we briefly investigate the converse situation in Sect. 3.4 in which the diffusion is homogeneous, but the convection varies throughout the domain. Particularly, we look at an example in which the variation of the convection coefficient causes the formation of an internal layer.

The purpose of studying these convection-dominated problem is to test the intrinsic (automatic) stability property of the AVS-FE discretizations, which we attained by using the DPG philosophy in the construction of our test functions. We are particularly interested to see if we indeed: (1) obtain automatic stability for any choice of mesh, (2) avoid overly diffused solutions for initial meshes, which is a commonly encountered impediment of LSFEM solution, and (3) avoid solutions with high oscillations at boundary and internal layers that do not tend to attenuate upon mesh refinements, as encountered in classical FE analyses of such problems.

3.1 Asymptotic Convergence Study

To ascertain the asymptotic convergence rates in terms of the $L^2(\Omega)$, $H^1(\Omega)$, and $\|\cdot\|_{U(\Omega)}$ norms of the error, we consider a scenario of our model convection diffusion problem (21) in which the diffusion coefficient $D = 1/Pe$, where we refer to $Pe \in \mathbb{R}^+$ as the Peclet number, and $\mathbf{b} = \{1, 1\}^T$. We choose $Pe = 10$ and the source function f such that the exact theoretical solution is given by:

$$u(x, y) = \left[x + \frac{e^{Pe \cdot x} - 1}{1 - e^{Pe}} \right] \left[y + \frac{e^{Pe \cdot y} - 1}{1 - e^{Pe}} \right].$$

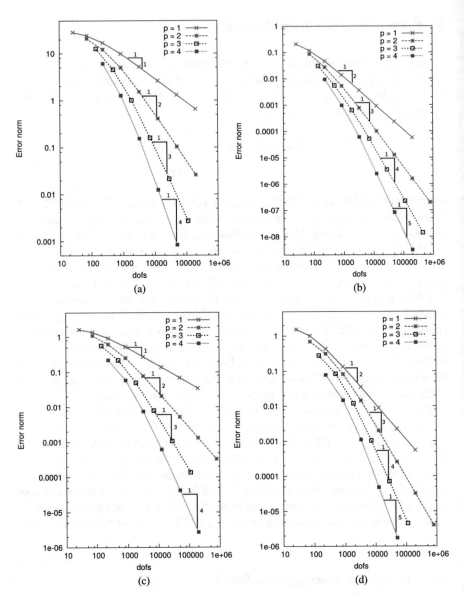

Fig. 2 Error convergence results for uniform h-refinements; $\Delta p = 0$. (**a**) $\|(u, \mathbf{q}) - (u^h, \mathbf{q}^h)\|_{U^h(\Omega)}$. (**b**) $\|u - u^h\|_{L^2(\Omega)}$. (**c**) $\|u - u^h\|_{H^1(\Omega)}$. (**d**) $\|\mathbf{q} - \mathbf{q}^h\|_{L^2(\Omega)}$

This solution exhibits a boundary layer along the boundaries $x = 1$ and $y = 1$, but since there is a moderate level of diffusion (due to the relatively low value of the Peclet number), these layers are not sharp.

In Fig. 2, we show error convergence results for uniform h-refinements in terms of various error norms. For each h-refinement study a uniform p-level has been

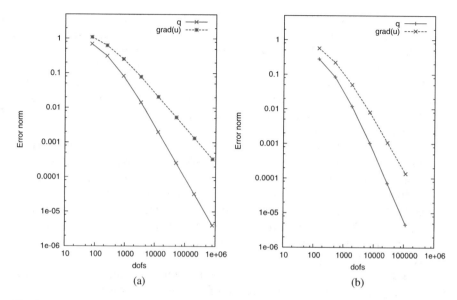

Fig. 3 Numerical accuracy comparison of **q** versus ∇u; $\Delta p = 0$. (**a**) $p = 2$. (**b**) $p = 3$

applied, ranging from $p = 1$ to $p = 4$. The test functions have been computed at the same local polynomial degree of approximation as their corresponding trial functions (i.e., $\Delta p = 0$). The plots in Fig. 2b and c clearly show that both the $L^2(\Omega)$ and $H^1(\Omega)$ norms of the error in the primal variable, $u - u^h$, exhibit optimal convergence rates of order $p + 1$ and p, respectively. Similarly, the $L^2(\Omega)$ norm of the error in the flux, $\mathbf{q} - \mathbf{q}^h$, has an optimal convergence rate of $p + 1$, as shown in Fig. 2d. The convergence rates in terms of the error norm $\|(u, \mathbf{q}) - (u^h, \mathbf{q}^u)\|_{U(\Omega)}$, presented in Fig. 2a, are also optimal at a rate of p.

These results are representative of extensive convergence studies we have conducted. In all these experiments, the observed asymptotic convergence rates have been optimal. The corresponding a priori estimates of these convergence rates, and their proofs, are to be presented in [19].

Lastly, we show a comparison of the $L^2(\Omega)$ norm of the error in $\mathbf{q} - \mathbf{q}^h$ versus $\nabla u - \nabla u^h$ in Fig. 3, for $p = 2$ and $p = 3$. These results are again representative of extensive numerical experiments, in which consistently a significantly higher accuracy is observed in the prediction of the flux variable versus the gradient of the primal variable.

3.2 Convection Dominated Diffusion: Homogeneous Coefficients

As mentioned in the introduction, we are particularly interested in diffusion problems in which convection plays a dominant role. We start here with the case in which the problem coefficients D and $\mathbf{b}$ in (21) are constant. For our numerical study, we enforce convection in the diagonal direction, i.e., the convection coefficient $\mathbf{b} = \{1, 1\}^T$. The source function is set at $f(\mathbf{x}) = 1$ and the diffusion coefficient again at $D = 1/Pe$. However, the Peclet number is now set at a high value of $Pe = 10^6$ to ensure the convection term is dominant in (21). With this choice of parameters in place, the distribution of the primal variable exhibits strong convection in the diagonal direction and a sharp boundary layer of width $1/Pe$ along the boundaries at $x = 1$ and $y = 1$.

For a graded regular mesh of only 2×2 elements, as illustrated in Fig. 4a, and a uniform $p = 2$ and $\Delta p = 0$, the corresponding AVS-FE approximation u^h is shown in Fig. 4b. The FE approximation, at just 75 dofs, is stable and does not exhibit any overly diffused behavior but captures the boundary layer well. Apparently, applying an identical local polynomial degree of approximation for the test functions in (17) (i.e. $\Delta p = 0$), suffices to capture the boundary layer with a relatively good accuracy. Results for $\Delta p = 1, 2, 3$, which are not presented here, do not show any significant difference with the results shown here.

If we apply several additional uniform refinements the solutions remain stable and converge. In Fig. 5, results for the AVS-FE approximation are provided for the fifth refinement (i.e., at 12,675 dofs). A zoomed-in plot of the distribution of u^h along the diagonal and in the vicinity of the corner at $y = x = 1$, do not show any oscillations, which are commonly observed in solutions obtained via classical FE methods or LSFEM. The resolution of the boundary layer is not distorted by any oscillations and continuously sharpens as the mesh is refined.

To demonstrate that the AVS-FE method also produces sequences of stable numerical solutions for unstructured meshes, we present results in Fig. 6 for $Pe = 400$. As depicted in Fig. 6a, the initial coarse mesh is unstructured and does not resolve the length scale of the boundary layer along $x = 1$ and $y = 1$. The

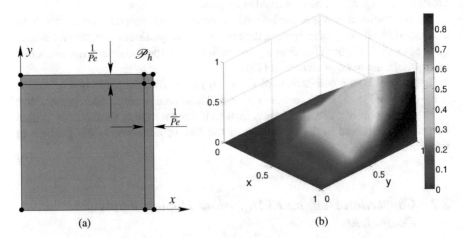

Fig. 4 AVS-FE results for homogeneous coefficients, a 2×2 graded mesh (75 dofs); $Pe = 10^6$, $p = 2$, and $\Delta p = 0$. (a) Initial 2×2 FE mesh. (b) Distribution of u^h

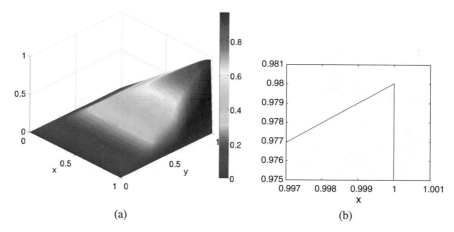

(b)

Fig. 5 AVS-FE results or homogeneous coefficients, a refined mesh (12,675 dofs); $Pe = 10^6$, $p = 2$, and $\Delta p = 0$. (**a**) Distribution of u^h throughout $(0, 1) \times (0, 1)$. (**b**) Distribution of u^h along $y - x = 0$

corresponding numerical solution of u^h is shown in Fig. 6b for $p = 2$ and has poor numerical accuracy, as is expected for such a coarse mesh. However, the solution is stable and upon applying uniform refinements (see Fig. 6c for the first refinement), the solutions indicate the presence of the boundary layer. Hence, any subsequent hp-adaptive strategies can then be applied to fully resolve the boundary layer. Since in this work our focus is not on hp-adaptivity, we simply apply several uniform h-refinements to demonstrate that the solutions do converge for unstructured meshes, as shown in Fig. 6d.

3.3 Heterogeneous Diffusion

We continue by looking at a more challenging case in which the diffusion D is a discontinuous piecewise constant function. Specifically, D has a value of Pe or $1/Pe$ following a checker board pattern, as depicted in Fig. 7a. Both the source function and convection coefficient remain unchanged from the experiment conducted in Sect. 3.2, i.e., $f(\mathbf{x}) = 1$ and $\mathbf{b} = \{1, 1\}^T$. By choosing a high Peclet number of 10^4, we essentially establish a zero solution in the diffusion dominant quadrants of the domain, while strong convection is observed in the remaining two quadrants. Consequently, in the convective regions, sharp internal layers are formed at the interface with the diffusion dominant quadrants, with a width of approximately $1/Pe$. Additionally, sharp boundary layers are present in the convective quadrants along their boundaries that intersect with the outer boundaries at $x = 1$ and $y = 1$.

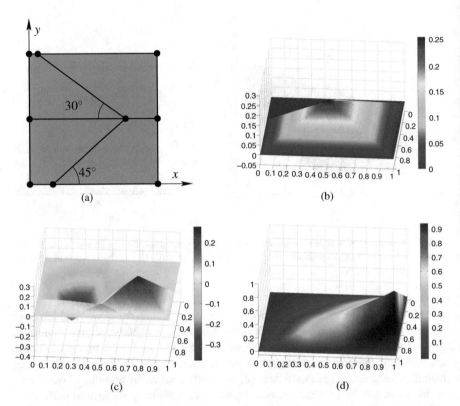

Fig. 6 AVS-FE results for homogeneous coefficients and unstructured meshes; $Pe = 400$, $p = 2$, and $\Delta p = 0$. (**a**) Initial unstructured mesh. (**b**) u^h for initial coarse mesh (27 dofs). (**c**) u^h after first refinement (75 dofs). (**d**) Converged u^h

For a graded regular mesh of only 4×4 elements (see Fig. 7b), $p = 2$, and $\Delta p = 0$, a contour plot of the of the distribution of the corresponding AVS-FE solution, u^h, throughout the unit square is depicted in Fig. 8a; whereas in Fig. 8b its distribution along the diagonal $y - x = 0$ is presented. Analogous to the results in Sect. 3.2, the numerical solution successfully captures the main features of the solution, i.e., the solution indeed vanishes in the diffusion dominant quadrants, strong convection is seen in the remaining regions, and the sharp internal and boundary layers are adequately captured. It is remarkable that with only 16 elements, and 243 dofs, the AVS-FE computation succeeds in resolving these features without any strong oscillations and without the need for any artificial stabilization. Again, using the same polynomial degree of approximation in solving the optimal test functions (17), does not appear to inhibit the corresponding AVS-FE computation to resolve the essential solution features.

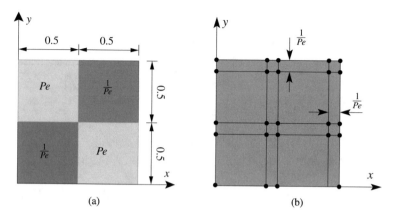

(a)   (b)

Fig. 7 Checker board problem. (**a**) Diffusion coefficient distribution. (**b**) Initial 4×4 graded mesh

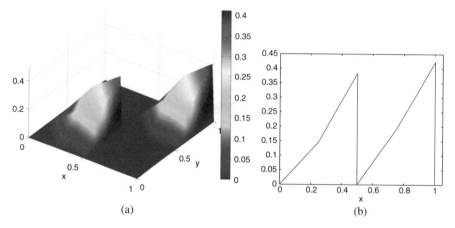

(a)   (b)

Fig. 8 AVS-FE results for heterogeneous diffusion, 4×4 graded mesh (243 dofs), $p = 2$, $\Delta p = 0$.
(**a**) Distribution of u^h throughout $(0, 1) \times (0, 1)$. (**b**) u^h along the diagonal $y - x = 0$

Subsequently applying uniform h-refinements results in a sequence of numerical solutions, in which the resolution of the internal and boundary continuously improves without inducing any oscillations. Results for the fourth h-refinement are given in Fig. 9.

3.4 Non-constant Convection

Lastly, let us now consider a case in which the convection coefficient, $\mathbf{b}$, rather than the diffusion coefficient, is non-constant, i.e., $\mathbf{b} = \{\frac{1}{2}(1 - 2x), 0\}^T$, i.e., we only have convection in the x-direction, which varies linearly throughout Ω and

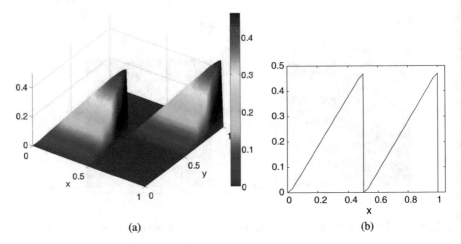

(a)   (b)

Fig. 9 AVS-FE results for heterogeneous D, a refined graded mesh (12,675 dofs), $p = 2$, $\Delta p = 0$. (**a**) Distribution of u^h throughout $(0, 1) \times (0, 1)$. (**b**) u^h along the diagonal $y - x = 0$

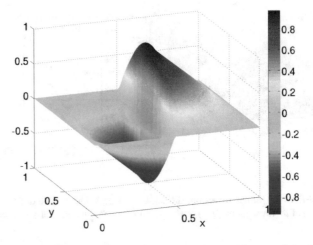

Fig. 10 AVS-FE results for non-constant convection, a refined uniform mesh ($\sim 790,000$ dofs), $Pe = 10^9$, $p = 1$, and $\Delta p = 0$

vanishes along the middle line segment $x = \frac{1}{2}$. By choosing the Peclet number at an extremely high level, $Pe = 10^9$, we ensure that convection is heavily dominant away from the line segment $x = \frac{1}{2}$. Next, the source function is chosen to be:

$$f(x, y) = \frac{4x - 2}{Pe} + y(1 - y^2)(8x - 4).$$

Under these conditions, the solution exhibits a sharp internal layer along the middle line segment $x = \frac{1}{2}$, with a width of the order of $1/Pe$, i.e., 10^{-9}. Away from the internal layer, or 'shock', the solution is convective. In Fig. 10, we present the distribution of u^h for the case in which we started with a 2×2 uniform mesh, $p = 1$, and $\Delta p = 0$, and subsequently applied seven uniform h-refinements, arriving at a mesh with approximately 790k dofs. The numerical solutions do not show any oscillatory behavior close to the shock and continuously provide sharper resolutions of the internal layer as the mesh is refined, while converging to a bounded amplitude. It is striking that the results are automatically stable for a staggering value of a billion for the Peclet number.

4 Concluding Remarks

We constructed a variationally stable finite element discretization. This hybrid continuous-discontinuous Petrov-Galerkin method uses solution (trial) functions that are piecewise continuous over the whole domain. That is, these functions correspond to standard finite element partitions. We then use as weight (test) functions a piecewise discontinuous basis. This broken test space allows us to extend the DPG approach to compute optimal test functions automatically and with these to establish numerically stable FE approximations. Important features of this discretization are as follows. The support of each discontinuous test function is identical to its corresponding continuous trial function. The local test-function contribution computed locally on an element by element fashion(i.e. decoupled). This has a linear cost with respect to the problem size and can be thought as an alternative assembly process, where not only inner products, but the functions themselves need to be computed on the fly. Additionally, our experience indicates that the computation of the optimal test functions is achieved with sufficient accuracy by using the same polynomial order of approximation, p, as that used in the trial function. As in every other DPG formulation, the resulting algebraic system is symmetric and positive definite, allowing us to use simple iterative strategies to compute the numerical solution. Our future work will include developing variationally stable discretizations based on isogeometric analysis (IGA) both in Galerkin as well as in collocation form. Our preliminary results indicate that these methods are very promising by delivering robust and efficient discretizations exploiting the smoothness of IGA basis functions to deliver intrinsically stable discretizations that are symmetric and positive definite for arbitrary partial differential equations.

We are confident in the impact this methodology will have. Thus, we are partnering with the development communities around FireDrake, Fenics-HPC and Camellia as well as our traditional partners who develop PetIGA and PetIGA-MF to release portable parallel implementations of this methodology.

Acknowledgements The support of the authors, Albert Romkes and Eirik Valseth, by the NSF CBET Program, under NSF Grant titled *Sustainable System for Mineral Beneficiation*, NSF Grant No. 1805550, is gratefully acknowledged. This publication was also made possible in part by the CSIRO Professorial Chair in Computational Geoscience at Curtin University and the Deep Earth Imaging Enterprise Future Science Platforms of the Commonwealth Scientific Industrial Research

Organisation, CSIRO, of Australia. Additional support was provided by the European Union's Horizon 2020 Research and Innovation Program of the Marie Skłodowska-Curie grant agreement No. 777778, and the Mega-grant of the Russian Federation Government (N 14.Y26.31.0013). Additional, support was provided at Curtin University by The Institute for Geoscience Research (TIGeR) and by the Curtin Institute for Computation. The J. Tinsley Oden Faculty Fellowship Research Program at the Institute for Computational Engineering and Sciences (ICES) of the University of Texas at Austin has partially supported the visits of author, Victor M. Calo, to ICES where he worked closely with Professor Leszek F. Demkowicz.

References

1. Ahmadi, A., Surana, K.S., Maduri, R.K., Romkes, A.: Higher order global differentiability local approximations for 2D distorted quadrilateral elements. Int. J. Comput. Methods Eng. Sci. Mech. **10**, 1–19 (2009)
2. Ahmadi, A., Surana, K.S., Maduri, R.K., Romkes, A.: Higher order global differentiability local approximations for 2D distorted triangular elements. Int. J. Comput. Methods Eng. Sci. Mech. **10**, 20–26 (2009)
3. Bazilevs, Y., Akkerman, I.: Large eddy simulation of turbulent Taylor–Couette flow using isogeometric analysis and the residual-based variational multiscale method. J. Comput. Phys. **229**(9), 3402–3414 (2010)
4. Bazilevs, Y., Beirao da Veiga, L., Cottrell, J.A., Hughes, T.J., Sangalli, G.: Isogeometric analysis: approximation, stability and error estimates for h-refined meshes. Math. Models Methods Appl. Sci. **16**(07), 1031–1090 (2006)
5. Bazilevs, Y., Calo, V.M., Cottrell, J.A., Hughes, T.J.R., Reali, A., Scovazzi, G.: Variational multiscale residual-based turbulence modeling for large eddy simulation of incompressible flows. Comput. Methods Appl. Mech. Eng. **197**(1–4), 173–201 (2007)
6. Bazilevs, Y., Michler, C., Calo, V.M., Hughes, T.J.R.: Weak Dirichlet boundary conditions for wall-bounded turbulent flows. Comput. Methods Appl. Mech. Eng. **196**(49–52), 4853–4862 (2007)
7. Bazilevs, Y., Calo, V.M., Hughes, T.J.R., Zhang, Y.: Isogeometric fluid-structure interaction: theory, algorithms, and computations. Comput. Mech. **43**(1), 3–37 (2008)
8. Bazilevs, Y., Gohean, J., Hughes, T., Moser, R., Zhang, Y.: Patient-specific isogeometric fluid–structure interaction analysis of thoracic aortic blood flow due to implantation of the Jarvik 2000 left ventricular assist device. Comput. Methods Appl. Mech. Eng. **198**(45–46), 3534–3550 (2009)
9. Bazilevs, Y., Calo, V.M., Cottrell, J.A., Evans, J.A., Hughes, T.J.R., Lipton, S., Scott, M.A., Sederberg, T.W.: Isogeometric analysis using T-splines. Comput. Methods Appl. Mech. Eng. **199**(5–8), 229–263 (2010)
10. Bazilevs, Y., Michler, C., Calo, V.M., Hughes, T.J.R.: Isogeometric variational multiscale modeling of wall-bounded turbulent flows with weakly enforced boundary conditions on unstretched meshes. Comput. Methods Appl. Mech. Eng. **199**(13–16), 780–790 (2010)
11. Bazilevs, Y., Hsu, M.C., Akkerman, I., Wright, S., Takizawa, K., Henicke, B., Spielman, T., Tezduyar, T.: 3D simulation of wind turbine rotors at full scale. Part I: Geometry modeling and aerodynamics. Int. J. Numer. Methods Fluids **65**(1–3), 207–235 (2011)
12. Behr, M.A., Franca, L.P., Tezduyar, T.E.: Stabilized finite element methods for the velocity-pressure-stress formulation of incompressible flows. Comput. Methods Appl. Mech. Eng. **104**, 31–38 (1993)
13. Bochev, P.B., Gunzburger, M.D.: Least-Squares Finite Element Methods, vol. 166. Springer Science & Business Media, New York (2009)
14. Brezzi, F., Fortin, M.: Mixed and Hybrid Finite Element Methods, vol. 15. Springer, New York (1991)

15. Brooks, A.N., Hughes, T.J.R.: Streamline upwind / Petrov-Galerkin formulations for convection dominated flows with particular emphasis on the incompressible Navier-Stokes equations. Comput. Methods Appl. Mech. Eng. **32**, 199–259 (1982)

16. Bubnov, I.G.: Reports on the works of professor Timoshenko which were awarded the Zhuranskyi Prize. Collection of Works of the Engineers Institute of Putey Soobshcheniya Imperatora Alexandra I **81**, 1–40 (1913; in Russian)

17. Calo, V.M., Brasher, N.F., Bazilevs, Y., Hughes, T.J.R.: Multiphysics model for blood flow and drug transport with application to patient-specific coronary artery flow. Comput. Mech. **43**(1), 161–177 (2008)

18. Calo, V.M., Chung, E.T., Efendiev, Y., Leung, W.T.: Multiscale stabilization for convection-dominated diffusion in heterogeneous media. Comput. Methods Appl. Mech. Eng. **304**, 359–377 (2016)

19. Calo, V.M., Romkes, A., Valseth, E., Kirby, R.C.: Automatic variationally stable analysis for FE computations: theory and the DPG framework. Comput. Methods Appl. Mech. Eng. (2020, in preparation)

20. Carstensen, C., Demkowicz, L., Gopalakrishnan, J.: A posteriori error control for DPG methods. SIAM J. Numer. Anal. **52**(3), 1335–1353 (2014)

21. Chang, K., Hughes, T.J.R., Calo, V.M.: Isogeometric variational multiscale large-eddy simulation of fully-developed turbulent flow over a wavy wall. Comput. Fluids **68**, 94–104 (2012)

22. Codina, R.: Comparison of some finite element methods for solving the diffusion-convection-reaction equation. Comput. Methods Appl. Mech. Eng. **156**, 185–210 (1998)

23. Codina, R.: On stabilized finite element methods for linear systems of convection-diffusion-reaction equations. Comput. Methods Appl. Mech. Eng. **188**, 61–82 (2000)

24. Collier, N., Dalcin, L., Calo, V.M.: On the computational efficiency of isogeometric methods for smooth elliptic problems using direct solvers. Int. J. Numer. Methods Eng. **100**(8), 620–632 (2014)

25. Côrtes, A.M.A., Coutinho, A.L.G.A., Dalcin, L., Calo, V.M.: Performance evaluation of block-diagonal preconditioners for the divergence-conforming B-spline discretization of the Stokes system. J. Comput. Sci. **11**, 123–136 (2015)

26. Cottrell, J.A., Reali, A., Bazilevs, Y., Hughes, T.J.: Isogeometric analysis of structural vibrations. Comput. Methods Appl. Mech. Eng. **195**(41–43), 5257–5296 (2006)

27. Cottrell, J., Hughes, T., Reali, A.: Studies of refinement and continuity in isogeometric structural analysis. Comput. Methods Appl. Mech. Eng. **196**(41–44), 4160–4183 (2007)

28. Cottrell, J.A., Hughes, T.J., Bazilevs, Y.: Isogeometric Analysis: Toward Integration of CAD and FEA. Wiley, New York (2009)

29. Demkowicz, L., Gopalakrishnan, J.: A class of discontinuous Petrov-Galerkin methods. Part I: the transport equation. Comput. Methods Appl. Mech. Eng. **199**(23), 1558–1572 (2010)

30. Demkowicz, L., Gopalakrishnan, J.: Analysis of the DPG method for the Poisson equation. SIAM J. Numer. Anal. **49**(5), 1788–1809 (2011)

31. Demkowicz, L., Gopalakrishnan, J.: A class of discontinuous Petrov-Galerkin methods. II. Optimal test functions. Numer. Methods Partial Differ. Equ. **27**(1), 70–105 (2011)

32. Demkowicz, L., Gopalakrishnan, J.: A class of discontinuous Petrov-Galerkin methods. Part III: adaptivity. Appl. Numer. Math. **62**(4), 396–427 (2012)

33. Demkowicz, L., Gopalakrishnan, J.: Discontinuous Petrov-Galerkin (DPG) method. Tech. rep., The Institute for Computational Engineering and Sciences, The University of Texas at Austin (2015)

34. Duddu, R., Lavier, L.L., Hughes, T.J.R., Calo, V.M.: A finite strain Eulerian formulation for compressible and nearly incompressible hyperelasticity using high-order B-spline finite elements. Int. J. Numer. Methods Eng. **89**(6), 762–785 (2012)

35. Elguedj, T., Bazilevs, Y., Calo, V.M., Hughes, T.J.R.: B and F projection methods for nearly incompressible linear and non-linear elasticity and plasticity using higher-order NURBS elements. Comput. Methods Appl. Mech. Eng. **197**(33–40), 2732–2762 (2008)

36. Espath, L.F.R., Sarmiento, A.F., Vignal, P., Varga, B.O.N., Cortes, A.M.A., Dalcin, L., Calo, V.M.: Energy exchange analysis in droplet dynamics via the Navier–Stokes–Cahn–Hilliard model. J. Fluid Mech. **797**, 389–430 (2016)

37. Franca, L.P., Frey, S.L.: Stabilized finite element methods: II. The incompressible Navier-Stokes equations. Comput. Methods Appl. Mech. Eng. **99**, 209–233 (1992)
38. Franca, L.P., Hughes, T.J.R.: Convergence analyses of Galerkin/leastsquares methods for symmetric advective-diffusive forms of the Stokes and incompressible Navier-Stokes equations. Comput. Methods Appl. Mech. Eng. **105**, 285–298 (1993)
39. Franca, L.P., Frey, S.L., Hughes, T.J.R.: Stabilized finite element methods: I. Application to the advective-diffusive model. Comput. Methods Appl. Mech. Eng. **95**, 253–276 (1992)
40. Girault, V., Raviart, P.A.: Finite Element Methods for Navier-Stokes Equations; Theory And Algorithms. Springer Series in Computational Mathematics, vol. 5. Springer, New York(1986)
41. Gómez, H., Calo, V.M., Bazilevs, Y., Hughes, T.J.R.: Isogeometric analysis of the Cahn–Hilliard phase-field model. Comput. Methods Appl. Mech. Eng. **197**(49–50), 4333–4352 (2008)
42. Hauke, G., Hughes, T.J.R.: A unified approach to compressible and incompressible flows. Comput. Methods Appl. Mech. Eng. **113**, 389–396 (1994)
43. Hsu, M.C., Bazilevs, Y.: Fluid–structure interaction modeling of wind turbines: simulating the full machine. Comput. Mech. **50**(6), 821–833 (2012)
44. Hughes, T.J.R.: Multiscale phenomena: Green's functions, the Dirichlet-to-Neumann formulation, subgrid-scale models, bubbles, and the origins of stabilized methods. Comput. Methods Appl. Mech. Eng. **127**, 387–401 (1995)
45. Hughes, T.J.R.: The finite element method: linear static and dynamic finite element analysis. Courier Corporation, North Chelmsford (2000)
46. Hughes, T.J.R., Mallet, M.: A new finite element formulation for fluid dynamics: III. The generalized streamline operator for multidimensional advective-diffusive systems. Comput. Methods Appl. Mech. Eng. **58**, 305–328 (1986)
47. Hughes, T.J.R., Stewart, J.R.: A space-time formulation for multiscale phenomena. J. Comput. Appl. Math. **74**, 217–229 (1996)
48. Hughes, T.J.R., Franca, L.P., Balestra, M.: A new finite element formulation for fluid dynamics: V. A stable Petrov-Galerkin formulation of the Stokes problem accommodating equal-order interpolations. Comput. Methods Appl. Mech. Eng. **59**, 85–99 (1986)
49. Hughes, T.J.R., Franca, L.P., Mallet, M.: A new finite element formulation for fluid dynamics: VI. Convergence analysis of the generalized SUPG formulation for linear time-dependent multidimensional advective-diffusive systems. Comput. Methods Appl. Mech. Eng. **63**, 97–112 (1987)
50. Hughes, T.J.R., Franca, L.P., Hulbert, G.M.: A new finite element formulation for fluid dynamics: VIII. The Galerkin/least-squares method for advective-diffusive equations. Comput. Methods Appl. Mech. Eng. **73**, 173–189 (1989)
51. Hughes, T.J.R., Feijóo, G., Mazzei, L., Quincy, J.B.: The variational multiscale method a paradigm for computational mechanics. Comput. Methods Appl. Mech. Eng. **166**, 2–24 (1998)
52. Hughes, T.J.R., Scovazzi, G., Franca, L.P.: Multiscale and stabilized methods. In: Encyclopedia of Computational Mechanics. Wiley, New York (2004)
53. Hughes, T.J., Cottrell, J.A., Bazilevs, Y.: Isogeometric analysis: CAD, finite elements, NURBS, exact geometry and mesh refinement. Comput. Methods Appl. Mech. Eng. **194**(39–41), 4135–4195 (2005)
54. Jansen, K.E., Collis, S.S., Whiting, C., Shakib, F.: A better consistency for low-order stabilized finite element methods. Comput. Methods Appl. Mech. Eng. **174**, 153–170 (1999)
55. Juanes, R., Patzek, T.W.: Multiscale-stabilized solutions to one-dimensional systems of conservation laws. Comput. Methods Appl. Mech. Eng. **194**, 25–26:2781–2805 (2005)
56. Niemi, A.H., Collier, N.O., Calo, V.M.: Automatically stable discontinuous Petrov-Galerkin methods for stationary transport problems: Quasi-optimal test space norm. Comput. Math. Appl. **66**(10), 2096–2113 (2013). ICNC-FSKD 2012
57. Oden, J.T., Reddy, J.N.: An Introduction to the Mathematical Theory of Finite Elements. Courier Corporation, North Chelmsford (2012)
58. Petrov, G.: Application of the method of Galerkin to a problem involving the stationary flow of a viscous fluid. Prikl. Matem. Mekh **4**(3), 3–12 (1940)

59. Puzyrev, V., Deng, Q., Calo, V.M.: Dispersion-optimized quadrature rules for isogeometric analysis: modified inner products, their dispersion properties, and optimally blended schemes. Comput. Methods Appl. Mech. Eng. **320**, 421–443 (2017)
60. Raviart, P.A., Thomas, J.M.: A Mixed Finite Element Method For Second Order Elliptic Problems. Springer, New York (1977)
61. Reddy, J.N.: An Introduction to the Finite Element Method, vol. 2. McGraw-Hill, New York (1993)
62. Shakib, F., Hughes, T.J.R.: A new finite element formulation for computational fluid dynamics: IX. Fourier analysis of space-time Galerkin/least-squares algorithms. Comput. Methods Appl. Mech. Eng. **87**, 35–58 (1991)
63. Shakib, F., Hughes, T.J.R., Johan, Z.: A new finite element formulation for computational fluid dynamics: X. The compressible Euler and Navier-Stokes equations. Comput. Methods Appl. Mech. Eng. **89**, 141–219 (1991)
64. Surana, K.S., Ahmadi, A.R., Reddy, J.N.: The k-version of finite element method for non-self-adjoint operators in BVP. Int. J. Comput. Eng. Sci. **4**(4), 737–812 (2003)
65. Surana, K.S., Reddy, J., Romkes, A.: h, p, k Mathematical and computational finite element framework for boundary value and initial value problems. Acta Mech. Solida Sin. **23**, 12–25 (2010)

Practical Contributions on the Fictitious Domain Method for a Fluid–Structure Interaction Problem

Guillaume Delay and Michel Fournié

Abstract The present study deals with the numerical simulation of a fluid–structure interaction problem. The fluid is represented by the incompressible Navier–Stokes equations and the structure is described by an ODE depending on two degrees of freedom. A recent fictitious domain method on a fixed mesh is considered. For that choice, we provide several tricks to meet the difficulties arising from the fluid–structure interaction. All developed tools can be applied to very general geometries and deformations of the structure. Finally, numerical simulations are conducted in a realistic aeronautics configuration.

1 Introduction

The interaction between a fluid and a deformable structure, which appears in a very large field of industrial problems, has recently received increasing attention from the scientific community. Numerical simulations of such problems remain a challenging task.

A first difficulty comes from the fact that the whole system is the assembling of two subsystems of different natures. We can then either consider the full system as a whole part and write a variational formulation that comprises the fluid and the structure equations or we can use appropriate solvers for each of these subsystems. The first approach is called monolithic approach [22]. In the sequel, we considered the second approach called partitioned approach [12]. We solve separately the

G. Delay (✉)
Institut de Mathématiques de Toulouse, Université Paul Sabatier III & CNRS, Toulouse Cedex, France
e-mail: guillaume.delay@math.univ-toulouse.fr; guillaume.delay@enpc.fr

M. Fournié
Institut de Mathématiques de Toulouse, Toulouse, France
e-mail: michel.fournie@math.univ-toulouse.fr

© Springer Nature Switzerland AG 2020
G. R. Barrenechea, J. Mackenzie (eds.), *Boundary and Interior Layers, Computational and Asymptotic Methods BAIL 2018*, Lecture Notes in Computational Science and Engineering 135,
https://doi.org/10.1007/978-3-030-41800-7_3

structure and the fluid equations at each time step. A possibility would have been, inside a time step, to iterate between the structure and the fluid solvers to reach an equilibrium between those states. This would have given a strong coupling at an increased computational cost. However, for small enough time steps, a weak coupling is sufficient [10, 15].

Another challenge is to handle the fluid domain that changes over time. This is indeed a difficulty as most of the numerical methods use fitted meshes, i.e. meshes such that the physical boundary is composed of cell faces. Hence, if we want to use such a fitted mesh, we need to change it at every time step to fit the moving boundary. Some algorithms such as the Arbitrary Lagrangian Eulerian (ALE) approach make it possible with small computation costs for small deformations of the fluid domain [13]. However, for large deformations of the domain, a complete remeshing is needed, which is highly expensive.

In order to avoid any change of the mesh during the computation, we consider a fictitious domain approach, i.e. the boundary of the physical domain can arbitrary cut the mesh. Several methods are included in this framework, for instance the immersed boundary method [8, 19, 20] or the penalization method [2, 16, 18].

This kind of method has already been used for fluid–structure interaction in [1, 14, 17]. A recent review can be found in [4].

In the present work, we use a XFEM type method that can also be found under the name of cutFEM. This method has been developed in the context of crack propagation in fatigue mechanics [9]. The main characteristic of this method is the use of a level-set function to locate an interface in the domain and an enrichment of the finite element basis with functions depending on the position of the interface. In the fluid–structure context, the interface will be the boundary between the fluid and the structure. We adapt and investigate a recent method of that type that is theoretically analyzed for the Stokes problem only in [11]. We focus our attention on determinant implementation stages that are non-standard.

We present the considered model in Sect. 2, the partitioned process in Sect. 3, the XFEM in Sect. 4 and the handling of the moving domain in Sect. 5. Numerical simulations are given in Sect. 6. Finally, Sect. 7 is a summary of the article.

2   The Problem

The system we are interested in corresponds to the 2D interaction of a fluid and a deformable structure that can be assimilated to a steering gear depending on two parameters that we denote $\theta = (\theta_1, \theta_2) \in \mathbb{R}^2$.

The whole system is enclosed in a box representing a wind tunnel Ω. The boundary of that box can be decomposed as $\partial\Omega = \Gamma_i \cup \Gamma_w \cup \Gamma_N$, where Γ_i, Γ_w and Γ_N are parts of the boundary where inflow Dirichlet boundary conditions, homogeneous Dirichlet boundary conditions and Neumann-like boundary conditions are respectively imposed (see Fig. 1).

Fig. 1 The domain configuration: the whole domain (a wind tunnel) is decomposed as $\Omega = S(\theta) \cup \mathscr{F}(\theta)$ (the structure and the fluid)

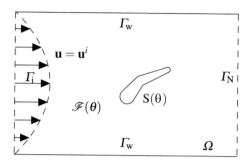

We consider the incompressible Navier–Stokes equations for the fluid and a virtual work principle for the structure, see [6, 7]. The structure and fluid domains depend on the parameters θ, we denote them respectively $S(\theta)$ and $\mathscr{F}(\theta)$. We denote $Q_\theta = \cup_{t \in (0,T)} \{t\} \times \mathscr{F}(\theta(t))$, $\Sigma_\theta = \cup_{t \in (0,T)} \{t\} \times \partial S(\theta(t))$, $\Sigma_i = (0,T) \times \Gamma_i$, $\Sigma_w = (0,T) \times \Gamma_w$ and $\Sigma_N = (0,T) \times \Gamma_N$.

The equations considered for the fluid are the following ones

$$\frac{\partial \mathbf{u}}{\partial t} + (\mathbf{u} \cdot \nabla)\mathbf{u} + \nu \Delta \mathbf{u} - \nabla p = \mathbf{f}_{\mathscr{F}} \quad \text{in} \quad Q_\theta, \tag{1}$$

$$\operatorname{div} \mathbf{u} = 0 \quad \text{in} \quad Q_\theta, \tag{2}$$

$$\mathbf{u} = \mathbf{u}^i \quad \text{on} \quad \Sigma_i, \tag{3}$$

$$\mathbf{u} = 0 \quad \text{on} \quad \Sigma_w, \tag{4}$$

$$\sigma_F(\mathbf{u}, p)\mathbf{n} = 0 \quad \text{on} \quad \Sigma_N, \tag{5}$$

$$\mathbf{u} = \mathbf{u}_S \quad \text{on} \quad \Sigma_\theta, \tag{6}$$

this system is completed by the following equation for the structure

$$M_\theta \ddot{\theta} = M_I(\theta, \dot{\theta}) + M_A(\theta, -\sigma_F(\mathbf{u}, p)\mathbf{n}_\theta) + \mathbf{f}_S \quad \text{on} \quad (0, T), \tag{7}$$

and we consider the initial conditions

$$\mathbf{u}(0) = \mathbf{u}_0 \quad \text{in} \quad \mathscr{F}(\theta_0), \tag{8}$$

$$\theta(0) = \theta_0 = (\theta_{1,0}, \theta_{2,0}), \qquad \dot{\theta}(0) = \omega_0 = (\omega_{1,0}, \omega_{2,0}). \tag{9}$$

In the previous equations, we denoted $\mathbf{u}$ the velocity field of the fluid, p the pressure field, ν the viscosity of the fluid, $\mathbf{f}_{\mathscr{F}}$ a source term acting as a force per unit volume, $\mathbf{u}^i$ a Dirichlet datum on the inflow boundary Γ_i, $\sigma_F(\mathbf{u}, p) = \nu(\nabla \mathbf{u} + (\nabla \mathbf{u})^T) - pI$ the stress tensor of the fluid, $\mathbf{n}$ the unit outward normal to Ω, $\mathbf{u}_S$ the velocity field of the structure, $\mathbf{f}_S$ a source term on the structure equation. These equations have to be completed with suitable expressions for M_θ, M_I, M_A and $\mathbf{u}_S$.

The matrix $M_\theta \in \mathbb{R}^{2 \times 2}$, which is invertible, and the vectors M_I, $M_A \in \mathbb{R}^2$ depend respectively on θ; on θ, $\dot{\theta}$; on θ and the force exerted by the fluid on the structure, i.e. $\sigma_F(\mathbf{u}, p)\mathbf{n}_\theta$, where $\mathbf{n}_\theta$ is the unit normal to $\partial S(\theta)$ pointing inward the structure. For a more general presentation of the numerical scheme, we do not specify those terms. The numerical simulations will be led with the expressions given in Sect. 6.1. For those terms, the well-posedness of Eqs. (1)–(9) has been proven in [7]. The reader can find there further information.

3   The Time-Marching Process

We describe in this section, the time-marching process which is of partitioned type. This means that we solve the structure and the fluid systems separately and one after the other.

The fluid step at time t^{n+1} will be obtained by discretizing with a semi-implicit scheme the variational formulation: Find $(\mathbf{u}^{n+1}, p^{n+1}, \boldsymbol{\lambda}^{n+1}) \in \big(\mathbf{H}^1(\mathscr{F}(\theta^{n+1})) \times (L^2(\mathscr{F}(\theta^{n+1}))) \times (\mathbf{H}^{-1/2}(\partial S(\theta^{n+1})) \times \mathbf{H}^{-1/2}(\Gamma_i \cup \Gamma_w))$ such that

$$
\begin{cases}
\displaystyle\int_{\mathscr{F}(\theta^{n+1})} \frac{\mathbf{u}^{n+1} - \mathbf{u}^n}{\Delta t} \cdot \mathbf{v} + (\mathbf{u}^n \cdot \nabla)\mathbf{u}^{n+1} \cdot \mathbf{v} + \frac{\nu}{2}(\nabla \mathbf{u}^{n+1} + (\nabla \mathbf{u}^{n+1})^T) : (\nabla \mathbf{v} + \nabla \mathbf{v}^T) \\
\qquad\qquad - p^{n+1}\mathrm{div}\,\mathbf{v} + \displaystyle\int_{\Gamma_i \cup \Gamma_w \cup \partial S(\theta^{n+1})} \boldsymbol{\lambda}^{n+1} \cdot \mathbf{v} = \int_{\mathscr{F}(\theta^{n+1})} \mathbf{f}_{\mathscr{F}}(t^{n+1}) \cdot \mathbf{v}, \\[2mm]
\displaystyle\int_{\mathscr{F}(\theta^{n+1})} q\,\mathrm{div}\,\mathbf{u}^{n+1} = 0, \\[2mm]
\displaystyle\int_{\Gamma_i \cup \Gamma_w \cup \partial S(\theta^{n+1})} \mathbf{u}^{n+1} \cdot \boldsymbol{\mu} = \int_{\Gamma_i} \mathbf{u}^i(t^{n+1}) \cdot \boldsymbol{\mu} + \int_{\partial S(\theta^{n+1})} \mathbf{u}_S(t^{n+1}) \cdot \boldsymbol{\mu},
\end{cases}
\tag{10}
$$

$\forall(\mathbf{v}, q, \boldsymbol{\mu}) \in \big(\mathbf{H}^1(\mathscr{F}(\theta^{n+1})) \times (L^2(\mathscr{F}(\theta^{n+1}))) \times (\mathbf{H}^{-1/2}(\partial S(\theta^{n+1})) \times \mathbf{H}^{-1/2}(\Gamma_i \cup \Gamma_w))$.

A Euler scheme semi-discretizes the partial derivative w.r.t. time. It requires $\mathbf{u}^n$ to be integrated over the domain $\mathscr{F}(\theta^{n+1})$ (while it is defined on $\mathscr{F}(\theta^n)$ only). We then need an extension procedure that we expose in Sect. 4.4. Note that the Dirichlet boundary conditions have been imposed in a weak way by the use of Lagrange multipliers $\boldsymbol{\lambda}^{n+1} = (\boldsymbol{\lambda}^{n+1}_{\partial S(\theta^{n+1})}, \boldsymbol{\lambda}^{n+1}_{\Gamma_i \cup \Gamma_w})$. This induces the computation of the additional variable $\boldsymbol{\lambda}^{n+1}_{\partial S(\theta^{n+1})} = -\sigma_F(\mathbf{u}^{n+1}, p^{n+1})\mathbf{n}_\theta$ which represents the fluid forces acting on the structure at time t^{n+1} and will be approximated by $\boldsymbol{\lambda}^{n+1}_h$. The index $\partial S(\theta^{n+1})$ is dropped in the sequel. The structure evolution is computed by the following finite difference method:

$$
\begin{cases}
\theta^{n+1} = 2\theta^n - \theta^{n-1} + (\Delta t_{n+1})^2 M_\theta^{-1}(M_A(\theta^n, \boldsymbol{\lambda}^n_h) + M_I(\theta^n, \omega^n)), & (11) \\[2mm]
\omega^{n+1} = \omega^n + \Delta t_{n+1} M_\theta^{-1}(M_A(\theta^n, \boldsymbol{\lambda}^n_h) + M_I(\theta^n, \omega^n)), & (12)
\end{cases}
$$

where we have denoted $\omega^n = (\omega_1^n, \omega_2^n)$ the approximation of $\dot{\theta}(t^n)$. Note that we use λ_h^n in the structure equations.

At each time step, we use the procedure described in Algorithm 1. The main difficulties are to compute the fluid step and to adapt the fluid domain, they are tackled respectively in Sects. 4 and 5.

Algorithm 1 The splitting scheme

Require: $(\mathbf{u}_h^n, p_h^n, \lambda_h^n, \theta^n, \theta^{n-1}, \omega^n, \Delta t_{n+1})$.
1 Compute $(\theta^{n+1}, \omega^{n+1})$ with the structure step (11)–(12).
2 Update $\mathscr{F}(\theta^{n+1})$.
3 Compute $(\mathbf{u}_h^{n+1}, p_h^{n+1}, \lambda_h^{n+1})$ with the fluid step.
4 Compute the next time step Δt_{n+2} with the CFL condition (15) (presented in Sect. 4.3).

---

4 The Discretization Method for the Fluid Equations

In this section we detail the way we approximate the evolution of the fluid state.

4.1 The Finite Element Method

We define a background mesh $\mathbb{T}_h$ covering the whole domain $\Omega = \mathscr{F}(\theta) \cup S(\theta)$. This mesh will not be modified during the simulation. The mesh cells can then be cut arbitrarily by the interface $\partial S(\theta)$. This means that the cells are either entirely contained in the fluid or the structure domain, or are shared between those two domains. For each variable, we define a finite element method on the whole triangulation $\mathbb{T}_h$. We choose the use of Taylor–Hood elements, i.e. $\mathbb{P}_2$ elements for $\mathbf{u}_h$, $\mathbb{P}_1$ elements for p_h and for λ_h.

Then, the approximated functions are the trace of these polynomials on the physical domain, $\mathscr{F}(\theta)$ for $\mathbf{u}_h$ and p_h, $\partial S(\theta)$ for λ_h. We depict the functional basis associated to this method in Fig. 2.

Away from the interface, we have the usual finite elements associated to usual degrees of freedom (dof). Near the interface, the elements are cut. Note that some dof of the fluid are located in the structure domain, we call them *fictitious degrees of freedom* in the sequel. They correspond to actual dof of the method, however, their value do not have any physical meaning. This is why we need to handle those dof carefully (see Sect. 4.4). In the structure, away from the interface, all degrees of freedom are discarded.

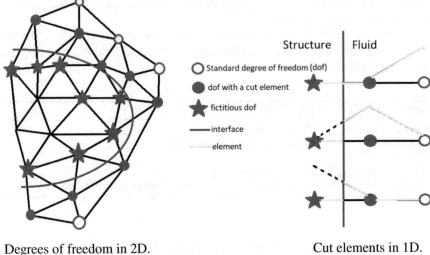

Degrees of freedom in 2D.   Cut elements in 1D.

Fig. 2 The degrees of freedom for $\mathbb{P}_1$ finite elements

4.2   The Stabilization Term

Classical XFEM applied to the Navier–Stokes equations does not ensure optimal convergence rate for the Lagrange multipliers. However, to simulate accurately the dynamics of the structure, we want λ_h to be a good approximation of λ.

In order to recover the optimal convergence rate, we add the following stabilization term to the variational formulation of the fluid problem (10)

$$- \gamma_0 h \int_{\partial S(\theta)} (\lambda_h + \sigma_F(\hat{\mathbf{u}}_h, \hat{p}_h)\mathbf{n}_\theta) \cdot (\mu_h + \sigma_F(\hat{\mathbf{v}}_h, \hat{q}_h)\mathbf{n}_\theta) \, \mathrm{d}x, \qquad (13)$$

with a mesh–independent constant $\gamma_0 > 0$. It corresponds to an augmented Lagrangian approach [5].

If we choose $\hat{\mathbf{u}}_h = \mathbf{u}_h$, $\hat{p}_h = p_h$, $\hat{\mathbf{v}}_h = \mathbf{v}_h$ and $\hat{q}_h = q_h$, then we get optimal convergence rate for λ_h if the mesh does not have any *bad triangle*. We call bad triangle a triangle that is cut by the interface and has only a tiny part of its surface in the fluid domain. For instance, in Fig. 3, T is a bad triangle.

The coefficients of the matrices are computed by an integration of the basis functions on the fluid domain. Then, such bad elements with only a tiny part in the fluid domain induce tiny coefficients in the matrices that are then ill-conditioned.

Since we consider a fixed mesh that can be arbitrarily cut by the interface, this stabilization is not enough. For such bad triangles, we take the velocity and pressure terms $\hat{\mathbf{u}}_h$, $\hat{p}_h$, $\hat{\mathbf{v}}_h$ and $\hat{q}_h$ in (13) as the extrapolation of the values of these variables in a good neighbor T' of T. Doing that way, we obtain optimal convergence for λ_h even if some triangles are badly cut in the mesh, see [11].

Fig. 3 A bad element T and a good neighbor T'. The fluid domain $\mathscr{F}(\theta)$ is on the right of the interface represented in blue

4.3 Algebraic Formulation

We denote with U, P and Λ the coordinate vectors of the velocity, the pressure and the multiplier into the $\mathbb{P}_2$, $\mathbb{P}_1$ and $\mathbb{P}_1$ Finite Element (FE) basis respectively. The FE method can be rewritten under the following matrix form:

$$\frac{1}{\Delta t}\begin{pmatrix} M_{\mathbf{uu}} & 0 & 0 \\ 0 & 0 & 0 \\ 0 & 0 & 0 \end{pmatrix}\begin{pmatrix} U^{n+1} - U^n \\ P^{n+1} - P^n \\ \Lambda^{n+1} - \Lambda^n \end{pmatrix} + \begin{pmatrix} A_{\mathbf{uu}} & A_{\mathbf{u}p} & A_{\mathbf{u}\lambda} \\ A_{\mathbf{u}p}^T & A_{pp} & A_{p\lambda} \\ A_{\mathbf{u}\lambda}^T & A_{p\lambda}^T & A_{\lambda\lambda} \end{pmatrix}\begin{pmatrix} U^{n+1} \\ P^{n+1} \\ \Lambda^{n+1} \end{pmatrix} = \begin{pmatrix} F^{n+1} \\ 0 \\ G^{n+1} \end{pmatrix}. \tag{14}$$

All the matrices are given by the discretization of the terms induced by (10) and (13). The matrices $A_{\mathbf{uu}}$, $A_{\mathbf{u}p}$, $A_{\mathbf{u}\lambda}$, $A_{\mathbf{u}p}^T$ and $A_{\mathbf{u}\lambda}^T$ correspond respectively, up to a perturbation induced by the stabilization term, to $\nu\Delta\mathbf{u}$, ∇p, λ, the incompressibility constraint and the Dirichlet boundary conditions, $M_{\mathbf{uu}}$ is the mass matrix. The vectors F^{n+1} and G^{n+1} correspond to the source term and to the Dirichlet data at t^{n+1}. The matrices A_{pp}, $A_{p\lambda}$, $A_{p\lambda}^T$ and $A_{\lambda\lambda}$ would have been null if there were no stabilization terms. Since the fluid domain changes over time, we discard and compute all the matrices at every time step. Some optimizations can be made to update only the coefficients linked to the cut cells.

In order to have a stable scheme, we use the following CFL condition:

$$\Delta t_{n+1} = \min\left(\text{cfl} \times \frac{h}{V_{\max}^n}, \Delta t_{\max}\right), \tag{15}$$

where $\text{cfl} \in (0, 1)$, $V_{\max}^n$ is the maximum velocity of the fluid in $\mathscr{F}(\theta^n)$ and $\Delta t_{\max}$ is the maximum time step allowed.

4.4 Initialization of the Fictitious Velocity Values

The time derivative has been discretized in (14) by a finite difference method. This is a natural way of discretizing it. However, it induces some difficulties. Indeed, at every time step, in order to compute $\mathbf{u}_h^{n+1}$, we need to provide the value of $\mathbf{u}_h^n$ at

Fig. 4 Initialization of the
velocity dof. In order to tackle
difficulties, all dof that are in
$S(\theta^n)$ (the fictitious and the
discarded ones) are given by
the velocity of the structure at
time t^n. Since we consider
adherence conditions between
the fluid and the structure, the
velocity of the structure near
the interface is close to the
one of the fluid. This justifies
the meaning of the value
given to the nodes in $S(\theta^n)$

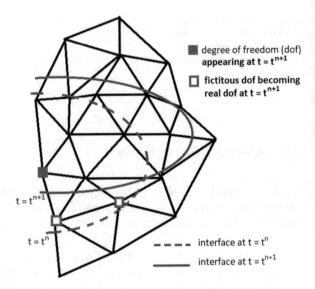

■ degree of freedom (dof)
appearing at $t = t^{n+1}$

☐ fictitous dof becoming
real dof at $t = t^{n+1}$

$t = t^{n+1}$

$t = t^n$

- - - - interface at $t = t^n$

———— interface at $t = t^{n+1}$

every dof considered for $\mathscr{F}(\theta^{n+1})$. The difficulty is due to the fact that such a value
is not available when that degree of freedom was discarded in $\mathscr{F}(\theta^n)$ (see Fig. 4).

Moreover, as exposed above, the values of the fictitious degrees of freedom have
to be carefully used since they do not have any physical meaning. Hence, a dof
that is fictitious in $\mathscr{F}(\theta^n)$ and becomes real in $\mathscr{F}(\theta^{n+1})$ cannot be straightforwardly
used. We then impose the velocity of the structure at t^n to all dof that are in $S(\theta^n)$.

5 The Evolution of the Fluid Domain

In this section, we describe the method used to adapt the fluid domain at every time
step. We first describe the integration over the cut cells, then we present the way we
locate the interface between the fluid and the structure.

5.1 Integration over the Cut Cells

All matrices in (14) are computed with an integration on the fluid domain only. We
then need to lead integrations on the mesh, in particular over cut cells. In order to
make these integrations possible, we divide every cut cell in sub-triangles that are
taken fitted to the interface. We then only have to integrate over the sub-triangles that
are in the fluid domain (see Fig. 5). This step is implemented with the *qhull* library
[3]. The method is classical, we recall it here for completeness of the presentation.

Fig. 5 Subdivision of the cut cells. Note that the operation consists in sub-dividing cut cells and not remeshing. Hence, the sub-cells are used only for integration and no new degrees of freedom are defined

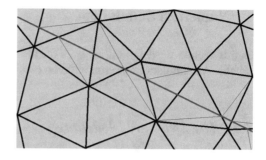

This process requires a level-set function to locate the interface (as the null level of this function). We describe in Sect. 5.2 the way that we compute this function.

5.2   Position of the Interface

We locate the interface as the zero of a level-set function. In the literature, most of the time, the level-set function is computed as the solution of a PDE representing its evolution [23]. This generates undesirable effects such as numerical diffusion.

In our setting, the structure position and then the interface depends only on the two parameters $\theta = (\theta_1, \theta_2)$. At each time step, we then compute a precise approximation of the level-set function associated to the parameters (θ_1^n, θ_2^n) (see Fig. 6).

We can tune the number of points considered to gain more precision. The drawback of this method is that it is time expensive. In order to mitigate the run time, we compute the level-set function only for the required nodes, i.e. the ones near the interface. A smart function chooses those nodes.

Fig. 6 Computation of the level-set function. The interface is represented by a list of points (computed analytically). The distance is computed with the polygon formed by those points

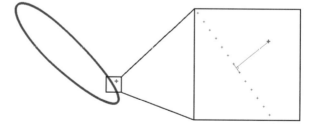

6 Numerical Simulations

In this section we detail our choices for the numerical simulations and show some results. All simulations are run with the *GetFEM++* library [21].

6.1 The Structure Modelling

To represent the deformations of the structure, we introduce a diffeomorphism $\mathbf{X}(\theta_1, \theta_2, .)$ that transforms a reference configuration corresponding to $\theta_{\mathrm{ref}} = (0, 0)$ into a given configuration corresponding to $\theta = (\theta_1, \theta_2)$ (see Fig. 7).

More precisely, we can define this diffeomorphism by

$$\mathbf{X}(\theta_1, \theta_2, \mathbf{x}) = \begin{pmatrix} \left(g_1(x_1) + x_2 \dfrac{N_1(x_1)}{|\mathbf{N}(x_1)|} \right) \cos(\theta_1) - \left(g_2(x_1) + x_2 \dfrac{N_2(x_1)}{|\mathbf{N}(x_1)|} \right) \sin(\theta_1) \\ \left(g_1(x_1) + x_2 \dfrac{N_1(x_1)}{|\mathbf{N}(x_1)|} \right) \sin(\theta_1) + \left(g_2(x_1) + x_2 \dfrac{N_2(x_1)}{|\mathbf{N}(x_1)|} \right) \cos(\theta_1) \end{pmatrix},$$

where $\mathbf{N}(x_1) = (N_1(x_1), N_2(x_1)) = (-g_1'(x_1), g_2'(x_1))$,

$$g_1(\ell) = \begin{cases} \ell & \text{if } \ell \le a, \\ a + (\ell - a)\cos(\theta_2/2) - f(\ell)\sin(\theta_2/2) & \text{if } \ell \in (a, b), \\ x_{B'} + (\ell - b)\cos(\theta_2) & \text{if } \ell \ge b, \end{cases}$$

$$g_2(\ell) = \begin{cases} 0 & \text{if } \ell \le a, \\ (\ell - b)\sin(\theta_2/2) + f(\ell)\cos(\theta_2/2) & \text{if } \ell \in (a, b), \\ y_{B'} + (\ell - b)\sin(\theta_2) & \text{if } \ell \ge b, \end{cases}$$

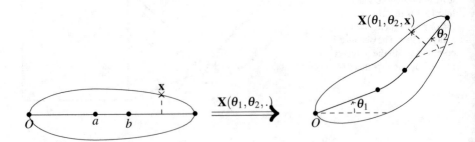

Fig. 7 The diffeomorphism $\mathbf{X}$

for $x_{B'} = b + (b-a)\cos(\theta_2/2)$, $y_{B'} = (b-a)\sin(\theta_2/2)$ and

$$f(\ell) = \frac{\tan(\theta_2/2)}{b-a}(\ell - (a+b)/2)^2 - \tan(\theta_2/2)\frac{b-a}{4}.$$

We use the values $a = 0.4$ and $b = 0.6$ in the sequel. To complete (1)–(9), we define

$$\mathbf{u}_S(t,\mathbf{x}) = \dot{\theta}_1\partial_{\theta_1}\mathbf{X}(\theta_1,\theta_2,\mathbf{x}) + \dot{\theta}_2\partial_{\theta_2}\mathbf{X}(\theta_1,\theta_2,\mathbf{x}), \quad (16)$$

$$(M_\theta)_{ij} = (\partial_{\theta_i}\mathbf{X}(\theta_1,\theta_2,\cdot), \partial_{\theta_j}\mathbf{X}(\theta_1,\theta_2,\cdot))_S, \quad (17)$$

$$(M_I)_j = -\left(\dot{\theta}_1^2\partial_{\theta_1}^2\mathbf{X}(\theta_1,\theta_2) + 2\dot{\theta}_1\dot{\theta}_2\partial_{\theta_1\theta_2}\mathbf{X}(\theta_1,\theta_2) + \dot{\theta}_2^2\partial_{\theta_2}^2\mathbf{X}(\theta_1,\theta_2), \partial_{\theta_j}\mathbf{X}(\theta_1,\theta_2)\right)_S, \quad (18)$$

$$(M_A(\theta,\mathbf{f}))_j = \int_{\partial S(\theta)} \mathbf{f}\cdot\partial_{\theta_j}\mathbf{X}(\theta_1,\theta_2,\mathbf{Y}(\theta_1,\theta_2,\mathbf{x})), \quad (19)$$

where $(\mathbf{f},\mathbf{g})_S = \rho\int_{\partial S(0)}\mathbf{f}\cdot\mathbf{g}$ and $\mathbf{Y}(\theta_1,\theta_2,\cdot)$ is the inverse diffeomorphism of $\mathbf{X}(\theta_1,\theta_2,\cdot)$.

We have proven well-posedness of the problem (1)–(9) with (16)–(19) in [7]. The reader will find there further information about this model.

6.2 Numerical Results

The whole domain $\Omega = (-1.0, 8.0) \times (y_{min}, y_{max})$ with $y_{min} = -2.5$ and $y_{max} = 2.1$ is discretized by a triangular mesh of 35,731 cells. It is locally refined near its boundary, near the corners, in the zone where lies the structure and in the wake behind the structure. We do not consider any forces in the fluid $\mathbf{f}_{\mathscr{F}} = 0$. The inflow condition is a perturbed Poiseuille profile $\mathbf{u}^i(t,x_2) = \frac{6U_m}{(y_{max}-y_{min})^2}(-x_2^2 + (y_{max}+y_{min})x_2 - y_{min}y_{max}) - z_p(x_2)e^{-(t-0.5)^2}$, where $U_m = 1.0$ and $z_p(x_2) = 0.8\sin(2\pi(x_2 + 0.75)/1.5)$ if $x_2 \in [-0.75, 0.75]$ and $z_p(x_2) = 0$ elsewhere is a profile chosen to perturb the stationary configuration.

We use the parameters $\nu = 1/120$ (Reynolds number of 120), $\rho = 5$, $\gamma_0 = 0.05$, cfl $= 0.8$ and $\Delta t_{max} = 5\cdot 10^{-4}$.

The initial parameters for the structure are $\theta_0 = (-20°, 0)$ and $\omega_0 = (0,0)$. The initial velocity profile $\mathbf{u}_0$ is obtained by solving the stationary Navier–Stokes equations in the initial configuration. The source on the structure $\mathbf{f}_S$ is chosen to compensate the forces of the fluid in the initial configuration and then enforce the initial state to be a stationary state.

The evolution of the system is reported in Figs. 8 and 9. The stationary setting is perturbed by the inflow condition. The perturbation propagates in the fluid and destabilizes the structure. Von Kármán vortex street appears in the wake behind the structure.

Fig. 8 The velocity magnitude profile (red: high, blue: low), left: whole domain, right: zoom on the structure. The pictures have been captured at $t = 0s$, $t = 1.5s$ and $t = 6s$

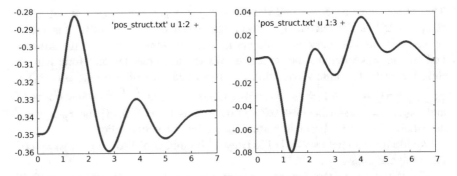

Fig. 9 The evolution of the structure parameters; left: θ_1, right: θ_2

7 Summary

In the present article, we proposed a method to compute numerically the evolution of a fluid–structure system with a fixed mesh. It is based on a finite difference method w.r.t. time and on XFEM w.r.t. space. The structure and fluid steps are partitioned. We presented some solutions to implement this method. We finally ran numerical simulations with small deformations of the structure. Since the advantage of using XFEM over ALE is to tackle large deformations, more complex test cases should be considered to fully validate this approach.

Acknowledgements The first author has been supported by IFSMACS ANR-15-CE40-0010.

References

1. Alauzet, F., Fabrèges, B., Fernández, M.A., Landajuela, M.: Nitsche–XFEM for the coupling of an incompressible fluid with immersed thin–walled structures. Comput. Methods Appl. Mech. Eng. **301**, 300–335 (2016)
2. Angot, P., Bruneau, C.-H., Fabrie, P.: A penalization method to take into account obstacles in incompressible viscous flows. Numer. Math. **81**(4), 497–520 (1999)
3. Barber, C.B., Dobkin, D.P., Huhdanpaa, H.: The quickhull algorithm for convex hulls. ACM Trans. Math. Softw. **22**(4), 469–483 (1996)
4. Bordas, S.P.A., Burman, E., Larson, M.G., Olshanskii, M.A.: Geometrically Unfitted Finite Element Methods and Applications. Springer, Berlin (2017)
5. Court, S., Fournié, M., Lozinski, A.: A fictitious domain approach for the Stokes problem based on the extended finite element method. Int. J. Numer. Methods Fluids **74**(2), 73–99 (2014)
6. Delay, G.: Étude d'un problème d'interaction fluide–structure: Modélisation, Analyse, Stabilisation et Simulations numériques. Ph.D. thesis. Université de Toulouse (2018)
7. Delay, G.: Existence of strong solutions to a fluid–structure system with a structure given by a finite number of parameters. Available on https://www.math.univ-toulouse.fr/~gdelay/ (2020)
8. Dillon, R.H., Li, Z.: An introduction to the immersed boundary and the immersed interface methods — from Interfaces problems and methods in biological and physical flows. Lect. Notes Ser. Inst. Math. Sci. Natl. Univ. Singap. **17**, 1–67 (2009)
9. Dolbow, J., Moës, N., Belytschko, T.: An extended finite element method for modeling crack growth with frictional contact. Comput. Methods Appl. Mech. Eng. **190**(51–52), 6825–6846 (2001)
10. Fernández, M.A., Mullaert, J., Vidrascu, M.: Generalized Robin–Neumann explicit coupling schemes for incompressible fluid–structure interaction: stability analysis and numerics. Int. J. Numer. Methods Eng. **101**(3), 199–229 (2014)
11. Fournié, M., Lozinski, A.: Stability and optimal convergence of unfitted extended finite element methods with Lagrange multipliers for the stokes equations. In: Bordas, S.P.A., Burman, E., Larson, M.G., Olshanskii, M.A. (eds.) Geometrically Unfitted Finite Element Methods and Applications, pp. 143–182. Springer, Berlin (2017)
12. Hou, G., Wang, J., Layton, A.: Numerical methods for fluid–structure interaction – a review. Commun. Comput. Phys. **12**(2), 337–377 (2012)
13. Hu, H.H., Patankar, N.A., Zhu, M.Y.: Direct numerical simulations of fluid–solid systems using arbitrary Lagrangian Eulerian technique. J. Comput. Phys. **169**, 427–462 (2001)
14. Kamensky, D., Hsu, M.-C., Schillinger, D., Evans, J.A., Aggarwal, A., Bazilevs, Y., Sacks, M.S., Hughes, T.J.R.: An immersogeometric variational framework for fluid–structure interaction: Application to bioprosthetic heart valves. Comput. Methods Appl. Mech. Eng. **284**, 1005–1053 (2015)
15. Landajuela, M., Vidrascu, M., Chapelle, D., Fernández, M.A.: Coupling schemes for the FSI forward prediction challenge: comparative study and validation. Int. J. Numer. Methods Biomed. Eng. **33**(4), pp.e02813 (2017)
16. Lefebvre, A.: Numerical simulation of gluey particles. M2AN Math. Model. Numer. Anal. **43**(1), 53–80 (2009)

17. Massing, A., Larson, M.G., Logg, A., Rognes, M.E.: A Nitsche–based cut finite element method for a fluid–structure interaction problem. Commun. Appl. Math. Comput. Sci. **10**(2), 97–120 (2015)
18. Maury, B.: Numerical analysis of a finite element/volume penalty method. SIAM Numer. Anal. **47**(2), 1126–1148 (2009)
19. Mittal, R., Iaccarino, G.: Immersed boundary methods. Annu. Rev. Fluid Mech. **37**, 239–261 (2005)
20. Peskin, C.S.: The immersed boundary method. Acta Numer. **11**, 476–517 (2002)
21. Renard, Y., Pommier, J.: Getfem finite element library. http://home.gna.org/getfem/
22. Richter, T.: A monolitic geometric multigrid solver for fluid–structure interactions in ALE formulation. Int. J. Numer. Methods Eng. **104**(5), 372–390 (2015)
23. Sethian, J.A.: Level Set Methods – From Volume 3 of Cambridge Monographs on Applied and Computational Mathematics. Cambridge University Press, Cambridge (1999)

Implicit–Explicit Multistep Methods for Nonlinear Convection–Diffusion Equations

Georgios Akrivis and Emmanuil H. Georgoulis

Abstract We construct and analyze implicit–explicit multistep schemes for nonlinear evolution convection–diffusion partial differential equations. We establish optimal order a priori error estimates. We are particularly interested in the dependence of the stability constants on the ratio between the convection and diffusion coefficients, the so-called Péclet number, and on the diffusion coefficient ε itself. In particular, we show that the second order implicit–explicit backward differentiation formula (BDF) admits stability constant independent of the Péclet number.

1 Introduction

This work is concerned with the design and analysis of implicit–explicit multistep methods for parabolic semilinear convection-diffusion partial differential equation (p.d.e.) problems. In order to focus on the time-stepping issues, only the time-discrete schemes are discussed.

Previous works on multistep methods for evolution p.d.e. problems include [1–8, 11, 12]. For the discretization of convection–diffusion equations by Runge–Kutta

G. Akrivis
Department of Computer Science & Engineering, University of Ioannina, Ioannina, Greece

Institute of Applied and Computational Mathematics, FORTH, Heraklion, Crete, Greece
e-mail: akrivis@cse.uoi.gr

E. H. Georgoulis (✉)
Department of Mathematics, University of Leicester, Leicester, UK

School of Applied Mathematical and Physical Sciences, National Technical University of Athens, Zografou, Greece

Institute of Applied and Computational Mathematics, FORTH, Heraklion, Crete, Greece
e-mail: Emmanuil.Georgoulis@le.ac.uk

© Springer Nature Switzerland AG 2020
G. R. Barrenechea, J. Mackenzie (eds.), *Boundary and Interior Layers,*
Computational and Asymptotic Methods BAIL 2018, Lecture Notes in
Computational Science and Engineering 135,
https://doi.org/10.1007/978-3-030-41800-7_4

methods we refer to [9, 10]. The standard monograph for numerical methods for parabolic equations is [15].

Multistep methods can be computationally attractive, as they do not require the calculation of intermediate stages (in contrast to, e.g., Runge–Kutta time-stepping schemes) to achieve high order convergence rates in time. The use of carefully constructed implicit–explicit schemes can further reduce the computational cost by requiring the solution of one linear equation at each time step.

More specifically, we shall construct and analyze implicit–explicit multistep methods for the following initial and boundary value problem: seek a function $u : \bar{\Omega} \times [0, T] \to \mathbb{R}$ satisfying

$$\begin{cases} u_t - \varepsilon \Delta u + \nabla \cdot \big(ub(x, t)\big) + c(x, t)u = f(u, x, t) & \text{in } \Omega \times (0, T), \\ u = 0 & \text{on } \partial\Omega \times (0, T), \quad (1.1) \\ u(\cdot, 0) = u^0 & \text{in } \Omega. \end{cases}$$

Here $\Omega \subset \mathbb{R}^d$ is a bounded domain, $\bar{\Omega}$ and $\partial\Omega$ are the closure and the boundary of Ω, respectively, T and ε are given positive numbers, and

$$\nabla \cdot (ub) := \sum_{i=1}^{d} (b_i u)_{x_i}.$$

The convective coefficient b, the coefficient c, the initial value u^0 and the forcing term f are given functions. We assume the vector-valued function $b : \bar{\Omega} \times [0, T] \to \mathbb{R}^d$ is continuously differentiable, $c : \bar{\Omega} \times [0, T] \to \mathbb{R}$ is continuous, $u^0 \in L^2(\Omega)$ and $f : \mathbb{R} \times \bar{\Omega} \times [0, T] \to \mathbb{R}$ is globally Lipschitz continuous in its first argument, uniformly with respect to its second and third arguments,

$$\exists L \geqslant 0 \; \forall x \in \bar{\Omega} \; \forall t \in [0, T] \; \forall y_1, y_2 \in \mathbb{R} \quad |f(y_1, x, t) - f(y_2, x, t)| \leqslant L|y_1 - y_2|, \tag{1.2}$$

and such that $f(0, \cdot, t) \in L^2(\Omega)$, for all $t \in [0, T]$. It is then easily seen that $f(v, \cdot, t) \in L^2(\Omega)$, for any $v \in L^2(\Omega)$ and all $t \in [0, T]$; indeed, using the Lipschitz condition (1.2) and elementary inequalities, we have

$$\big|f\big(v(x), x, t\big)\big|^2 \leqslant 2\big|f\big(v(x), x, t\big) - f(0, x, t)\big|^2 + 2|f(0, x, t)|^2$$
$$\leqslant 2L^2|v(x)|^2 + 2|f(0, x, t)|^2,$$

and our claim is evident. Additional hypotheses on the data will be imposed below.

We assume that the initial and boundary value problem (1.1) admits a sufficiently smooth solution u.

Let now (α, β) be a strongly $A(0)$-stable q-step scheme and (α, γ) be an explicit q-step scheme, characterized by three polynomials α, β and γ,

$$\alpha(\zeta) = \sum_{i=0}^{q} \alpha_i \zeta^i, \quad \beta(\zeta) = \sum_{i=0}^{q} \beta_i \zeta^i, \quad \gamma(\zeta) = \sum_{i=0}^{q-1} \gamma_i \zeta^i.$$

For simplicity, we assume that the order of both q-step schemes, the implicit (α, β) and the explicit (α, γ), is p, i.e.,

$$\sum_{i=0}^{q} i^\ell \alpha_i = \ell \sum_{i=0}^{q} i^{\ell-1} \beta_i = \ell \sum_{i=0}^{q-1} i^{\ell-1} \gamma_i, \quad \ell = 0, 1, \ldots, p. \tag{1.3}$$

As an example of schemes satisfying our assumptions we mention the implicit–explicit BDF methods, described by the polynomials

$$\alpha(\zeta) = \sum_{j=1}^{q} \frac{1}{j} \zeta^{q-j} (\zeta - 1)^j, \quad \beta(\zeta) = \zeta^q, \quad \gamma(\zeta) = \zeta^q - (\zeta - 1)^q. \tag{1.4}$$

The corresponding implicit (α, β)-schemes are the well-known BDF methods, which are strongly $A(0)$-stable for $q = 1, \ldots, 6$; their order is $p = q$. For a given α, the scheme (α, γ) is the unique explicit q-step scheme of order $p = q$. Thus, the implicit–explicit BDF methods satisfy the order conditions (1.3) with $p = q$.

Let $N \in \mathbb{N}$, $k := T/N$ be the constant time step, and $t^n := nk, n = 0, \ldots, N$, be a uniform partition of the interval $[0, T]$. We assume that starting approximations $U^0, \ldots, U^{q-1}$ are given, as we shall be concerned with q-step schemes for (1.1).

We first write the differential equation in (1.1) in the form

$$u_t + Au + C(t)u = B(t, u(t)), \quad t \in (0, T), \tag{1.5}$$

with appropriate for our purposes operators A, B and C. The numerical scheme depends on the particular choice of the operators A, B and C in (1.5). We shall discuss specific choices later on. We discretize the operators A and C implicitly, by the implicit scheme (α, β), and the operator B explicitly, by the explicit scheme (α, γ). Thus, we define approximations U^m to the nodal values $u^m := u(\cdot, t^m)$ of the exact solution by

$$\sum_{i=0}^{q} \left[\alpha_i I + k\beta_i \big(A + C(t^{n+i}) \big) \right] U^{n+i} = k \sum_{i=0}^{q-1} \gamma_i B(t^{n+i}, U^{n+i}), \tag{1.6}$$

for $n = 0, \ldots, N - q$.

1.1 Consistency Error

We shall now discuss a suitable representation of the consistency error of the implicit–explicit scheme (1.6), which will later be used to derive optimal order consistency estimates; see also [1]. We assume that the order of both schemes (α, β) and (α, γ) is p; cf. (1.3).

The consistency error E^n of the scheme (1.6) for the solution u of (1.1), i.e., the amount by which the exact solution fails to satisfy (1.6), is given by

$$kE^n = \sum_{i=0}^{q} \left[\alpha_i I + k\beta_i \left(A + C(t^{n+i})\right)\right] u^{n+i} - k \sum_{i=0}^{q-1} \gamma_i B(t^{n+i}, u^{n+i}), \qquad (1.7)$$

$n = 0, \ldots, N - q$. First, letting

$$E_1^n := \sum_{i=0}^{q} \left[\alpha_i u^{n+i} - k\beta_i u_t(t^{n+i})\right], \quad E_2^n := k \sum_{i=0}^{q} (\beta_i - \gamma_i) B(t^{n+i}, u^{n+i}),$$

with $\gamma_q := 0$, and using the differential equation in (1.5), we infer that

$$kE^n = E_1^n + E_2^n. \qquad (1.8)$$

Furthermore, via a Taylor expansion about t^n, we see that, due to the order conditions of the implicit (α, β)-scheme (i.e., the first equality in (1.3)) and the second equality in (1.3), respectively, leading terms of order up to $p - 1$ cancel, and we obtain

$$\begin{cases} E_1^n = \dfrac{1}{p!} \sum_{i=1}^{q} \int_{t^n}^{t^{n+i}} (t^{n+i} - s)^{p-1} \left[\alpha_i (t^{n+i} - s) - pk\beta_i\right] \dfrac{\partial^{p+1} u}{\partial t^{p+1}}(s)\, ds, \\[4mm] E_2^n = \dfrac{k}{(p-1)!} \sum_{i=1}^{q} (\beta_i - \gamma_i) \int_{t^n}^{t^{n+i}} (t^{n+i} - s)^{p-1} \dfrac{d^p}{dt^p} B(s, u(s))\, ds. \end{cases} \qquad (1.9)$$

This representation of the consistency error will allow us to derive optimal order consistency estimates in suitable norms, under reasonable regularity assumptions.

The remainder of this work is structured as follows. In Sect. 2, we study linearly implicit numerical schemes, characterised by the same discrete operator for all time levels, and we derive optimal order error estimates. We show that, under such general setting, the constant in the stability estimate for this family of methods depends on the Péclet number. Moreover, the constant in the error estimate depends also on the diffusion parameter ε implicitly, through high order Sobolev norms of the exact solution u. In such general setting one does not expect better dependence on the singular perturbation parameter and the results should be treated as an indication of the numerical challenges for this class of methods. At the other end of the spectrum, in Sect. 3, we focus on first and second order schemes of BDF type, requiring again one linear solve at every time level to advance in time. Crucially, however, we shall make use of possibly time-dependent *nonsymmetric*, in general, linear discrete operators, stemming from possible respective time-dependence of the convective and/or reaction coefficients b and c. This somewhat nonstandard choice will be justified below. Using the A-stability of these low order schemes, we improve crucially on the estimates of Sect. 2: the stability constant is now *independent* of the diffusion parameter ε, while the constant in the error estimate depends on it only implicitly, through appropriate norms of the exact solution u. To test the potential of the proposed method, in Sect. 4, we present a series of numerical experiments demonstrating the performance of the second order BDF method in the discretization of semilinear convection-diffusion equations, with nonlinearities admitting non-Lipschitz growth. Although the latter result is somewhat in the folklore of this class of methods, we were not able to locate a proof under the same assumptions.

2 Error Estimates with Constants Depending on the Péclet Number

Let $c^\star$ be a fixed positive number, let the operator C in (1.5) vanish, and choose the operators A and B as $Av := -\varepsilon \Delta v + c^\star v$ and $B(t, v) := f\big(v(\cdot), \cdot, t\big) - \nabla \cdot \big(vb(t)\big) + \big(c^\star - c(t)\big)v$. We can then, obviously, write the p.d.e. in (1.1) in the form (1.5). For simplicity, we suppressed the dependence on x; we shall follow this convention also below.

With this splitting, the scheme (1.6) takes the form

$$\sum_{i=0}^{q} (\alpha_i I + k\beta_i A) U^{n+i} = k \sum_{i=0}^{q-1} \gamma_i B(t^{n+i}, U^{n+i}), \qquad (2.1)$$

$n = 0, \ldots, N - q$.

To advance with (2.1) in time, i.e., to compute the unknown U^{n+q}, we need to solve one linear equation, with the same operator for all time levels. As we shall see below, this operator is self-adjoint and positive definite; in particular, the approximate solutions are well defined.

We shall follow the analysis in [4, 7] and [1] to study the implicit–explicit numerical scheme (1.6).

2.1 Notation and Lipschitz Conditions

We let $H := L^2(\Omega)$, denote by $(\cdot, \cdot)$ and $\| \cdot \|$ its inner product and norm, respectively, and we recall the standard Sobolev spaces. Evidently, the operator $A : \mathscr{D}(A) = H^2(\Omega) \cap H_0^1(\Omega) \to H$ is linear, self-adjoint and positive definite; the domain $V := \mathscr{D}(A^{1/2})$ of $A^{1/2}$ is $V = H_0^1(\Omega)$. We denote by V' the dual of V, with respect to the pivot space H, i.e., $V' = H^{-1}(\Omega)$, and we introduce the norms $\| \cdot \|$ and $\| \cdot \|_\star$ in V and V', respectively, by

$$\|v\| := \|A^{1/2}v\| = \left(-\varepsilon(\Delta v, v) + c^\star(v, v)\right)^{1/2} = \left(\varepsilon\|\nabla v\|^2 + c^\star\|v\|^2\right)^{1/2}$$

and

$$\|v\|_\star := \|A^{-1/2}v\| = \left(v, (-\varepsilon\Delta + c^\star I)^{-1}v\right)^{1/2}.$$

In a standard fashion, A can be extended to an operator from V onto V'; the notation $(\cdot, \cdot)$ will additionally signify the duality pairing between V' and V. The operator $B(t, \cdot) : \mathscr{D}(A) \to H$ can also be viewed as an operator from V into V'.

For notational convenience, we split the operator B into two parts, $B = B_1 + B_2$, with

$$B_1(t)v := -\nabla \cdot \left(vb(t)\right) \quad \text{and} \quad B_2(t, v) := f\left(v(\cdot), \cdot, t\right) + \left(c^\star - c(t)\right)v.$$

Useful Estimates for $\|v\|_\star$ We have

$$\|v\|_\star = \sup_{u \in V \setminus \{0\}} \frac{(v, u)}{\left(\varepsilon\|\nabla u\|^2 + c^\star\|u\|^2\right)^{1/2}},$$

which leads to the estimate

$$\|v\|_\star \leqslant \min\left\{\frac{1}{\sqrt{c^\star}}\|v\|, \frac{1}{\sqrt{\varepsilon}}\|v\|_{H^{-1}}\right\}. \tag{2.2}$$

Indeed, first

$$\|v\|_\star \leqslant \sup_{u \in V \setminus \{0\}} \frac{(v, u)}{\sqrt{c^\star}\|u\|} \leqslant \sup_{u \in V \setminus \{0\}} \frac{\|v\|\,\|u\|}{\sqrt{c^\star}\|u\|},$$

whence

$$\|v\|_\star \leqslant \frac{1}{\sqrt{c^\star}}\|v\|; \tag{2.3}$$

furthermore,

$$\|v\|_\star \leqslant \sup_{u\in V\setminus\{0\}} \frac{(v,u)}{\sqrt{\varepsilon}\|\nabla u\|} \leqslant \sup_{u\in V\setminus\{0\}} \frac{\|v\|_{H^{-1}}\|\nabla u\|}{\sqrt{\varepsilon}\|\nabla u\|} \leqslant \frac{1}{\sqrt{\varepsilon}}\|v\|_{H^{-1}}.$$

Lipschitz Conditions First, for $v, \tilde{v} \in V$, we have

$$(B_1(t,v),\tilde{v}) = \int_\Omega vb(\cdot,t)\cdot\nabla\tilde{v}\,dx = \sum_{i=1}^d \int_\Omega b_i(\cdot,t)v(\tilde{v})_{x_i}\,dx,$$

whence, with

$$\hat{b} := \max_{1\leqslant i\leqslant d}\ \max_{\substack{x\in\bar{\Omega}\\0\leqslant t\leqslant T}} |b_i(x,t)|,$$

we obtain

$$\left|(B_1(t,v),\tilde{v})\right| \leqslant \hat{b}\|v\|\sum_{i=1}^d \|(\tilde{v})_{x_i}\| \leqslant \hat{b}\sqrt{d}\|v\|\Big(\sum_{i=1}^d \|(\tilde{v})_{x_i}\|^2\Big)^{1/2} = \hat{b}\sqrt{d}\|v\|\,\|\nabla\tilde{v}\|.$$

Therefore, since $\sqrt{\varepsilon}\|\nabla\tilde{v}\| \leqslant \|\tilde{v}\|$,

$$\forall t\in[0,T]\ \forall v\in V \quad \|B_1(t,v)\|_\star \leqslant \mu_1\|v\| \quad \text{with}\quad \mu_1 := \frac{\hat{b}}{\sqrt{\varepsilon}}\sqrt{d}. \tag{2.4}$$

Furthermore, with

$$\hat{c} := \max_{\substack{x\in\bar{\Omega}\\0\leqslant t\leqslant T}} |c^\star - c(x,t)|,$$

in view of (1.2), for $v, \tilde{v} \in V$, we have

$$\|B_2(t,v) - B_2(t,\tilde{v})\|^2 \leqslant \int_\Omega \left[(L+\hat{c})|v(x)-\tilde{v}(x)|\right]^2 dx = (L+\hat{c})^2\|v-\tilde{v}\|^2,$$

whence, according to (2.3),

$$\|B_2(t,v) - B_2(t,\tilde{v})\|_\star \leqslant \frac{1}{\sqrt{c^\star}}\|B_2(t,v) - B_2(t,\tilde{v})\| \leq \frac{L+\hat{c}}{\sqrt{c^\star}}\|v-\tilde{v}\|,$$

i.e.,

$$\forall t\in[0,T]\ \forall v,\tilde{v}\in V \quad \|B_2(t,v) - B_2(t,\tilde{v})\|_\star \leqslant \mu_2\|v-\tilde{v}\| \quad \text{with}\quad \mu_2 := \frac{L+\hat{c}}{\sqrt{c^\star}}.$$

$$\tag{2.5}$$

From (2.4) and (2.5) we obtain the desired global Lipschitz condition

$$\forall t \in [0, T] \, \forall v, \tilde{v} \in V \quad \|B(t, v) - B(t, \tilde{v})\|_\star \leqslant \mu \|v - \tilde{v}\|, \tag{2.6}$$

with Lipschitz constant $\mu := \mu_1 + \mu_2$,

$$\mu = \frac{\hat{b}}{\sqrt{\varepsilon}} \sqrt{d} + \frac{L + \hat{c}}{\sqrt{c^\star}}. \tag{2.7}$$

Notice that the Lipschitz constant μ is bounded for uniformly bounded Péclet numbers $\hat{b}/\sqrt{\varepsilon}$.

2.2 Consistency

From the representations (1.8) and (1.9) of the consistency error, we immediately obtain, under obvious regularity requirements, the desired optimal order consistency estimate

$$\max_{0 \leqslant n \leqslant N-q} \|E^n\|_\star \leqslant ck^p; \tag{2.8}$$

the coefficient c depends on the diffusion parameter ε, through appropriate norms of the exact solution u.

2.3 Stability

Let $U^m, V^m \in V, m = 0, \ldots, N$, satisfy (2.1) and

$$\sum_{i=0}^{q} (\alpha_i I + k\beta_i A) V^{n+i} = k \sum_{i=0}^{q-1} \gamma_i B(t^{n+i}, V^{n+i}), \tag{2.9}$$

$n = 0, \ldots, N - q$, respectively. Then, with $\vartheta^m := U^m - V^m, m = 0, \ldots, N$, we have the stability estimate

$$\|\vartheta^n\|^2 + k \sum_{\ell=0}^{n} \|\vartheta^\ell\|^2 \leqslant c \sum_{j=0}^{q-1} (\|\vartheta^j\|^2 + k\|\vartheta^j\|^2), \quad n = q - 1, \ldots, N, \tag{2.10}$$

with a constant c independent of U^m, V^m and k; see [7, Theorem 2.1]. Actually, since the Lipschitz constant μ in the Lipschitz condition (2.6) depends on ε only through the Péclet number, the stability constant c in (2.10) depends on the diffusion parameter ε also only through the Péclet number; indeed, it is bounded, for bounded Péclet numbers.

2.4   Error Estimates

According to [7, Theorem 2.1], we have the following error estimate:

Theorem 2.1 (Error Estimate) *Let (α, β) be a strongly $A(0)$-stable q-step scheme and (α, γ) be an explicit q-step scheme. Let the order of both schemes (α, β) and (α, γ) be p. Assume we are given starting approximations $U^0, U^1, \ldots, U^{q-1} \in V$ to $u^0, \ldots, u^{q-1}$ such that*

$$\max_{0 \leqslant j \leqslant q-1} \left(\|u^j - U^j\| + k^{1/2} \|u^j - U^j\| \right) \leqslant ck^p. \tag{2.11}$$

Let $U^n \in V, n = q, \ldots, N$, be recursively defined by (2.1). Let $e^n = u^n - U^n, n = 0, \ldots, N$, be the approximation error. Then, there exists a constant c, independent of k and n, depending exponentially on $\hat{b}^2/\varepsilon$, such that

$$\|e^n\|^2 + k \sum_{\ell=0}^{n} \|e^\ell\|^2 \leqslant c \left\{ \sum_{j=0}^{q-1} \left(\|e^j\|^2 + k\|e^j\|^2 \right) + k \sum_{\ell=0}^{n-q} \|E^\ell\|_\star^2 \right\}, \tag{2.12}$$

$n = q - 1, \ldots, N$, whence, in view of (2.11) and (2.8),

$$\max_{0 \leqslant n \leqslant N} \|u(t^n) - U^n\| \leqslant ck^p. \tag{2.13}$$

In the estimate (2.12) E^ℓ is the consistency error of the scheme; see (1.7) with $C(t) = 0$.   □

Remark 2.1 (BDF Schemes) We focus here on the case that the method (2.1) is the implicit–explicit q-step BDF scheme. Since in the Lipschitz condition (2.6) the norm $\|\|\cdot\|\|$ does not enter on the right-hand side, the stability constants λ in the notation of [7] vanishes; therefore, Remark 7.2 of [4] applies, and we can relax condition (2.11) on the starting approximations $U^0, \ldots, U^{q-1}$. More precisely, the statement of Theorem 2.1 is valid in this case under the assumption

$$\max_{0 \leqslant j \leqslant q-1} \|u^j - U^j\| \leqslant ck^p.$$

□

Remark 2.2 (A Wider Class of Linearly Implicit Methods) The splitting of the p.d.e. in (1.1) that we used in the scheme (2.1) satisfies also the assumptions imposed in [4]. Therefore, with this specific splitting, the initial and boundary value problem

(1.1) can be discretized by the wider class of linearly implicit methods discussed in [4]. Thus, the abstract results of [4] apply and lead to error estimates with error constants depending exponentially on $\hat{b}^2/\varepsilon$, as is the case in Theorem 2.1. □

2.5   Alternative Forms of the Implicit–Explicit Schemes

Various possibilities of splitting the p.d.e. in (1.1) before discretizing by implicit–explicit multistep schemes are possible; here, we comment on two such alternatives.

2.5.1   First Choice

Consider now the operators $\tilde{A}$ and $\tilde{B}$ defined as $\tilde{A}v := -\varepsilon \Delta v$, and $\tilde{B}(t, v) :=$ $f(v(\cdot), \cdot, t) - \nabla \cdot (vb(t)) - c(t)v$. Notice that the only difference to the splitting we used in (2.1) is that here we set $c^\star = 0$. With this splitting, the scheme (1.6) takes the form

$$\sum_{i=0}^{q}(\alpha_i I + k\beta_i \tilde{A})U^{n+i} = k\sum_{i=0}^{q-1}\gamma_i \tilde{B}(t^{n+i}, U^{n+i}), \qquad (2.14)$$

$n = 0, \ldots, N - q$. In this case,

$$\|v\| := \|\tilde{A}^{1/2}v\| = \sqrt{\varepsilon}\|\nabla v\| \quad \text{and} \quad \|v\|_\star := \|\tilde{A}^{-1/2}v\| = \frac{1}{\sqrt{\varepsilon}}\|(-\Delta)^{-1/2}v\|.$$

and, with $\tilde{B}_2(t, v) := f(v(\cdot), \cdot, t)$, we have

$$\|\tilde{B}_2(t, v) - \tilde{B}_2(t, \tilde{v})\|^2 = \int_\Omega \left|f(v(x), x, t) - f(\tilde{v}(x), x, t)\right|^2 dx$$

$$\leqslant L^2 \int_\Omega |v(x) - \tilde{v}(x)|^2 dx = L^2\|v - \tilde{v}\|^2.$$

Furthermore, using the Poincaré–Friedrichs inequality $\|w\| \leqslant C_{PF}\|\nabla w\|$, for $w \in V$, we easily see that $\|v\|_\star \leqslant \frac{1}{\sqrt{\varepsilon}}C_{PF}\|v\|$, for $v \in H$. Thus, we infer that

$$\forall t \in [0, T] \, \forall v, \tilde{v} \in V \quad \|\tilde{B}_2(t, v) - \tilde{B}_2(t, \tilde{v})\|_\star \leqslant \tilde{\mu}_2\|v - \tilde{v}\| \qquad (2.15)$$

with $\tilde{\mu}_2 := \frac{1}{\sqrt{\varepsilon}}LC_{PF}$. Therefore, a straightforward application of the analysis of [7] leads to a stability estimate with constant depending exponentially on $1/\varepsilon$, rather than on the Péclet number. However, the scheme (2.14) can, obviously, be equivalently written in the form

$$\sum_{i=0}^{q}(\alpha_i I + k\beta_i A)U^{n+i} - kc^{\star}\sum_{i=0}^{q}\beta_i U^{n+i} = k\sum_{i=0}^{q-1}\gamma_i \tilde{B}(t^{n+i}, U^{n+i}), \qquad (2.16)$$

$n = 0, \ldots, N - q$. Applying here the analysis of [1], rather than the one of [7], we can then easily see that the results of both Theorem 2.1 and Remark 2.1 remain valid also for the scheme (2.14).

2.5.2 Second Choice

We discuss here the discretization of the linear part of Eq. (1.1) implicitly, by the implicit scheme (α, β), and of the nonlinear part explicitly, by the explicit scheme (α, γ). With the operator A used above, $C(t)v := \nabla \cdot (vb(t)) + (c(t) - c^{\star})v$, and $B(t, v) := f(v(\cdot), \cdot, t)$, the scheme can be written in the form (2.1). Applying the analysis of [1] (with the operator A rather than with $\tilde{A}$) we easily see that the results of Theorem 2.1 remain valid in this case as well. Furthermore, in the case of BDF schemes the requirement on the starting approximations can be relaxed as in Remark 2.1. Notice that $c^{\star}$ is used in the analysis of the schemes only; the schemes themselves are independent of $c^{\star}$.

Let us emphasize that, in contrast to the schemes (2.1) and (2.14), in this case the operator of the linear equations is in general time dependent, and so varies from one time level to the next, if b and/or c are/is time dependent. Also, the numerical approximations are well defined, provided the time step k is sufficiently small. More precisely, for a given $w \in V'$, it suffices to show that equation

$$\alpha_q v + k\beta_q\big[-\varepsilon\Delta v + \nabla \cdot (vb(t^{n+q})) + c(t^{n+q})v\big] = w, \qquad (2.17)$$

possesses a unique solution $v \in V$. A well-known property of $A(0)$-stable multistep schemes (α, β) is that the product $\alpha_q\beta_q$ is positive. Assume, without loss of generality, that α_q is positive. According to the Lax–Milgram Lemma, it obviously suffices to show that the bilinear form $a : V \times V \to \mathbb{R}$,

$$a(v, \tilde{v}) := \alpha_q(v, \tilde{v}) + k\beta_q\varepsilon(\nabla v, \nabla\tilde{v}) + k\beta_q\big(\nabla \cdot (vb(t^{n+q})), \tilde{v}\big) + k\beta_q\big(c(t^{n+q})v, \tilde{v}\big)$$

is coercive and continuous. First, for $v \in V$, using an elementary inequality, we have

$$a(v, v) = \alpha_q\|v\|^2 + k\beta_q\varepsilon\|\nabla v\|^2 - k\beta_q\big(vb(t^{n+q}), \nabla v\big) + k\beta_q\big(c(t^{n+q})v, v\big)$$

$$= \alpha_q\|v\|^2 + k\beta_q\varepsilon\|\nabla v\|^2 + \frac{1}{2}k\beta_q\big(v\nabla \cdot b(t^{n+q}), v\big)k\beta_q\big(c(t^{n+q})v, v\big)$$

$$\geqslant \big(\alpha_q - \frac{1}{2}\beta_q k\|\nabla \cdot b(t^{n+q})\|_{L^{\infty}(\Omega)} - \beta_q k\|c(t^{n+q})\|_{L^{\infty}(\Omega)}\big)\|v\|^2 + k\beta_q\varepsilon\|\nabla v\|^2,$$

and we infer that a is indeed coercive for sufficiently small k, independent of ε. Similarly, for $v, \tilde{v} \in V$, we have

$$|a(v, \tilde{v})| \leqslant \alpha_q \|v\| \|\tilde{v}\| + k\beta_q \varepsilon \|\nabla v\| \|\nabla \tilde{v}\| + k\beta_q b^\star \|v\| \|\nabla \tilde{v}\| + k\beta_q c^\star \|v\| \|\tilde{v}\|,$$

and see that a is also continuous.

It is evident from the above discussion that, Péclet-number independent stability analysis, in such general setting, is an essential challenge and may potentially not be true for schemes that are not A-stable. Indeed, as we shall see below, upon employing non-standard and specialised techniques, we are able to recover Péclet-number independent stability bounds for implicit–explicit Euler method (a known result, but nevertheless, included for completeness of the presentation) and for the classical implicit–explicit second order BDF method; the latter is an improvement upon the results presented in [12].

3 Low Order Schemes

Focusing, now, on low order time stepping schemes, our goal is to establish error estimates via the energy technique with stability constants independent of ε.

We consider, again, the initial and boundary value problem (1.1) and assume that

$$c(x, t) + \frac{1}{2} \nabla \cdot b(x, t) \geqslant \underline{b}^2 \quad \forall x \in \bar{\Omega} \ t \in [0, T], \tag{3.1}$$

for some positive $\underline{b}$; notice that this can always be achieved by either adding, if necessary, to both sides of the differential equation a term of the form au with a sufficiently large coefficient a, or by the change of variables $\tilde{u} := e^{-at} u$. Notice that this affects the constant in the error estimate. Indeed, in the first case the Lipschitz constant changes from L to $L + a$. In the second case $\tilde{f}(\tilde{u}, x, t) = e^{-at} f(e^{at} \tilde{u}, x, t)$ satisfies the Lipschitz condition with constant L, but we need to multiply the approximations $\tilde{U}^n$ of $\tilde{u}^n$ by e^{at^n} to obtain approximations U^n of u^n and thus $u^n - U^n = e^{at^n}(\tilde{u}^n - \tilde{U}^n)$.

For convenience we introduce the operator $A(t) : H^2(\Omega) \cap H_0^1(\Omega) \to L^2(\Omega) =: H, t \in [0, T]$, by

$$A(t)v := -\varepsilon \Delta v + \nabla \cdot (vb(x, t)) + c(x, t)v.$$

Notice, however, that $A(t)$ is not self-adjoint and possibly time-dependent. Obviously, $A(t)$ can be extended to an operator from $V := H_0^1(\Omega)$ to $V' = H^{-1}(\Omega)$.

Taking the duality paring with v and integrating by parts, we have

$$(A(t)v, v) = \varepsilon \|\nabla v\|^2 + (\nabla \cdot (vb(x, t)) + c(x, t)v, v)$$

$$= \varepsilon \|\nabla v\|^2 + ((c + \frac{1}{2} \nabla \cdot b)v, v).$$

Thus, in view of (3.1), $A(t)$ is coercive:

$$(A(t)v, v) \geqslant \varepsilon \|\nabla v\|^2 + \underline{b}^2 \|v\|^2 \quad \forall v \in V = H_0^1(\Omega). \tag{3.2}$$

3.1   Implicit–Explicit Euler Method

For completeness, we begin by presenting the lowest order case of implicit–explicit Euler method, keeping careful track of the stability constants.

We recursively define a sequence of approximations U^m to the nodal values $u(t^m)$ of the solution u of the initial and boundary value problem (1.1) by the implicit–explicit Euler method,

$$U^{n+1} + kA(t^{n+1})U^{n+1} = U^n + kf(U^n, \cdot, t^n), \quad n = 0, \dots, N - 1, \tag{3.3}$$

with starting approximation $U^0 := u^0$. In view of the coercivity (3.2), it can be easily seen that the numerical approximations are well defined.

3.1.1   Consistency

The consistency error E^n of the implicit–explicit Euler scheme (3.3) for the solution u of (1.1),

$$kE^n = u^{n+1} + kA(t^{n+1})u^{n+1} - u^n - kf(u^n, \cdot, t^n), \quad n = 0, \dots, N - 1, \tag{3.4}$$

can be written in the form $kE^n = E_1^n + E_2^n$ with

$$E_1^n = \int_{t^n}^{t^{n+1}} (t^n - s)u_{tt}(s)\, ds, \quad E_2^n = k \int_{t^n}^{t^{n+1}} \frac{d}{dt} f(u(s), \cdot, s)\, ds; \tag{3.5}$$

cf. (1.9). Therefore, under obvious regularity assumptions, we derive the desired optimal order consistency estimate

$$\max_{0 \leqslant n \leqslant N-1} \|E^n\| \leqslant ck \tag{3.6}$$

with a suitable positive constant c. (Of course, c depends implicitly on ε, since the solution u depends on ε.)

3.1.2 Stability

Let $U^m, V^m \in V, m = 0, \ldots, N$, satisfy (3.3) and

$$V^{n+1} + kA(t^{n+1})V^{n+1} = V^n + kf(V^n, \cdot, t^n), \quad n = 0, \ldots, N-1, \tag{3.7}$$

respectively. Then, $\vartheta^m := U^m - V^m, m = 0, \ldots, N$, satisfy the relation

$$\vartheta^{n+1} + kA(t^{n+1})\vartheta^{n+1} = \vartheta^n + k\big[f(U^n, \cdot, t^n) - f(V^n, \cdot, t^n)\big], \tag{3.8}$$

$n = 0, \ldots, N-1$. Taking in (3.8) the inner product with ϑ^{n+1} and utilizing (3.2), we obtain

$$\|\vartheta^{n+1}\|^2 + k\varepsilon\|\nabla\vartheta^{n+1}\|^2 + k\underline{b}^2\|\vartheta^{n+1}\|^2 \leqslant (\vartheta^n, \vartheta^{n+1}) + k(f(U^n, \cdot, t^n) - f(V^n, \cdot, t^n), \vartheta^{n+1}).$$

Therefore,

$$\|\vartheta^{n+1}\|^2 + k\varepsilon\|\nabla\vartheta^{n+1}\|^2 + k\underline{b}^2\|\vartheta^{n+1}\|^2 \leqslant \frac{1}{2}\|\vartheta^n\|^2 + \frac{1}{2}\|\vartheta^{n+1}\|^2$$
$$+ k\|f(U^n, \cdot, t^n) - f(V^n, \cdot, t^n)\| \, \|\vartheta^{n+1}\|$$
$$\leqslant \frac{1}{2}\|\vartheta^n\|^2 + \frac{1}{2}\|\vartheta^{n+1}\|^2 + \frac{1}{2}Lk\|\vartheta^n\|^2 + \frac{1}{2}Lk\|\vartheta^{n+1}\|^2,$$

whence, by multiplying by 2,

$$\big[1 - (L - 2\underline{b}^2)k\big]\|\vartheta^{n+1}\|^2 \leq (1 + Lk)\|\vartheta^n\|^2. \tag{3.9}$$

Now, for sufficiently small k,

$$\|\vartheta^{n+1}\|^2 \leqslant (1 + c_\star k)\|\vartheta^n\|^2, \tag{3.10}$$

with a suitable constant $c_\star$. This is obviously valid in the case $2\underline{b}^2 \geqslant L$, with $c_\star = L$. In the case $L > 2\underline{b}^2$, (3.9) yields, for $k < 1/(L - 2\underline{b}^2)$,

$$\|\vartheta^{n+1}\|^2 \leq \frac{1 + Lk}{1 - (L - 2\underline{b}^2)k}\|\vartheta^n\|^2. \tag{3.11}$$

For any fixed $c_\star$ (strictly) larger than $2(L - \underline{b}^2)$, it is easily seen that

$$\frac{1 + Lk}{1 - (L - 2\underline{b}^2)k} \leqslant 1 + c_\star k, \tag{3.12}$$

provided that k is sufficiently small,

$$k \leqslant \frac{c_\star - 2(L - \underline{b}^2)}{c_\star(L - 2\underline{b}^2)}, \tag{3.13}$$

and (3.10) follows from (3.11) and (3.12).

Now, from (3.10) we obtain

$$\|\vartheta^n\|^2 \leqslant (1 + c_\star k)^n \|\vartheta^0\|^2, \quad n = 0, \ldots, N,$$

and thus

$$\|\vartheta^n\|^2 \leqslant e^{c_\star nk} \|\vartheta^0\|^2, \quad n = 0, \ldots, N.$$

Hence, we arrive at the desired stability estimate

$$\max_{1 \leqslant n \leqslant N} \|\vartheta^n\| \leqslant e^{c_\star T/2} \|\vartheta^0\|. \tag{3.14}$$

Crucially, the above stability constant is *independent* of the diffusion coefficient ε.

3.1.3 Error Estimate

Let $e^n := u^n - U^n, n = 0, \ldots, N$. Subtracting (3.3) from (3.4), we obtain the error equation

$$e^{n+1} + kA(t^{n+1})e^{n+1} = e^n + k\big[f(u^n, \cdot, t^n) - f(U^n, \cdot, t^n)\big] + kE^n, \tag{3.15}$$

$n = 0, \ldots, N - 1$. Taking here the inner product with e^{n+1}, proceeding as in the stability proof, and utilizing the consistency estimate (3.6) as well as the fact that e^0 vanishes, we easily derive the desired error estimate

$$\max_{1 \leqslant n \leqslant N} \|e^n\| \leqslant ck. \tag{3.16}$$

The constant c on the right-hand side of (3.16) depends on ε only implicitly through Sobolev norms of the solution u (see (3.5) and (3.6)).

3.2 Implicit–Explicit Two-Step BDF Method

Here we present a robust error analysis for the IMEX using a two-step BDF method. Although we present the time-discrete analysis only, the result can be used to improve fully discrete a priori error bounds for fully discrete BDF-IMEX schemes for various stable spatial discretizations, e.g., [12]; in particular, the exponential dependence of the a priori error bound constant on the Péclet number from [12] can be avoided.

With starting approximation $U^0 := u^0$, we first perform one step of the implicit–explicit Euler scheme to compute U^1, i.e., we let U^1 be given by

$$U^1 + kA(t^1)U^1 = U^0 + kf(U^0, \cdot, t^0), \tag{3.17}$$

and then let the approximations $U^2, \ldots, U^N$ be given by the implicit–explicit two-step BDF scheme,

$$\frac{3}{2}U^{n+2} - 2U^{n+1} + \frac{1}{2}U^n + kA(t^{n+2})U^{n+2} = 2kf(U^{n+1}, \cdot, t^{n+1}) - kf(U^n, \cdot, t^n), \tag{3.18}$$

$n = 0, \ldots, N - 2$. Again, in view of (3.2), it can be easily seen that the numerical approximations are well defined.

3.2.1 Consistency

The consistency error E^n of the implicit–explicit BDF scheme (3.18),

$$kE^n = \frac{3}{2}u^{n+2} - 2u^{n+1} + \frac{1}{2}u^n + kA(t^{n+2})u^{n+2} \\ - 2kf(u^{n+1}, \cdot, t^{n+1}) + kf(u^n, \cdot, t^n), \tag{3.19}$$

$n = 0, \ldots, N - 2$, can be written in the form

$$kE^n = E_1^n + E_2^n$$

with

$$E_1^n = -\int_{t^n}^{t^{n+1}} (t^{n+1} - s)^2 u_{ttt}(s)\, ds + \frac{3}{4}\int_{t^n}^{t^{n+2}} (t^{n+2} - s)(t^{n+2} - s - \frac{4}{3}k)u_{ttt}(s)\, ds,$$

$$E_2^n = -2k\int_{t^n}^{t^{n+1}} (t^{n+1} - s)\frac{d^2}{dt^2}f(u(s), \cdot, s)\, ds + k\int_{t^n}^{t^{n+2}} (t^{n+2} - s)\frac{d^2}{dt^2}f(u(s), \cdot, s)\, ds;$$

cf. (1.9). Therefore, under the regularity assumptions

$$\|u_{ttt}(t)\|_\star \leqslant c_1 \quad \text{and} \quad \left\|\frac{d^2}{dt^2}f(u(t), \cdot, t)\right\| \leqslant c_2, \tag{3.20}$$

for all $t \in [0, T]$, we immediately conclude that

$$\max_{0 \leqslant n \leqslant N-2} \|E_1^n\| \leqslant 2c_1k^2 \quad \text{and} \quad \max_{0 \leqslant n \leqslant N-2} \|E_2^n\| \leqslant 2c_2k^2.$$

Thus, we obtain the desired estimate for the consistency error E^n,

$$\max_{0 \leqslant n \leqslant N-2} \|E^n\| \leqslant Ck^2. \tag{3.21}$$

Remark 3.1 (Regularity Requirement) Note that (3.20) can be replaced by slightly weaker $C^{2,1}$-requirements on u and f. Similar remark applies to (3.5) and (3.6).

3.2.2 Stability

Let $U^0, \ldots, U^N \in V$ satisfy (3.17) and (3.18), and $V^0, \ldots, V^N \in V$ satisfy

$$\frac{3}{2}V^{n+2} - 2V^{n+1} + \frac{1}{2}V^n + kA(t^{n+2})V^{n+2} = 2kf(V^{n+1}, \cdot, t^{n+1}) - kf(V^n, \cdot, t^n), \tag{3.22}$$

$n = 0, \ldots, N - 2$. Let

$$\vartheta^m := U^m - V^m \quad \text{and} \quad b^m := f(U^m, \cdot, t^m) - f(V^m, \cdot, t^m),$$

$m = 0, \ldots, N$. Subtracting (3.22) from (3.18), we obtain

$$\frac{3}{2}\vartheta^{n+2} - 2\vartheta^{n+1} + \frac{1}{2}\vartheta^n + kA(t^{n+2})\vartheta^{n+2} = 2kb^{n+1} - kb^n. \tag{3.23}$$

Now, we observe the identity

$$\left(\frac{3}{2}\vartheta^{n+2} - 2\vartheta^{n+1} + \frac{1}{2}\vartheta^n, \vartheta^{n+2}\right) = \frac{5}{4}\|\vartheta^{n+2}\|^2 - \|\vartheta^{n+1}\|^2 - \frac{1}{4}\|\vartheta^n\|^2$$

$$- \left((\vartheta^{n+2}, \vartheta^{n+1}) - (\vartheta^{n+1}, \vartheta^n)\right) + \frac{1}{4}\|\vartheta^{n+2} - 2\vartheta^{n+1} + \vartheta^n\|^2; \tag{3.24}$$

cf. [16]. We note that (3.24) stems from the G-stability of the BDF method (α, β) with the positive definite symmetric matrix G,

$$G = \frac{1}{4}\begin{pmatrix} 5 & -2 \\ -2 & 1 \end{pmatrix};$$

see [14, Example 6.5].

Taking the inner product with ϑ^{n+2} in (3.23), and using (3.24) and (3.2), we have

$$\frac{5}{4}\|\vartheta^{n+2}\|^2 - \|\vartheta^{n+1}\|^2 - \frac{1}{4}\|\vartheta^n\|^2 - \big((\vartheta^{n+2}, \vartheta^{n+1}) - (\vartheta^{n+1}, \vartheta^n)\big)$$
$$+ k\varepsilon\|\nabla\vartheta^{n+2}\|^2 + k\underline{b}^2\|\vartheta^{n+2}\|^2 \leqslant 2k\|b^{n+1}\|\,\|\vartheta^{n+2}\| + k\|b^n\|\,\|\vartheta^{n+2}\|.$$
$$(3.25)$$

Now, in view of the Lipschitz condition (1.2), we have

$$\|b^m\| \leqslant L\|\vartheta^m\|; \tag{3.26}$$

therefore, (3.25) yields

$$\frac{5}{4}\|\vartheta^{n+2}\|^2 - \|\vartheta^{n+1}\|^2 - \frac{1}{4}\|\vartheta^n\|^2 - \big((\vartheta^{n+2}, \vartheta^{n+1}) - (\vartheta^{n+1}, \vartheta^n)\big)$$
$$+ k\varepsilon\|\nabla\vartheta^{n+2}\|^2 + k\underline{b}^2\|\vartheta^{n+2}\|^2 \leqslant 2Lk\|\vartheta^{n+1}\|\,\|\vartheta^{n+2}\| + Lk\|\vartheta^n\|\,\|\vartheta^{n+2}\|,$$
$$(3.27)$$

and, using a standard Poincaré-Friedrichs inequality $\|v\|^2 \leqslant c_{PF}\|\nabla v\|^2$, we infer

$$\frac{5}{4}\big(\|\vartheta^{n+2}\|^2 - \|\vartheta^{n+1}\|^2\big) + \frac{1}{4}\big(\|\vartheta^{n+1}\|^2 - \|\vartheta^n\|^2\big)$$
$$-\big((\vartheta^{n+2}, \vartheta^{n+1}) - (\vartheta^{n+1}, \vartheta^n)\big) \leqslant \big(\frac{3L}{2} - \delta\big)k\|\vartheta^{n+2}\|^2 + Lk\|\vartheta^{n+1}\|^2 + \frac{Lk}{2}\|\vartheta^n\|^2,$$
$$(3.28)$$

with $\delta := c_{PF}\varepsilon + \underline{b}^2$. Summing in (3.28) from $n = 0$ to $n = \ell$, we obtain

$$\frac{5}{4}\big(\|\vartheta^{\ell+2}\|^2 - \|\vartheta^1\|^2\big) + \frac{1}{4}\big(\|\vartheta^{\ell+1}\|^2 - \|\vartheta^0\|^2\big) - (\vartheta^{\ell+2}, \vartheta^{\ell+1})$$
$$\leqslant (3L - \delta)k\sum_{n=0}^{\ell+2}\|\vartheta^n\|^2 - (\vartheta^1, \vartheta^0),$$

whence, easily,

$$\frac{1}{4}\|\vartheta^{\ell+2}\|^2 \leqslant (3L - \delta)k\sum_{n=0}^{\ell+2}\|\vartheta^n\|^2 + \frac{1}{2}\big(\|\vartheta^1\|^2 + \|\vartheta^0\|^2\big).$$

Therefore, we have,

$$\|\vartheta^\ell\|^2 \leqslant 4(3L - \delta)k\sum_{n=0}^{\ell}\|\vartheta^n\|^2 + 2\big(\|\vartheta^1\|^2 + \|\vartheta^0\|^2\big), \tag{3.29}$$

$\ell = 2, \ldots, N$. Now, for sufficiently small k, whose size depends adversely *only* on L, application of the discrete Gronwall inequality leads to the desired local stability estimate

$$\|\vartheta^n\|^2 \leqslant c\big(\|\vartheta^1\|^2 + \|\vartheta^0\|^2\big), \quad n = 1, \ldots, N. \tag{3.30}$$

Note that, in particular, the dependence of the stability constant c on ε is desirable, in that it diminishes as $\varepsilon \to 0$ and can even be beneficial for large ε.

Now, let V^1 and V^0 be related by

$$V^1 + kA(t^1)V^1 = V^0 + kf(V^0, \cdot, t^0), \tag{3.31}$$

i.e., starting with initial value V^0 we obtain V^1 by performing one step with the implicit–explicit Euler scheme to the differential equation in (1.1); see (3.17) and (3.18). Obviously, in view of the stability property (3.14) of the implicit–explicit Euler method, $\|\vartheta^1\| \leqslant c\|\vartheta^0\|$, which combined with (3.30) leads to our final stability estimate

$$\max_{1 \leqslant n \leqslant N} \|\vartheta^n\| \leqslant c\|\vartheta^0\|, \tag{3.32}$$

with a constant $c > 0$ independent of ε.

3.2.3 Error Estimates

Let the implicit–explicit BDF2 approximations $U^0, \ldots, U^N$ be given by (3.17) and (3.18). Assume that the solution u of (1.1) is sufficiently smooth, such that (3.21) and (3.6) be valid. Then, combining stability and consistency in the standard way we establish the following optimal order error estimate

$$\max_{0 \leqslant n \leqslant N} \|u(t^n) - U^n\| \leqslant ck^2. \tag{3.33}$$

Again, here the constant c in (3.32) depends only implicitly on ε, through its dependence on Sobolev norms of the exact solution u, via the consistency estimates (3.21) and (3.6).

Remark 3.2 (Energy Technique for Higher Order BDF Methods) Proceeding as in Sect. 2.1, we can see that

$$\big|(A(t)v, u)\big| \leqslant \big(\sqrt{\varepsilon}\|\nabla v\| + \mu\|v\|\big)\big(\varepsilon\|\nabla u\|^2 + \underline{b}^2\|u\|^2\big)^{1/2} \quad \forall v, u \in V = H_0^1(\Omega), \tag{3.34}$$

with a constant μ depending on d, $\max_{x,t} |c(x, t)|$ and the Péclet number $\hat{b}/\sqrt{\varepsilon}$. In view of the coercivity condition (3.2) and (3.34), as well as of the fact that the time interval $[0, T]$ is bounded, a slight modification of the stability analysis of [2] leads to optimal order error estimates in our case with constants depending on the Péclet number for BDF methods up to order five. Notice that (3.34) is a slight relaxation of the corresponding boundedness condition [2, (1.7)]. The energy stability analysis in [2] is based on the Nevanlinna–Odeh multiplier technique. Moreover, it is not clear if it is possible to improve upon these estimates to arrive to a Péclet-number independent stability analysis for $A(\alpha)$-stable BDF methods with $\alpha < \pi/2$.   □

The proposed implicit-explicit methods are somewhat non-standard in that they require the solution of a *non-symmetric* linear system per time-step. Indeed, it is a usual practice to treat convection explicitly also in an effort to arrive at symmetric linear systems instead. Such methods, however, require careful tuning of the discretization parameters, as hyperbolic-type CFL restrictions are introduced by the explicit treatment of the dominant convection term. The latter is, crucially, *not* the case for the low order schemes studied in this work.

We shall now argue that the implicit treatment of the predominantly skew-symmetric convection term does *not* hinder the computational efficiency of the proposed methods in any essential fashion. This is because nonsymmetric linear systems arising from the discretization of the convection-diffusion spatial operator through some stable finite elements (e.g., streamline upwinded Petrov-Galerkin methods, discontinuous Galerkin approaches, etc.) admit a number of special properties that can be exploited in the design of scalable preconditioning strategies. For instance, for discontinuous Galerkin methods for steady convection-diffusion problems it has been shown in [13] that, preconditioning by the symmetric part of the convection-diffusion stiffness matrix for the interior penalty discontinuous Galerkin method within a preconditioned GMREs iteration, provides a *three-step recurrence* Krylov method that converges independently of the spatial mesh-size. This means that, up to the cost of inversion of a symmetric preconditioner, the complexity of the preconditioned GMREs is comparable to that of a standard Conjugate Gradient iteration that would normally be used for the respective symmetric linear system of the classical diffusion-only implicit IMEX scheme. At the same time the step of inverting the symmetric part of the stiffness matrix can be efficiently tackled by standard multilevel approaches, whose convergence is further aided by the presence of a strong reaction coefficient stemming from the time discretization.

4   Numerical Experiments

We present some numerical experiments investigating the convergence rates for the implicit-explicit second order BDF (BDF2) method, described in Sect. 3.2, as well as its robustness with respect to the Péclet number. To fully asses the applicability of the proposed method, our numerical investigations are not confined to globally Lipschitz nonlinearities.

Table 1 Example 1: error and convergence rates

$\varepsilon = 1$			$\varepsilon = 10^{-1}$		
k	$\|e\|_{L^\infty(L^2)}$	Rate	k	$\|e\|_{L^\infty(L^2)}$	Rate
1.0e−1	2.519e−4	–	1.0e−1	1.646e−4	–
5.0e−2	6.603e−5	1.93	5.0e−2	4.289e−5	1.94
1.0e−2	2.674e−6	1.99	1.0e−2	2.221e−6	1.84
5.0e−3	6.687e−7	1.99	5.0e−3	5.832e−7	1.93
1.0e−3	2.682e−8	1.99	1.0e−3	2.451e−8	1.97
$\varepsilon = 10^{-3}$			$\varepsilon = 10^{-5}$		
k	$\|e\|_{L^\infty(L^2)}$	Rate	k	$\|e\|_{L^\infty(L^2)}$	Rate
1.0e−1	3.075e−4	–	1.0e−1	3.790e−4	–
5.0e−2	8.958e−5	1.78	5.0e−2	1.102e−4	1.78
1.0e−2	3.874e−6	1.95	1.0e−2	4.786e−6	1.95
5.0e−3	9.731e−7	1.99	5.0e−3	1.203e−6	1.99
1.0e−3	3.897e−8	2.00	1.0e−3	4.572e−8	2.03

4.1 Example 1

We begin with considering the semilinear convection-diffusion problem for $\Omega = [0, 1]^2$, $T = 1$, $b = (1, 1)^T$, $c = 0$ and $f(u, x, t) = -u^2 + g(x, t)$, with g such that the exact solution of the problem is given by

$$u(x, t) := (1 - \exp(-t))x_1 x_2(1 - x_1)(1 - x_2), \quad x := (x_1, x_2)^T.$$

The implicit-explicit BDF2 method is implemented for $\varepsilon = 1, 10^{-1}, 10^{-3}, 10^{-5}$, using the finite element library FEniCS, with spatial discretization via conforming finite elements on a 32×32 triangular mesh. The mesh is fine enough to ensure that the time-discretization error dominates the spatial error, which is, generally, non-zero as cubic conforming elements on triangular meshes are used for all computations but one, namely for $k = 10^{-3}$, $\varepsilon = 10^{-5}$, where quadric elements are used.

The errors and the convergence rates are given in Table 1, where k is the time-step size, $\|e\|_{L^\infty(L^2)} := \max_{0 \leqslant n \leqslant N} \|u(t^n) - U^n\|$, and 'rate' is the respective convergence rate between two consecutive time-step sizes. As predicted by the theory, second order convergence with respect to k is observed in all cases.

4.2 Example 2

Next, we consider a convective variant of the classical Fisher equation, namely problem (1.1) with $\Omega = [0, 1]^2$, $T = 1$, $b = (1, 1)^T$, $c = 0$ and $f(u, x, t) = 10u(1 - u)$. We apply the implicit-explicit BDF2 method for $\varepsilon = 10^{-1}, 10^{-2}$, with spatial discretization via conforming quadratic finite elements on a 64×64

Table 2 Example 2: error and convergence rates

	$\varepsilon = 10^{-1}$			$\varepsilon = 10^{-2}$		
k	$\|e\|_{L^\infty(L^2)}$	Rate	k	$\|e\|_{L^\infty(L^2)}$	Rate
1.0e−1	4.982e−2	–	5.0e−2	6.743e−1	–
5.0e−2	1.647e−2	1.58	2.5e−2	2.653e−1	1.34
2.5e−2	4.348e−3	1.87	1.25e−2	2.382e−2	3.48
1.25e−2	8.677e−4	2.12	6.125e−3	6.793e−3	1.81

triangular mesh. The finite element space is accurate enough to ensure that the time-discretization error dominates the spatial error and that the boundary layers are, therefore, sufficiently resolved.

As no analytical solution is available, the time-discretization error is computed by comparing the numerical solution to a much finer reference numerical solution $\tilde{U}^n$, $n = 0, 1, \ldots, N$. The reference numerical solution is computed using the implicit-explicit Euler method from Sect. 3.1, with $k = 2.5 \cdot 10^{-4}$ and cubic conforming finite elements on the same meshes as the numerical solution. The errors and the convergence rates are given in Table 2, where k is the time-step size, $\|e\|_{L^\infty(L^2)} := \max_{0 \leqslant n \leqslant N} \|\tilde{U}^n - U^n\|$, and 'rate' is the respective convergence rate between two consecutive time-step sizes. Approximately second order convergence with respect to k is observed in this case also.

References

1. Akrivis, G.: Implicit–explicit multistep methods for nonlinear parabolic equations. Math. Comput. **82**, 45–68 (2013)
2. Akrivis, G.: Stability of implicit–explicit backward difference formulas for nonlinear parabolic equations. SIAM J. Numer. Anal. **53**, 464–484 (2015)
3. Akrivis, G.: Stability of implicit and implicit–explicit multistep methods for nonlinear parabolic equations. IMA J. Numer. Anal. **38**, 1768–1796 (2018)
4. Akrivis, G., Crouzeix, M.: Linearly implicit methods for nonlinear parabolic equations. Math. Comput. **73**, 613–635 (2004)
5. Akrivis, G., Lubich, C.: Fully implicit, linearly implicit and implicit–explicit backward difference formulae for quasi-linear parabolic equations. Numer. Math. **131**, 713–735 (2015)
6. Akrivis, G., Crouzeix, M., Makridakis, Ch.: Implicit–explicit multistep finite element methods for nonlinear parabolic problems. Math. Comput. **67**, 457–477 (1998)
7. Akrivis, G., Crouzeix, M., Makridakis, Ch.: Implicit–explicit multistep methods for quasilinear parabolic equations. Numer. Math. **82**, 521–541 (1999)
8. Banjai, L., Peterseim, D.: Parallel multistep methods for linear evolution problems. IMA J. Numer. Anal. **32**, 1217–1240 (2011)
9. Burman, E., Ern, A.: Implicit-explicit Runge-Kutta schemes and finite elements with symmetric stabilization for advection-diffusion equations. ESAIM Math. Model. Numer. Anal. **46**, 681–707 (2012)
10. Burman, E., Ern, A., Fernández, M.A.: Explicit Runge-Kutta schemes and finite elements with symmetric stabilization for first-order linear PDE systems. SIAM J. Numer. Anal. **48**, 2019–2042 (2010)

11. Crouzeix, M.: Une méthode multipas implicite–explicite pour l'approximation des équations d'évolution paraboliques. Numer. Math. **35**, 257–276 (1980)
12. Dolejší, V., Vlasák, M.: Analysis of a BDF–DGFE scheme for nonlinear convection–diffusion problems. Numer. Math. **110**, 405–447 (2008)
13. Georgoulis, E.H., Loghin, D.: Norm preconditioners for discontinuous Galerkin hp-finite element methods. SIAM J. Sci. Comput. **30**, 2447–2465 (2008)
14. Hairer, E., Wanner, G.: Solving Ordinary Differential Equations II: Stiff and Differential–Algebraic Problems. Springer Series in Computational Mathematics v. 14, 2nd revised edn. Springer, Berlin (2002)
15. Thomée, V.: Galerkin Finite Element Methods for Parabolic Problems, 2nd edn. Springer, Berlin (2006)
16. Zlámal, M.: Finite element methods for nonlinear parabolic equations. RAIRO **11**, 93–107 (1977)

H(div)-Conforming Spaces Based on General Meshes, with Interface Constraints: Accuracy Enhancement, Multiscale, and *hp*-Adaptividy

Philippe R. B. Devloo, Omar Durán, and Sônia M. Gomes

Abstract The importance of **H**(div)-conforming approximations is well recognized for conservative mixed formulations of multiphysics systems. There exists in the literature a variety of such approximation spaces, which are usually restricted to standard element geometry: triangular, quadrilateral, hexahahedral, tetrahedral, or prismatic meshes. We describe the principles in the construction of more general **H**(div)-conforming contexts, the meshes allowing non-convex polygonal (polyhedral) local subdomains. Given a finite dimensional normal flux space Λ_c, piecewise defined over a partition of the mesh skeleton, **H**(div)-conforming approximation spaces $\mathbf{V}_c$ keep fixed the face flux components constrained by Λ_c, but the internal flux components and the potential approximations inside the subdomains may be enriched in different extents: with respect to internal mesh size, internal polynomial degree, or both. The definition and implementation of such constrained space configurations are possible due to an available hierarchy of vector shape functions for classic polynomial spaces of arbitrary degree defined on usual master elements.

1 Introduction

There is a growing interest on numerical methods for partial differential equations using approximations based on general polygonal/polyhedral meshes due to their appeal in various domains in computational practice (e.g. HDG [5], HHO [12], virtual finite element methods [1], and references there in). On this direction, the focus of the current paper is on the construction of **H**(div)-conforming spaces that can be used in mixed formulations based on such kinds of meshes.

P. R. B. Devloo · O. Durán
FEC-Universidade Estadual de Campinas, Campinas, SP, Brazil
e-mail: phil@fec.unicamp.br; oduran@unicamp.br

S. M. Gomes (✉)
IMECC-Unicamp, Campinas, SP, Brazil
e-mail: soniag@ime.unicamp.br

© Springer Nature Switzerland AG 2020
G. R. Barrenechea, J. Mackenzie (eds.), *Boundary and Interior Layers,*
Computational and Asymptotic Methods BAIL 2018, Lecture Notes in
Computational Science and Engineering 135,
https://doi.org/10.1007/978-3-030-41800-7_5

Darcys's flows in a domain Ω consider flux fields of the form $\boldsymbol{\sigma} = -\mathbf{A}\nabla u$, where u is the fluid pressure, verifying $\nabla \cdot \boldsymbol{\sigma} = f$, where a positive definite tensor $\mathbf{A}$, $f \in L^2(\Omega)$, Dirichlet and Neumann boundary data u_D and g are given. As studied in [2], discrete variational mixed formulations of such problems search for approximations in finite dimensional spaces $u \in U^{\mathscr{T}} \subset L^2(\Omega)$ and $\boldsymbol{\sigma} \in \mathbf{V}^{\mathscr{T}} \subset \mathbf{H}(\mathrm{div}, \Omega)$, with $\boldsymbol{\sigma} \cdot \mathbf{n} = g$ on $\partial\Omega_N$, such that, for all $\mathbf{q} \in \mathbf{V}^{\mathscr{T}}$, with $\mathbf{q} \cdot \mathbf{n}|_{\partial\Omega_N} = 0$, and $v \in U^{\mathscr{T}}$,

$$\int_\Omega \mathbf{A}^{-1}\boldsymbol{\sigma} \cdot \mathbf{q} \, d\Omega - \int_\Omega u \, \nabla.\mathbf{q} \, d\Omega = -\int_{\partial\Omega_D} u_D \mathbf{q} \cdot \mathbf{n} \, ds, \qquad (1.1)$$

$$\int_\Omega \nabla \cdot \boldsymbol{\sigma} \, v \, d\Omega \qquad = \int_\Omega f v d\Omega. \qquad (1.2)$$

There exist in the literature a variety of such approximation spaces, which are usually restricted to partitions $\mathscr{T}$ of the computational domain formed by standard element geometry: triangular, quadrangular, hexahedral, tetrahedral, or prismatic meshes. The functions $\mathbf{q} \in \mathbf{V}^{\mathscr{T}}$ must have continuous normal traces over element interfaces, but for $v \in U^{\mathscr{T}}$ no interface continuity is required. Some principles for the construction of more general $\mathbf{H}(\mathrm{div})$-conforming contexts shall be described in Sect. 3, which can be based on polygonal/polyhedral meshes (even non-convex or with curved sides). Given a finite dimensional normal flux space Λ_c, piecewise defined over a partition of the mesh skeleton, $\mathbf{H}(\mathrm{div})$-conforming approximation spaces $\mathbf{V}_c$ are such that the face flux components are determined by Λ_c, but the internal fluxes (with vanishing normal traces over the mesh skeleton) and the potential approximations inside the subdomains may have different refinement patterns: with respect to internal mesh size, internal polynomial degree, or both. The methodology is strongly based on an available hierarchy of vector shape functions for classic polynomial spaces of arbitrary degree defined on usual master elements, which are summarized in Sect. 2. In Sect. 4, three particular cases of constrained spaces configurations recently proposed are reviewed, which are verified for some test problems with known exact solutions. The first one refers to uniform internal polynomial degree enrichment, while the second one considers two-scale uniform refinements of local spaces by internal mesh size and polynomial enrichment. The third example considers non-uniform hp-adaptive frameworks.

2 Standard Constructions of Compatible Space Configurations

Space configurations $U^{\mathscr{T}}$, $\mathbf{V}^{\mathscr{T}}$ to be used for potential and flux approximations in mixed finite element methods are piecewise defined over the elements of $\mathscr{T}$. For stable simulations, they should be compatible (i.e., inf-sup condition holds).

Basic Characteristics There exists in the literature different classic families of compatible space configurations, and their constructions have some common basic characteristics. Master elements $\hat{K}$ are associated to the elements $K \in \mathscr{T}$ by a geometric diffeomorfism $F_K : \hat{K} \to K$. Usually, for two dimensional problems, $\hat{K}$ may be a triangle Tr or a square Sq, and a tetrahedron $\hat{K} = Te$, a cube $\hat{K} = He$, or a prism $\hat{K} = Pr$ are used for three dimensional cases. A pair of vector polynomial space $\mathbf{V}_k(\hat{K})$ and a scalar polynomial space $U_k(\hat{K})$ is assigned to the master element $\hat{K}$, k indicating the degree of the normal traces $\hat{\mathbf{q}} \cdot \hat{\mathbf{n}}|_{\partial \hat{K}}$. In all the cases, compatibility holds if

$$\nabla \cdot \mathbf{V}_k(\hat{K}) = U_k(\hat{K}). \qquad (2.1)$$

On the computational elements K, local approximation spaces $\mathbf{V}_k(K)$ and $U_k(K)$ are defined backtracking the polynomial spaces $\mathbf{V}_k(\hat{K})$ and $U_k(\hat{K})$. The Piola transformation $\mathbb{F}_K^{\text{div}}$ or the usual mapping of scalar functions $\mathbb{F}_K$ are used, both induced by F_K. Precisely, $U_k(K) = \mathbb{F}_K U_k(\hat{K}) = \{\varphi|\ \varphi \circ F_K \in U_k(\hat{K})\}$, and

$$\mathbf{V}_k(K) = \mathbb{F}_K^{\text{div}} \mathbf{V}_k(\hat{K}) = \left\{\mathbf{q}|\ \hat{\mathbf{q}} = J_K\ DF_K^{-1}\mathbf{q} \circ F_K \in \mathbf{V}_k(\hat{K})\right\},$$

where DF_K is the Jacobian of F_K, $J_K = det(DF_K) > 0$ is assumed, affine elements occurring when J_K is constant. For $\hat{\mathbf{q}} \in \mathbf{V}_k(\hat{K})$, and $\hat{\mathbf{x}} \in \hat{K}$, $\nabla \cdot \hat{\mathbf{q}} = J_K(\hat{\mathbf{x}})\nabla \cdot \mathbf{q}$ holds.

Hierarchy of High Order Vector Shape Functions The applications of higher order mixed formulations can lead to complications with enforcing the appropriate **H**(div)-conformity, namely the continuity of normal traces over element interfaces. Therefore, the definition and implementation of globally defined hierarchical vector shape functions of arbitrary order are very helpful, particularly if elements of non-uniform local order are employed. Recently, increasing efforts have been put into the development and/or implementation of suitable sets of basis functions $\mathbf{B}_k^{\hat{K}}$ for classic vector spaces $\mathbf{V}_k(\hat{K})$. The methodology is described in [15] for Tr and Sq, and in [4] for Te, He, and Pr, following a common sequence of steps:

1. Scalar polynomial spaces P_n, and hierarchic basis functions $\hat{\varphi}$ of $P_n(\hat{K})$ are provided. For Tr and Te, the polynomials in $P_n(\hat{K})$ have total degree n, for Sq and He they have maximum degree n in each coordinate, and for Pr, $P_n(\hat{K})$ is formed by polynomials of total degree n for the coordinates over the planar triangular sections, and of maximum degree n for the orthogonal coordinate. They are classified according to the goemetric components of the element: vertex, edge, face or internal types, being non-zero at the corresponding geometric component, vanishing at the other ones of same dimension, and at all other components of smaller dimensions. These H^1 hierarchical scalar bases have been constructed in [7].

2. Constant vector fields $\hat{\mathbf{v}}$ are defined over $\hat{K}$, which are classified as being of face (or of edges for two dimensional elements) or internal type. A field associated to a given face is incident to it, with compatible normal component there. There are

internal fields connected to the interior of the master element and internal fields which are tangent to an edge or to a face.

3. Vector shape function $\hat{\boldsymbol{\Phi}} \in \mathbf{B}_k^{\hat{K}}$ are defined by the multiplication of a vector field $\hat{\mathbf{v}}$ by a scalar basic function $\hat{\varphi}$, getting $\hat{\boldsymbol{\Phi}} = \hat{\varphi}\hat{\mathbf{v}}$. There are vector shape functions $\hat{\boldsymbol{\Phi}}$ of interior type, with vanishing normal traces over all element faces $\hat{F} \subset \partial\hat{K}$. Otherwise, $\hat{\boldsymbol{\Phi}}$ is classified as of face type. The normal trace $\hat{\boldsymbol{\Phi}} \cdot \hat{\mathbf{n}}|_{\partial\hat{K}}$ of a shape function associated to a face F vanishes over all other faces, and coincides over F with the trace $\hat{\varphi}|_F$ of the scalar shape function used in its definition, having degree $\leq k$. For the internal shape functions $\hat{\boldsymbol{\Phi}} = \hat{\varphi}\hat{\mathbf{v}}$, $\hat{\varphi}$ may have higher degree.

4. Thus, the bases are decomposed as $\mathbf{B}_k^{\hat{K}} = \mathbf{B}_k^{\partial\hat{K}} \cup \mathring{\mathbf{B}}_k^{\hat{K}}$. Accordingly, a direct factorization $\mathbf{V}_k(\hat{K}) = \mathbf{V}_k^{\partial}(\hat{K}) \oplus \mathring{\mathbf{V}}_k(\hat{K})$ holds, in terms of a face component $\mathbf{V}_k^{\partial}(\hat{K})$ spanned by $\mathbf{B}_k^{\partial\hat{K}}$, and an internal component $\mathring{\mathbf{V}}_k(\hat{K})$ spanned by $\mathring{\mathbf{B}}_k^{\hat{K}}$.

5. Vector-valued basis $\mathbf{B}_k^K$ for $\mathbf{V}(K)$ are defined after the application of $\mathbb{F}_K^{\mathrm{div}}$ to $\mathbf{B}_k^{\hat{K}}$. Since zero normal traces are preserved by the action of $\mathbb{F}_K^{\mathrm{div}}$, shape functions in $\mathbf{B}_k^K$ keep the classification in terms of face or internal types, according to their counterpart ones in the master element $\hat{K}$.

Assembly of Approximation Spaces $\mathbf{V}^{\mathcal{T}} \subset \mathbf{H}(\mathbf{div}, \Omega)$ Associated to $\mathcal{T}$, let the mesh skeleton $\mathcal{E}$ be the union of all element faces $F \subset \partial K$. All over $\mathcal{E}$ the normal traces $\mathbf{q} \cdot \mathbf{n}$ across an interface $F = K_\ell \cap K_j$ between two neighbouring elements K_ℓ and K_j should coincide. For the enforcement of this crucial property, the local expansions of $\mathbf{q}$ in terms of the hierarchic shape functions described above are fundamental, recalling that only the face functions associated to F on each side play a role, requiring the constraint of the coefficients multiplying them. Furthermore, the particular definition of the face shape functions facilitate this task. In fact, recall that a face shape function $\boldsymbol{\Phi}^\ell$ associated to F on the side K_ℓ is defined backtracking a face shape function $\hat{\boldsymbol{\Phi}} = \hat{\varphi}\hat{\mathbf{v}}$, such that $\boldsymbol{\Phi}^\ell = \mathbb{F}_{K_\ell}^{\mathrm{div}}\hat{\boldsymbol{\Phi}} = \varphi^\ell\mathbf{b}^\ell$, where $\varphi^\ell = \mathbb{F}_{K_\ell}(\hat{\varphi})$ and $\mathbf{b}^\ell = \mathbb{F}_{K_\ell}^{\mathrm{div}}\hat{\mathbf{v}}$. Thus, $\boldsymbol{\Phi}^\ell \cdot \mathbf{n}|_F = \varphi|_F \, \mathbf{b}^\ell \cdot \mathbf{n}|_F$. Using the fact that the face vector field $\hat{\mathbf{v}}$ is chosen to have compatible normal component over $\hat{F}$, then $\mathbf{b}^\ell \cdot \mathbf{n}|_F = \mathbf{b}^j \cdot \mathbf{n}|_F$ holds. Using this statement, the continuous assembly of the normal traces of $\boldsymbol{\Phi}^\ell$ and $\boldsymbol{\Phi}^j$ over F reduces to the continuous assembly of the corresponding scalar functions φ^ℓ and φ^j (which is described in [3] for meshes without limitations on hanging sides and distribution of approximation orders). Consequently, the normal trace continuity of $\mathbf{q}$ requires that the coefficients multiplying the face functions $\boldsymbol{\Phi}^\ell$ and $\boldsymbol{\Phi}^j$ associated to F in the expansions of $\mathbf{q}|_{K_\ell}$ and $\mathbf{q}^{K_j}$, respectively, should coincide.

3   General Space Configurations with Interface Constraints

For the general mixed finite element formulations we have in mind, the construction of finite-dimensional approximation spaces require a (macro) partition $\mathcal{T}_H = \{\Omega_e\}$ of the region Ω by subdomains Ω_e. Regular conforming local partitions $\mathcal{T}^e = \{K\}$

are defined for each subdomain Ω_e, and local approximation spaces $U^{\mathcal{T}^e} \subset L^2(\Omega_e)$, and $\mathbf{V}^{\mathcal{T}^e} \subset \mathbf{H}(\text{div}, \Omega_e)$ are defined for them, as described in the previous section.

The choice of the subdomains Ω_e may be quite general: they may have one of the usual element geometry (triangle, quadrilateral, tetrahedral, hexahedral or prismatic), but non-convex polygonal (polyhedral) domains are allowed. The elements $K \in \mathcal{T}^e$ are supposed to have one of the usual geometry, which may vary between them. Instead, for neighboring subdomains Ω_e and Ω_i, the meshes $\mathcal{T}^e$ and $\overset{\rightarrow}{\mathcal{T}}$ may be non-conforming over a face $F \subset \partial\Omega_e \cap \partial\Omega_i$. Associated to $\mathcal{T}_H$, let the mesh skeleton $\mathcal{E}_H$ be the union of all element faces $F \subset \partial\Omega_e$. For most of the examples discussed here, the elements $K \in \mathcal{T}^e$ are supposed to be affine, but general frameworks with non-planar element faces may be admitted as well.

Consider the lower dimension functional space

$$\Lambda = \left\{ \mu \in H^{-\frac{1}{2}}(\mathcal{E}_H); \ \exists \sigma \in \mathbf{H}(\text{div}, \Omega) \text{ such that } \mu = \sigma \cdot \mathbf{n}|_{\mathcal{E}_H} \right\},$$

and give a finite dimensional subspace $\Lambda_c \subset \Lambda$, of functions piecewise defined over a given skeleton partition $\Gamma^{\mathcal{E}_H} = \{E\}$. The index c (for coarse) indicates that:

Hypothesis 1 A micro face induced on $\partial\Omega_e$ by $\mathcal{T}^e$ should intersect only one of the elements E of $\Gamma^{\mathcal{E}_H}$. This means that the restriction of $\Gamma^{\mathcal{E}_H}$ over a face $F \subset \partial\Omega_e$ defines a partition of F coarser than (or equal to) the one induced on it by the internal mesh $\mathcal{T}^e$.

Hypothesis 2 The functions in $\Lambda_c|_F$ are embedded in the space formed by the normal components $\mathbf{v} \cdot \mathbf{n}|_F$ of functions $\mathbf{v} \in \mathbf{V}^{\mathcal{T}^e}$.

Figure 1 illustrates the kind of general meshes that can be used to form approximation space configurations with interface constraints for mixed formulations.

Based on Λ_c, and on the local spaces $\mathbf{V}^{\mathcal{T}^e}$, the $\mathbf{H}(\text{div})$-conforming space constrained to Λ_c is then defined as $\mathbf{V}_c^{\mathcal{T}_H} = \left\{ \mathbf{v} \in \mathbf{H}(\text{div}, \Omega); \ \mathbf{v}|_{\Omega_e} \in \mathbf{V}_c^{\mathcal{T}^e} \right\}$, where $\mathbf{V}_c^{\mathcal{T}^e} = \left\{ \mathbf{v} \in \mathbf{V}^{\mathcal{T}^e}, \ \mathbf{v} \cdot \mathbf{n}|_{\mathcal{E}_H} \in \Lambda_c \right\}$. Similarly, consider $U^{\mathcal{T}_H} = \left\{ \varphi \in L^2(\Omega); \ \varphi|_{\Omega_e} \in U^{\mathcal{T}^e} \right\}$. Furthermore, bases $\mathbf{B}_c^{\Omega_e} = \{\boldsymbol{\Phi}_c^{\partial\Omega_e}\} \cup \{\overset{\circ}{\boldsymbol{\Phi}}^{\Omega_e}\}$ can be constructed for $\mathbf{V}_c^{\mathcal{T}^e}$, formed by vector shape functions $\overset{\circ}{\boldsymbol{\Phi}}$ of internal type (with vanishing normal components over $\partial\Omega_e$), and face shape functions $\boldsymbol{\Phi}_c^{\partial\Omega_e}$ otherwise, inducing a direct factorization $\mathbf{V}_c^{\mathcal{T}^e} = \mathbf{V}_c^{\partial\Omega_e} \oplus \overset{\circ}{\mathbf{V}}^{\Omega_e}$.

It should be emphasized that applications of such constrained space configurations lead to stable mixed finite element simulations. In fact, as in standard cases, the compatibility condition (2.1) holds for all elements K of the global refined partition $\cup_e \mathcal{T}^e$ of Ω. Furthermore, error estimates for the approximated variables can be expressed in terms of the projection errors of the exact solution on the corresponding spaces, as for classic error estimates. In fact, accuracy estimations for approximations of sufficiently smooth vector fields in the constrained spaces $\mathbf{V}_c^{\mathcal{T}_H}$ can be obtained for bounded projections $\boldsymbol{\Pi} : \mathbf{H}^\alpha(\Omega) \to \mathbf{V}_c^{\mathcal{T}_H}$. They are piecewise

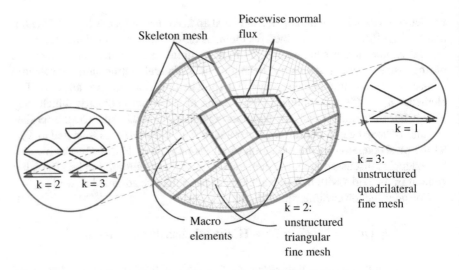

Fig. 1 Sketch of a general mesh $\mathcal{T}_H$, skeleton mesh, Λ_c, local meshes and spaces

constructed from local projections $(\boldsymbol{\Pi}\mathbf{q})|_{\Omega_e} = \boldsymbol{\pi}^e(\mathbf{q}|_{\Omega_e})$. Let $\Gamma : L^2(\Omega) \to U^{\mathcal{T}_H}$ be the standard L^2-projection, which can also be locally defined as $\Gamma\varphi|_{\Omega_e} = \gamma^e(\varphi|_{\Omega_e})$, in terms of the local L^2-projections γ^e over $U^{\mathcal{T}^e}$. The following property should be valid

$$\int_{\Omega_e} \nabla \cdot [\boldsymbol{\pi}^e\mathbf{q} - \mathbf{q}]\varphi \, d\Omega_e = 0, \quad \forall\varphi \in U^{\mathcal{T}^e}. \tag{3.1}$$

As suggested in [6], there is a general form to define projections verifying property (3.1), without requiring any specific geometric aspect. Inspired by the representation of the spaces $\mathbf{V}_c^{\mathcal{T}^e}$ by a direct sum of face and internal functions, the corresponding local projections $\boldsymbol{\pi}^e : \mathbf{H}^\alpha(\Omega_e) \to \mathbf{V}_c^{\mathcal{T}^e}$ are also defined by the combination of two similar contributions $\boldsymbol{\pi}_e\mathbf{q} = \boldsymbol{\pi}_e^\partial\mathbf{q} + \mathring{\boldsymbol{\pi}}_e(\mathbf{q} - \boldsymbol{\pi}_e^\partial\mathbf{q})$, such that: $\boldsymbol{\pi}_e^\partial\mathbf{q} \in \mathbf{V}_c^{\partial\Omega_e}$ is determined by

$$\int_{\partial\Omega_e} \boldsymbol{\pi}_e^\partial\mathbf{q} \cdot \mathbf{n} \, \mu \, ds = \int_{\partial\Omega_e} \mathbf{q} \cdot \boldsymbol{\eta}^{\Omega_e} \mu \, ds, \quad \forall\mu \in \Lambda_c|_{\partial\Omega_e}, \tag{3.2}$$

and $\mathring{\boldsymbol{\pi}}_e\mathbf{q} \in \mathring{\mathbf{V}}^{\Omega_e}$ is determined by the following constraints, valid for $\sigma \in \mathring{\mathbf{V}}^{\Omega_e}$,

$$\int_{\Omega_e} \nabla \cdot [\boldsymbol{\pi}_e^\partial\mathbf{q} + \mathring{\boldsymbol{\pi}}_e\mathbf{q}] \, \nabla \cdot \sigma \, d\Omega_e = \int_{\Omega_e} \nabla \cdot \mathbf{q} \, \nabla \cdot \sigma \, d\Omega_e, \tag{3.3}$$

$$\int_{\Omega_e} [\boldsymbol{\pi}_e^\partial\mathbf{q} + \mathring{\boldsymbol{\pi}}_e\mathbf{q}] \cdot \sigma \, d\Omega_e = \int_{\Omega_e} \mathbf{q} \cdot \sigma \, d\Omega_e, \text{ with } \nabla \cdot \sigma = 0. \tag{3.4}$$

As for the special cases considered in [8, 11, 13], which are summarized in Examples 1 and 2 of the next section, similar arguments can be applied for the general contexts described above to prove that the local projections $\pi_e \mathbf{q}$ are uniquely defined, and verify the required commutation property (3.1).

Concerning the implementation of mixed finite element simulations based on space configurations $U^{\mathscr{T}_H}$, $\mathbf{V}_c^{\mathscr{T}_H}$, standard static condensation, supported by purely algebraic procedures, can be applied to solve the associated linear systems, by selecting as primary variables the degrees-of-freedom associated to the flux face shape functions $\boldsymbol{\Phi}_c^{\partial\Omega}$, and one degree-of-freedom u_e per macro-element, and by condensing all the remaining ones (those associated to internal flux functions $\overset{\circ}{\boldsymbol{\Phi}}{}^{\Omega_e}$, and the degrees-of-freedom of $u|_{\Omega_e}$, excepting u_e). Thus, the number of equations to be solved in the condensed system is proportional to the dimension of the space Λ_c.

4 Examples of Space Configurations with Interface Constraints

Three different examples of mixed formulations based on space configurations with interface constraint are described in this section, and their performance are illustrated for applications to test problems with known solutions. As starting point of their definitions, a basic standard framework $U^{\mathscr{T}}$, $\mathbf{V}^{\mathscr{T}}$ is given, which is based on a conforming coarse partition $\mathscr{T} = \{K\}$, verifying the properties described in Sect. 2. Furthermore, the next assumptions are supposed to hold: (a) a partition $\mathscr{T}_H$ is formed by subregions Ω_e obtained by grouping some elements $K \in \mathscr{T}$, leading to local meshes $\mathscr{T}^e$; (b) a mesh skeleton $\mathscr{E}_H$ is created by the faces $E \subset \partial\Omega_e$ induced by $\mathscr{T}$, and local spaces $U^{\mathscr{T}^e}$, $\mathbf{V}^{\mathscr{T}^e}$ are defined by restricting the functions of $U^{\mathscr{T}}$ and $\mathbf{V}^{\mathscr{T}}$ to Ω_e; (c) new local vector and scalar spaces defined in Ω_e are obtained by uniform h or p refinement of the local spaces $U(K)$, $\mathbf{V}(K)$ of the elements $K \in \mathscr{T}_e$, but these refinements may differ from one subdomain to another; finally, (c) a partition $\Gamma^{\mathscr{E}_H}$ of $\mathscr{E}_H$, and a normal trace space Λ_c, piecewise defined over $\Gamma^{\mathscr{E}_H}$, are chosen such that Hypotheses 1 and 2 are fulfilled.

Example 1 (Enhanced Potential and Flux Divergence Approximations) Let us consider a framework as described above, for which the starting point is a basic standard framework $U_k^{\mathscr{T}} \subset L^2(\Omega)$ and $\mathbf{V}_k^{\mathscr{T}} \subset \mathbf{H}(\text{div}, \Omega)$, based on a partition $\mathscr{T}$, with local spaces $U_k(K)$, $\mathbf{V}_k(K)$, obtained backtracking a stable reference space configuration $\mathbf{V}_k(\hat{K})$ and $U_k(\hat{K})$, according to the description in Sect. 2. Identify each of the elements $K \in \mathscr{T}$ as a subdomain, i.e. $\mathscr{T}_H = \mathscr{T}$, and the mesh skeleton $\mathscr{E}_H$ as the union of the faces $F \subset \partial K$, $K \in \mathscr{T}$. Also set $\Gamma^{\mathscr{E}_H}$ as the partition of $\mathscr{E}_H$ formed by these element faces, and consider the subspace $\Lambda_{c,k} \subset \Lambda$ of functions μ piecewise defined over the mesh skeleton by $\mu|_F = \mathbf{v} \cdot \mathbf{n}|_F$, $F \subset \partial K$, of functions $\mathbf{v} \in \mathbf{V}_k(K)$. Define new local spaces $\mathbf{V}_{k+n}(K)$ and $U_{k+n}(K)$ by uniform p-refinement. Thus, Hypotheses 1 and 2 are clearly fulfilled. Let $U_{k+n}^{\mathscr{T}}$ be the scalar space based $\mathscr{T}$ associated to the enriched local spaces $U_{k+n}(K)$, and define the constrained vector space

$$\mathbf{V}_k^{\mathcal{T},+n} = \{\mathbf{v} \in \mathbf{H}(\mathrm{div}, \Omega); \ \mathbf{v}|_K \in \mathbf{V}_{k+n}(K); \ \mathbf{v} \cdot \mathbf{n}|_{\mathcal{E}_H} \in \Lambda_{c,k}\}.$$

These constrained frameworks have been analysed in [4, 8, 11] for stable simulations of mixed finite element formulations, where they have been interpreted under the equivalent opposite point of view of space enrichment. In fact, the spaces $\mathbf{V}_k^{\mathcal{T},+n}$ may also be defined by the addition to $\mathbf{V}_k(K)$ of internal shape functions of $\mathbf{V}_{k+n}(K)$, while keeping the original face flux space of $\mathbf{V}_k(K)$. If they are associated to shape-regular meshes $\mathcal{T}_h$ (which should also be non degenerated for 3D problems), h indicating mesh resolution, it has been proved that, for some classic basic standard frameworks (Raviart-Thomas RT_k for quadrilaterals and hexahera), Brezzi-Douglas-Fortin-Marini $BDFM_{k+1}$ for triangles and tetrahedra, and Nédélec spaces N_k for prisms), flux divergence can be obtained with arbitrary desired accuracy (of order h^s, with $s = k+n+1$ for affine meshes, $s = k+n$ for bilinearly mapped quadrilaterals, and $s = k+n-1$ for general hexahedra mapped by trilinear geometric transformations F_K). The potential accuracy can also be enhanced by increasing its convergence rate in one unit (to order $m = k+2$ for affine meshes or bilinearly mapped quadrilaterals, instead of $m = k+1$ otherwise). Flux accuracy are not affected with increasing n, keeping the convergence rate order obtained with the basic framework $U_k^{\mathcal{T}_h}$, $\mathbf{V}_k^{\mathcal{T}_h}$ (namely, order h^s, with $s = k+1$ for affine meshes, and bilinearly mapped quadrilaterals, and $s = k-1$ for trilinearly mapped hexahedra).

The constrained point of view is convenient for the implementations of the space configurations $U_k^{\mathcal{T},+n}$, $\mathbf{V}_k^{\mathcal{T},+n}$. Assuming that the full spaces $U_{k+n}(K)$, $\mathbf{V}_{k+n}(K)$ have been already implemented, for any polynomial degree $k + n$, and that their internal and face shape functions can be identified, then the assembly algorithm for vector functions in $\mathbf{V}_k^{\mathcal{T},+n} \subset \mathbf{V}_{k+n}^{\mathcal{T}}$ is defined by setting to zero the coefficients multiplying the face shape function with normal traces of degree $m, k < m \leq k+n$.

Test Problem 1 Consider an application of the constrained space configuration $U_k^{\mathcal{T},+n}$, $\mathbf{V}^{\mathcal{T},+n}$ to solve a Darcy's problem by the mixed formulation to a test model problem defined on the unit cube $\Omega = (0, 1)^3$, with $f = -\Delta u_{exact}$, and $u_D = u_{exact}|_{\partial\Omega}$, where the analytic solution u_{exact} is given by

$$u_{exact} = \frac{\pi}{2} - \tan^{-1}\left(5\left(\sqrt{(x-1.25)^2 + (y+0.25)^2 + (z+0.25)^2} - \frac{\pi}{3}\right)\right).$$

Uniform meshes, and the basic frameworks RT_k for hexahera, $BDFM_{k+1}$ for tetrahedra, and N_k for prisms, are adopted. Convergence histories for flux, flux divergence and pressure variables are shown in Fig. 2, taking $k = 2$, $n = 0$ and 3. The predicted convergence rates are verified: order 3 for the flux in all the cases, order 3 for u when the classic schemes are applied ($n = 0$), and order 4 for the enriched versions ($n = 3$). Divergence accuracy improves from order 3 when $n = 0$ to order 6 for $n = 3$.

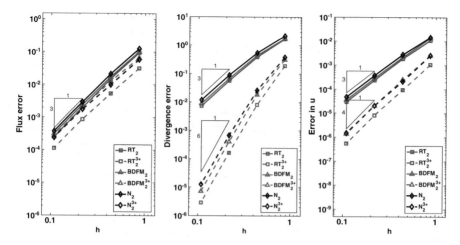

Fig. 2 Test Problem 1: convergence histories for different mixed approximation settings $U_k^{\mathcal{T},+n}$, $\mathbf{V}_k^{\mathcal{T},+n}$, for $k = 2$ and $n = 0, 3$, based on tetrahedral, hexahedral and prismatic meshes

Example 2 (Constrained Two-Scale Space Configurations) Now let us consider the framework described above, for which there is a standard basic one-scale stable framework $U_k^{\mathcal{T}_h} \subset L^2(\Omega)$ and $\mathbf{V}_k^{\mathcal{T}_h} \subset \mathbf{H}(\text{div}, \Omega)$, based on a partition $\mathcal{T}_h$ of Ω, as described in Sect. 2. A partition $\mathcal{T}_H$ is formed by subregions Ω_e composed by local partitions $\mathcal{T}_h$ inherited from $\mathcal{T}_h$. Therefore, neighboring local partitions $\mathcal{T}_h$ and $\dot{\mathcal{T}}_h$ match over a single partition $\mathcal{T}_h^F$ of each interface $F \subset \partial \Omega_e \cap \Omega_i$. Define $\Lambda_c := \Lambda_{h,k}$ by the normal traces of functions in $\mathbf{V}_k^{\mathcal{T}_h}$ over the faces included in $\partial \Omega_e$, and consider constrained two-scale spaces $\mathbf{V}_k^{\mathcal{T}_{\bar{h}},n+}$ by: (a) uniform subdivision of $\mathcal{T}_h$, given new internal refined partitions $\mathcal{T}_{\bar{h}}^e$ of Ω_e, $\bar{h} < h$, combined with (b) uniform polynomial degree increment to form local spaces $U_{\bar{h},k+n}(\Omega_e)$ and $\mathbf{V}_{\bar{h},k+n}(\Omega_e)$ based on $\mathcal{T}_{\bar{h}}^e$, and (c) constraint of face functions to those ones matching the normal traces to $\Lambda_{h,k}$. Precisely, define $U_k^{\mathcal{T}_{\bar{h}},n+}$ by local spaces $U_{\bar{h},k}^{n+}(\Omega_e) = U_{\bar{h},k+n}(\Omega_e)$, and

$$\mathbf{V}_k^{\mathcal{T}_{\bar{h}},n+} = \left\{ \mathbf{v} \in \mathbf{V}_{\bar{h},k+n}(\Omega_e); \ \mathbf{v}|_{\Omega_e} \in \mathbf{V}_{\bar{h},k}^{n+}(\Omega_e), \text{ and } \mathbf{v} \cdot \mathbf{n}|_{\mathcal{E}} \in \Lambda_{h,k} \right\}.$$

These kinds of constrained two-scale space configurations have beed used in [13] for applications of a new version of the Multiscaled Hybrid Mixed (MHM) method, denoted by MHM-**H**(div). As introducced in [14], MHM method refers to a numerical technique targeted to approximate systems of differential equations with strongly varying solutions. It follows the "divide-and-conquer" principle, for which the solutions are characterized in terms of boundary value problems locally set on each macrodomain Ω_e, which are assembled by using transmission conditions throughout the mesh skeleton, normal fluxes (multiplier) making the interelement connection. The multiplier and coarse piecewise constant potential approximations in each subdomain are computed (upscaling). Then, small details are resolved by local problems, using fine representations inside the subdomains, setting the multiplier as Neumann boundary conditions (downscaling). The MHM-$\mathbf{H}$(div) variant adopts mixed finite elements at the dowscaling stage, instead of continuous finite elements used in all previous publications of the MHM method. It has been shown in [13] that the MHM-$\mathbf{H}$(div) method can be interpreted as a mixed formulation based on constrained space configurations, as described in Sect. 3. Error estimations for sufficiently smooth exact solutions reveal that for two-scale space configurations for classic spaces RT_k, $BDFM_{k+1}$ and N_k based on shape-regular and non degenerate affine meshes of convex regions, convergence is of order h^{k+1} for the flux, $\bar{h}^{-k+n+1}$ for the flux divergence, and h^{k+2} for u.

Test Problem 2 Consider $\Omega = (0, 1)^2$, $f = -\Delta u_{exact}$, and $u_D = u_{exact}|_{\partial\Omega}$, such that $u_{exact} = \operatorname{arctg}(100(r - 0.5))(1 + 0.3\sin(10\pi x))(1 + 0.5\cos(10\pi y))$, $r = \sqrt{x^2 + y^2}$. Macro partitions $\mathscr{T}_H$ are formed by uniform square subdomains with side length $H = 2^{-j}$, $j = 2, \cdots, 6$. No skeleton subdivision is used ($h = H$), and one more refinement level inside the macro elements is adopted ($\bar{h} = H/2$). History of convergence for potential u and flux $\sigma = \nabla u$, measured by L^2-norms, and expressed in terms of macro element size H, are illustrated in Fig. 3 for the approximate solutions obtained with the space configurations $U_k^{\mathscr{T}_{\bar{h}},n+}$, $\mathbf{V}_k^{\mathscr{T}_{\bar{h}},n+}$, where $k = 3$ is used to define the normal traces over the mesh skeleton, and Raviart-Thomas elements RT_{3+n}, $n = 0, 1, 2$ are applied inside the subdomains. Observe that for $n = 0$ the errors have the standard behavior of the corresponding one-level schemes, with rates of order $k + 1$ for potential and flux variables, after the strong solution oscillations are properly resolved. Increasing the polynomial degree

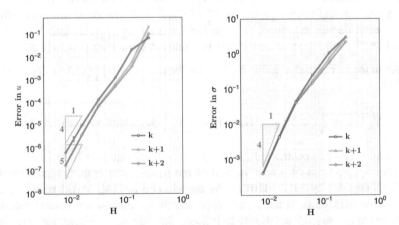

Fig. 3 Test Problem 2: L^2 error curves for u (left side) and ∇u (right side), expressed in terms of macro element size H for the approximate solutions obtained with space configurations $U_k^{\mathscr{T}_{\bar{h}},n+}$, $\mathbf{V}_k^{\mathscr{T}_{\bar{h}},n+}$, where $\bar{h} = H/2$, $k = 3$ and $n = 0, 1, 2$

to $k+n$, for potential and internal fluxes inside the macro elements, the error curves show enhanced convergence rate of order $k+2$ for the potential, reaching order $k+1$ for the flux.

Example 3 (Adaptive hp Space Configurations) The purpose here is to illustrate the construction of *hp*-adaptive space configurations under the point of view of interface constrained spaces, as described above. This means that, instead of h and/or p refinements uniformly defined over the subdomains $\Omega_e \in \mathcal{T}_H$, as adopted in Examples 1 and 2, these refinements may differ from one subdomain to another. Consequently, non-uniform patterns $\mathbf{k} = (k_e)$ and $\mathbf{h} = (h_e)$ may occur for the polynomial degree and mesh resolution of the local approximation spaces $U_{h_e,k_e}(\Omega_e)$ and $\mathbf{V}_{h_e,k_e}(\Omega_e)$ based on the refined local partitions $\mathcal{T}_{h_e}$. Assembly of such kinds of local spaces require the definition of a partition $\Gamma_H^{\mathcal{E}}$ of mesh skeleton $\mathcal{E}_H$, and of a normal flux space Λ_c verifying the Hypotheses 1 and 2. As described in [9, 10] for 2D and 3D *hp*-meshes, and similarly to the assembly in the case of conforming meshes (no hanging sides) and uniform polynomial order distribution discussed in Sect. 2, continuous normal traces over the mesh skeleton is obtained by using the special properties hold by the face vector fields, reducing the task to enforce continuity of function expansions in terms of the associated scalar shape functions, which is described in [3] for *hp*-adaptive meshes without restrictions concerning hanging faces and polynomial degree distributions.

Test Problem 3 Consider $u_{exact} = \frac{\pi}{2} - \arctan\left[\alpha\left(\sqrt{(x - 1.25)^2 + (y + 0.25)^2} - \frac{\pi}{3}\right)\right]$, and chose $f = -\Delta u_{exact}$, $\Omega = (0, 1)^2$, and boundary conditions accordingly. Strong gradients occur, with magnitude determined by the parameter $\alpha = 200$, in the proximity to the circumference centered at the point $(1.25, -0.25)$, with radius $\pi/3$. For this problem, let us consider a sequence of *hp*-adaptive meshes with triangular geometry, which are obtained from a conforming coarse (father) mesh composed of uniform triangles of width $h = 2^{-3}$ and polynomial degree $k = 2$. A first adaptive mesh is formed by refining once all the elements, and taking mesh size $h(K) = 2^{-4}$ and $k(K) = 3$ for the elements intersecting a layer of diameter $1/2$ around the singularity curve. The refinement process follows a sequence of steps $\ell = 2, 3$, and 4 by first increasing by one unit the approximation order of all elements of the previous step, and then by subdividing the elements intersecting a layer of diameter $2^{-\ell}$ around the singularity curve, and by further increasing their approximation order by one unit. In Fig. 4 (left side) a zoom in close to the singularity part illustrates the *hp* refinement at the final refinement level. L^2-error curves are shown for approximate flux and potential, in terms of the number of primary variables of the condensed systems, obtained with the triangular *hp*-adaptive space configurations for $BDFM_{k+1}^{+n}$ elements, with $n = 0$ (continuous curves) and $n = 1$ (dashed curves). The dashed-dotted curves correspond to simulations for uniform meshes and $BDFM_{k+1}$ configuration, with $k = 2$. For comparison, results for H^1-conforming formulation based on the same *hp*-meshes are also included (dotted curves). As expected, the application of *hp* refinement improves considerably the performance of the methods, with exponential

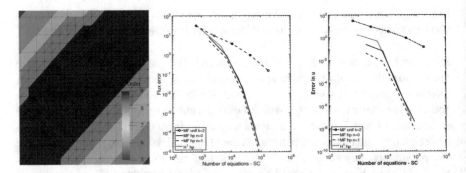

Fig. 4 Test Problem 3: zoom in close to the singularity region of the triangular hp-adaptive mesh at the final refinement step (left side), and L^2-error curves for the flux and potential variables versus the number of condensed equations (right side), using the hp-adaptive space configurations for the mixed (MF) and H^1-conforming formulations

convergence. Furthermore, accuracy in u improves when $BDFM_k^{+1}$ configuration is applied in the hp-adaptive mixed formulation.

References

1. Beirão da Veiga, L., Brezzi, F., Marini, L.D., Russo, A.: Virtual Element implementation for general elliptic equations. In: Barrenechea, G., Brezzi, F., Cangiani, A., Georgoulis, E. (eds.) Connections and Challenges in Modern Approaches to Numerical Partial Differential Equations. Lecture Notes in Computational Science and Engineering, vol. 114, pp. 39–71. Springer, New York (2016)
2. Brezzi, F., Fortin, M.: Mixed and Hybrid Finite Element Methods. Springer Series in Computational Mathematics, vol. 15. Springer, New York (1991)
3. Calle, J.L.D., Devloo, P.R.B., Gomes, S.M.: Implementation of continuous hp-adaptive finite element spaces without limitations on hanging sides and distribution of approximation orders. Comput. Math. Appl. **70**(5), 1051–1069 (2015)
4. Castro, D.A., Devloo, P.R.B., Farias, A.M., Gomes, S.M., de Siqueira, D., Durán, O.: Three dimensional hierarchical mixed finite element approximations with enhanced primal variable accuracy. Comput. Methods Appl. Mech. Eng. **306**, 479–502 (2016)
5. Cockburn, B.: Static condensation, hybridization, and the devising of the HDG methods. In: Barrenechea, G., Brezzi, F., Cangiani, A., Georgoulis, E. (eds.) Connections and Challenges in Modern Approaches to Numerical Partial Differential Equations. Lecture Notes in Computational Science and Engineering, vol. 114, pp. 129–177. Springer, New York (2016)
6. Demkowicz, L., Monk, P., Vardapetvan, L., Rachowicz, W.: De Rham diagram for hp finite element spaces. Comput. Math. Appl. **39**, 29–38 (2000)
7. Devloo, P.R.B., Bravo, C.M.A., Rylo, C.M.A.: Systematic and generic construction of shape functions for p-adaptive meshes of multidimensional finite elements. Comput. Methods Appl. Mech. Eng. **198**, 1716–1725 (2009)
8. Devloo, P.R.B., Durán, O., Farias, A.M., Gomes, S.M.: H(div) finite elements based on non-affine meshes for three dimensional mixed formulations of flow problems with arbitrary high order accuracy of the divergence of the flux. Int. J. Numer. Methods Eng. (2020). https://doi.org/10.1002/nme.6337

9. Devloo, P.R.B., Farias, A.M., Gomes, S.M., de Siqueira, D.: Two-dimensional hp adaptive finite element spaces for mixed formulations. Math. Comput. Simul. **126**, 104–122 (2016)

10. Devloo, P.R.B., Durán, O., Gomes, S.M., Shauer, N.: Mixed finite element approximations based on 3D *hp*-adaptive curved meshes with two types of H(div)-conforming spaces. Int. J. Numer. Methods Eng. **113**(7), 1045–1060 (2017)

11. Devloo, P.R.B., Farias, A.M., Gomes, S.M.: A remark concerning divergence accuracy order for H(div)-conforming finite element flux approximations. Comput. Math. Appl. **77**, 1864–1872 (2019)

12. Di Pietro, D.A., Ern, A., Lemaire, S.: A review of hybrid high- order methods: formulations, computational aspects, comparison with other methods. In: Barrenechea, G., Brezzi, F., Cangiani, A., Georgoulis, E. (eds.) Connections and Challenges in Modern Approaches to Numerical Partial Differential Equations. Lecture Notes in Computational Science and Engineering, vol. 114, , pp. 205–236. Springer, New York (2016)

13. Durán, O., Devloo, P.R.B., Gomes, S.M., Valentin, F.: A multiscale hybrid method for Darcy's problems using mixed finite element local solvers. Comput. Methods Appl. Mech. Eng. **354**, 213–244 (2019)

14. Harder, C., Paredes, D., Valentin, F.: A family of multiscale hybrid mixed finite element methods for the darcy equation with rough coefficients. J. Comput. Phys. **245**, 107–130 (2013)

15. Siqueira, D., Devloo, P.R.B., Gomes, S.M.: A new procedure for the construction of hierarchical high order Hdiv and Hcurl finite element spaces. J. Comput. Appl. Math. **240**, 204–214 (2013)

A Numerical Method for the Hemker Problem

Alan F. Hegarty and Eugene O'Riordan

Abstract We construct a new numerical method comprising upwind finite difference operators on asymptotically appropriate Shishkin meshes to obtain a numerical approximation to the solution of the Hemker problem. Numerical results indicate that the numerical approximations are computationally uniformly convergent with respect to the small parameter ε.

1 Introduction

A model problem proposed by Hemker [7] was to find a globally accurate numerical approximation to the solution of the linear elliptic problem:

$$-\varepsilon \Delta u + u_x = 0, \qquad (x, y) \in \mathbb{R}^2 \setminus \bar{\Omega}, \quad \bar{\Omega} := \{(x, y) | x^2 + y^2 \leq 1\}$$

$$u(x, y; \varepsilon) = 1 \quad \text{for} \quad x^2 + y^2 = 1; \quad u(x, y; \varepsilon) = 0 \quad \text{for} \quad x^2 + y^2 \to \infty,$$

whose pointwise accuracy was retained irrespective of the size of the singular perturbation parameter $\varepsilon \in (0, 1]$.

For those interested in realistic simulation, such as non-linear flow models on domains with complicated geometry, the problem can be seen as a simple academic problem to compare and modify numerical algorithms. From this exercise, the most promising algorithms would subsequently be utilised in more complicated simulation models. In this respect, one is reluctant to incorporate *a priori*

A. F. Hegarty (✉)
University of Limerick, Limerick, Ireland
e-mail: alan.hegarty@ul.ie

E. O'Riordan
Dublin City University, Dublin, Ireland
e-mail: eugene.oriordan@dcu.ie

© Springer Nature Switzerland AG 2020
G. R. Barrenechea, J. Mackenzie (eds.), *Boundary and Interior Layers, Computational and Asymptotic Methods BAIL 2018*, Lecture Notes in Computational Science and Engineering 135, https://doi.org/10.1007/978-3-030-41800-7_6

information into the numerical method that may be specific to the Hemker problem. Many different finite element/finite volume methods (applied to this particular test problem) were reviewed in [1], but none were seen to be satisfactory from the point of stability or accuracy.

An alternative perspective would be taken by numerical analysts, who are primarily interested in establishing theoretical parameter-uniform error bounds. At present, the Hemker problem is beyond the reach of current theory, as sharp *a priori* bounds on the partial derivatives of the solution are not available. Nevertheless, systematic computations that would indicate parameter-uniform convergence of the numerical approximations emanating from a carefully designed numerical algorithm would identify a potentially useful path for the theory to develop in its quest to develop robust and accurate numerical approximations for a wide class of singularly perturbed problems. From this perspective, one is typically willing to use as much *a priori* information about the continuous solution that is deemed necessary (and is available) to achieve parameter-uniform convergence.

In Han et al. [4], a stable numerical method was constructed on a quasi-uniform mesh, but there was a limited discussion of the accuracy of the numerical approximations. The focus in this *Tailored Finite Point Method* was to design a discrete scheme, which was stable for all values of the singular perturbation parameter. The method is related to the fitted operator schemes, developed by Il'in [8] and others.

In the design of a parameter-uniform numerical method, we also believe that uniform stability of the numerical method should be a necessary component of the numerical algorithm. We achieve this by using simple upwinding. However, our main focus is on the design of an appropriate fitted mesh, so that we guarantee that a significant proportion of the mesh point lie within the layers. To effect this, we design Shishkin meshes whose location and widths are identified using formal asymptotic analysis of the continuous problem. In addition, we emphasise that we shall estimate the global convergence of the interpolated computed approximations across the entire domain.

We are interested in parameter-uniform numerical methods [3] that satisfy the global error bound

$$\|\bar{U}^N - u\|_{\infty,\Omega} \leq CN^{-p}, \quad p > 0,$$

where C and p are independent of the singular perturbation parameter ε. We stress that we are using the pointwise L^∞ norm. Here u is the solution of the continuous problem approximated by the bilinear interpolated $\bar{U}^N$ and U^N is the solution of the discrete problem (where N denotes the number of elements used in any co-ordinate direction). In this paper, we construct a numerical algorithm for a bounded domain version of the Hemker problem, which is seen to perform computationally as a parameter-uniform numerical method.

2 Problem in an Annulus

In this paper, we confine our discussion to the solution of the Hemker problem posed on a finite domain: Find $\tilde{u}(x, y)$ such that

$$-\varepsilon\Delta\tilde{u} + \tilde{u}_x = 0, \quad \text{in} \quad A := \{(x, y)|R_1 < x^2 + y^2 < R_2\} \quad \text{(2.1a)}$$

$$\tilde{u} = 0, \quad \text{when} \quad x^2 + y^2 = R_2; \quad \tilde{u} = 1, \quad \text{when} \quad x^2 + y^2 = R_1. \quad \text{(2.1b)}$$

Note that the solution of this problem, unlike the original problem in the unbounded domain, has an exponential layer at the outer boundary for $x > 0$. The numerical solutions in this paper will all be for the particular values of $R_1 = 1$ and $R_2 = 4$. There is an obvious symmetry of $\tilde{u}(x, y) = \tilde{u}(x, -y)$ in this problem. To avoid discussions about the discretization of an artificial boundary at $y = 0$, we simply choose not to utilize this symmetry in our numerical algorithm.

It is of course natural in this annular domain [6] to consider the problem in polar coordinates: Find $u(r, \theta)$ such that

$$-\frac{\varepsilon}{r^2}u_{\theta,\theta} - \varepsilon u_{rr} + a_1 u_r + a_2 u_\theta = 0, \quad r \in (R_1, R_2), \theta \in [0, 2\pi]; \quad \text{(2.2a)}$$

$$\text{where} \quad a_1 := \cos(\theta) - \frac{\varepsilon}{r}, a_2 := -\frac{\sin(\theta)}{r}; \quad \text{(2.2b)}$$

$$\text{with} \quad u(R_1, \theta) = 1, \quad u(R_2, \theta) = 0, \quad 0 \le \theta \le 2\pi. \quad \text{(2.2c)}$$

To solve the problem (2.2) numerically we use the internal computational domain

$$\Omega^N := \{(r_i, \theta_j)|0 \le \theta_j \le 2\pi, R_1 < r_i < R_2\}_{i=1, j=0}^{N-1, N}$$

and a finite difference method with an upwind approximation to the first derivative to ensure inverse monotonicity:

$$-\frac{\varepsilon}{r_i^2}\delta_\theta^2 U - \varepsilon\delta_r^2 U + a_1 D_r^\pm U + a_2 D_\theta^\pm U = 0, \quad (r_i, \theta_j) \in \Omega^N;$$

$$U(R_1, \theta_j) = 1, \quad U(R_2, \theta_j) = 0, \quad 0 \le \theta_j \le 2\pi.$$

Here $\delta_\theta^2 u, \delta_r^2 u$ are the classical second order centered difference approximations [3] to $u_{\theta\theta}, u_{rr}$ respectively and $D_r^\pm u, D_\theta^\pm u$ are the standard upwind discretizations of u_r, u_θ, where all four finite difference operators are defined on an arbitrary mesh.

In [5] we examined the solution of this problem (2.1) using a simple Shishkin mesh with refinement near the inner and outer circles, but the numerical results were not parameter uniform, even for $x \le 0$. We cannot expect any numerical method to be parameter uniform in the full domain if we are taking no account of the asymptotic nature of the solution for $x > 0$. However, we can do this for $x \le 0$ if we employ the following mesh motivated by the asymptotic nature of the solution in the neighbourhood of the characteristic points $x = 0, y = \pm 1$. In the vicinity of the characteristic point $(0, 1)$, asymptotic analysis [8, pg. 163] suggests that the appropriate scalings for the layers are of the form

$$\frac{r - 1}{\varepsilon^{2/3}} \quad \text{and} \quad \frac{\pi/2 - \theta}{\varepsilon^{1/3}}.$$

On the annulus A we now define a mesh Ω_A^N, which incorporates this asymptotic information.

Shishkin Mesh Ω_A^N *In the radial direction, the interval $[R_1, R_2]$ is discretised using a Shishkin mesh with two transitions points*

$$\sigma_1 := \min\{\frac{R_2 - R_1}{4}, 2\varepsilon \ln N\}, \quad \sigma_2 := \min\{\frac{R_2 - R_1}{4}, \varepsilon^{2/3} \ln N\};$$

so that the interval $[R_1, R_2]$ is subdivided into the intervals

$$[R_1, R_1 + \sigma_1], [R_1 + \sigma_1, R_1 + \sigma_1 + \sigma_2], [R_1 + \sigma_1 + \sigma_2, R_2 - \sigma_1], [R_2 - \sigma_1, R_2].$$

We use a uniform mesh with $N/4$ elements in each of these four subintervals. Similarly the angular domain $[0, 2\pi]$ is discretised using a Shishkin mesh with the transition point

$$\tau := \min\{\frac{\pi}{4}, 2\varepsilon^{1/3} \ln N\}$$

so that $[0, 2\pi]$ is divided into the five intervals

$$[0, \frac{\pi}{2} - \tau], [\frac{\pi}{2} - \tau, \frac{\pi}{2} + \tau], [\frac{\pi}{2} + \tau, \frac{3\pi}{2} - \tau], [\frac{3\pi}{2} - \tau, \frac{3\pi}{2} + \tau], [\frac{3\pi}{2} + \tau, 2\pi],$$

on each of which a uniform mesh is used with, respectively, $N/8$, $N/4$, $N/4$, $N/4$ and $N/8$ subintervals. This completes the construction of the mesh Ω_A^N.

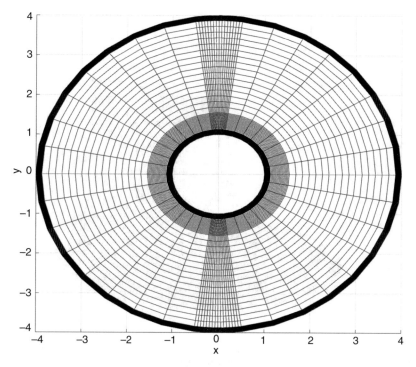

Fig. 1 A schematic image of the annular mesh Ω_A^N

A schematic image of this mesh is presented in Fig. 1. However, in practice, the refinement in the radial direction only becomes apparent to the user for very small values of ε.

A sample computed solution on this mesh Ω_A^N is presented in Fig. 2, which is free of oscillations (due to the choice of upwinding) and displays boundary and interior layers in the expected areas. However, the approximate global pointwise errors in this computed approximation are presented in Fig. 3. Clearly we do not obtain a good numerical solution on the entire computational domain. However, using computed global orders of two-mesh differences[1] [3, Chapter 8], Table 1 shows that we do in fact obtain a uniformly in ε accurate solution, not only for $x \leq 0$ but in fact for $x \leq \varepsilon^{1/3}$. This is a significant improvement over the corresponding

---

[1]Let U_ε^N be the computed solutions on certain meshes Ω_ε^N. Define the maximum two-mesh global differences by

$$D_\varepsilon^N := \|\bar{U}_\varepsilon^N - \bar{U}_\varepsilon^{2N}\|_{\Omega_\varepsilon^N \cup \Omega_\varepsilon^{2N}} \quad \text{and} \quad D^N := \max_\varepsilon D_\varepsilon^N, \tag{2.3a}$$

where $\bar{U}^N$ denotes the bilinear interpolation of the discrete solution U^N on the mesh Ω_ε^N. Then, for any particular value of ε and N, the computed orders of convergence are $\bar{p}_\varepsilon^N$ and, for any particular value of N and *all values of* ε, the **parameter-uniform** computed orders of convergence $\bar{p}^N$ are

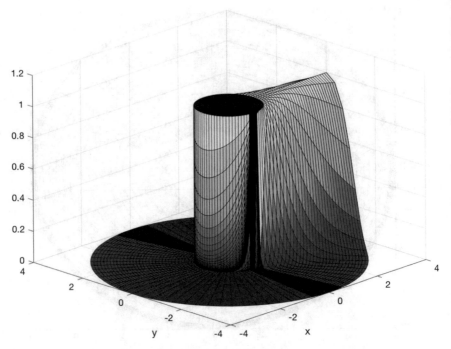

Fig. 2 Numerical solution generated on the mesh Ω_A^{128}, with $\varepsilon = 2^{-15}$

results in [5]. However, the convergence is lost beyond the vertical line $x = \varepsilon^{1/3}$ (which can be seen in Fig. 3), as there is no refinement in the vicinity of the internal parabolic layers near the half-lines $y = \pm 1, x > 0$.

Given this improvement in the vicinity of the characteristic points, we could now follow the approach taken in [5] and use the numerical solution generated on the annular mesh Ω_A^N as the left boundary condition at $x = L$ (where $0 \leq L \leq \varepsilon^{1/3}$ is to be determined) and solve the original problem (2.1) on a Cartesian mesh (with refinement near the internal parabolic layers) to obtain a numerical approximation for $x > 0$, downwind of the unit circle. However, even by starting the left boundary at $L = \varepsilon^{1/3}$, we observed that the numerical solutions are not convergent for all ε irrespective of the various different layer-adapted meshes we tried to discretize the rectangular domain $[L, R_2] \times [-R_2, R_2]$. The errors grew soon after the line $x = \varepsilon^{1/3}$.

defined, respectively, by

$$\bar{p}_\varepsilon^N := \log_2\left(\frac{D_\varepsilon^N}{D_\varepsilon^{2N}}\right) \quad \text{and} \quad \bar{p}^N := \log_2\left(\frac{D^N}{D^{2N}}\right). \tag{2.3b}$$

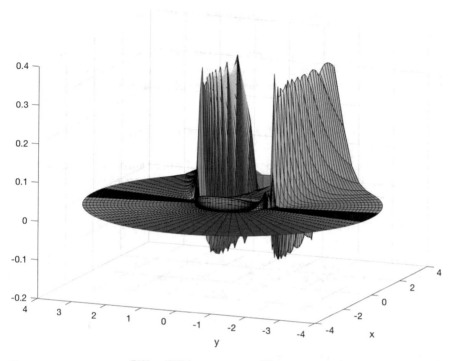

Fig. 3 Approximate error $\bar{U}^{128} - \bar{U}^{1024}$ on the mesh Ω_A^{128}, with $\varepsilon = 2^{-15}$

Table 1 Computed double-mesh orders $\bar{p}_\epsilon^N$ and parameter-uniform orders $\bar{p}^N$ of global convergence for the meshes Ω_A^N for $x \leq \epsilon^{1/3}$

ε	$\bar{p}_\epsilon^N$						
	8	16	32	64	128	256	512
1	1.0308	1.0221	1.0168	1.0104	1.0057	1.0029	1.0014
2^{-2}	0.9657	0.9264	0.9728	0.9915	0.9942	0.9971	0.9985
2^{-4}	-0.0467	0.6265	1.0062	0.9362	0.9822	0.9838	0.9913
2^{-6}	0.2668	0.4357	0.6837	0.8353	0.7666	0.7582	0.8260
2^{-8}	0.4093	0.4479	0.6905	0.8325	0.7653	0.7628	0.8242
2^{-10}	0.5854	0.4569	0.6920	0.8305	0.7662	0.7645	0.8239
2^{-12}	0.7471	0.4647	0.6926	0.8294	0.7667	0.7651	0.8242
2^{-14}	0.6091	0.7337	0.6929	0.8287	0.7671	0.7654	0.8244
2^{-16}	0.4341	0.7128	1.0516	0.8283	0.7674	0.7656	0.8245
2^{-18}	0.3254	0.5224	0.8344	1.1599	1.0866	0.7658	0.8246
2^{-20}	0.2363	0.3823	0.6625	0.9245	1.1442	1.1864	1.1055
$\bar{p}^N$	0.2363	0.3823	0.6625	0.9245	1.1442	1.1864	1.1055

3   In the Vicinity of the Characteristic Points

To retain convergence for $x > \varepsilon^{1/3}$, we next introduce a new patched region Q^+ in a neighbourhood of the point $(x, y) = (0, 1)$ (and a corresponding region Q^- near $(x, y) = (0, -1)$), whose dimensions are of the order $O(\varepsilon^{1/3} f(N)) \times O(1)$. Consider

$$Q^+ := \{(x, y) | y = t - x^2/2, \quad x_0 \le x \le x_1, \ 1 - 3\delta \le t \le 1 + 5\delta\}$$

where the dimensions of the region $[x_0, x_1]$ and δ will be specified in such a way so as to retain the stability of the discrete operator. A natural coordinate system [8] for this region is

$$s = x, \quad t = y + x^2/2, \quad (y = t - s^2/2)$$

which transforms the patch Q^+ to a rectangle $[x_0, x_1] \times [1 - 3\delta \le t \le 1 + 5\delta]$. Then let $\hat{u}(s, t) := u(x, y)$. Under this transformation the differential equation in (2.1) on the patched region Q^+ is of the form

$$-\varepsilon(\hat{u}_{ss} + 2s\hat{u}_{ts} + (1+s^2)\hat{u}_{tt}) + \hat{a}(s, t)\hat{u}_s + (\hat{a}(s, t)s - \varepsilon)\hat{u}_t = \hat{f}, \ (s, t) \in \hat{Q}^+ \quad (3.1)$$

Observe that there is a mixed second order derivative term $s\hat{u}_{ts}, \ s \in [x_0, x_1]$ present in this transformed differential equation. The influence of this mixed derivative term on the stability of any discretization will be moderated by limiting the maximum value of the coefficient. That is, x_1 will be limited in size. In this paper, we simply choose $x_0 = 0$ and $x_1 = L$, where the choice of L will be specified below.

Shishkin Mesh Ω_P^N   *We solve the transformed problem (3.1) in the s, t coordinate system, using a uniform mesh in the s direction and, in the t direction, a Shishkin mesh with two transition points*

$$\tau_1 := \min\{2\delta, \varepsilon^{2/3} \ln M\}, \qquad \tau_2 := \min\{2\delta, \sqrt{\varepsilon} \ln M\}$$

located above and below the line $t = 1$. The choice of $\sqrt{\varepsilon}$ in the scaling comes again from asymptotic analysis, which indicates that interior layers of width $O(\sqrt{\varepsilon})$ will evolve around the horizontal lines $y = \pm 1$ downwind of the unit circle, where $x > 0$. The vertical strip $[1 - 3\delta, 1 + 5\delta]$ is split into the following five sub-regions

$$[1 - 3\delta, 1 - \tau_1] \cup [1 - \tau_1, 1] \cup [1, 1 + \tau_1] \cup [1 + \tau_1, 1 + (\tau_2 + \tau_1)] \cup [1 + (\tau_2 + \tau_1), 1 + 5\delta]$$

and these sub-intervals are subdivided in the ratio $M/8 : M/4 : M/4 : M/4 : M/8$. This completes the construction of the mesh Ω_P^N.

Let $\bar{U}_1$ be the solution of the Hemker problem (2.1) on the annular mesh Ω_A^N, which we use as boundary values to solve on the region $\hat{Q}^+$. Hence we wish to solve the problem: Find $\hat{u}(s, t)$ such that

$$- \varepsilon(\hat{u}_{ss} + 2s\hat{u}_{ts} + (1+s^2)\hat{u}_{tt}) + \hat{a}(s, t)\hat{u}_s + (\hat{a}(s, t)s - \varepsilon)\hat{u}_t = \hat{f}, \quad (s, t) \in \hat{Q}^+$$

$$\hat{u}(s, t) = 1, \quad \text{if} \quad s^2 + (t - s^2/2)^2 \le 1, (\text{ inside inner circle})$$

$$\hat{u}(s, 1 + 5\delta) = \bar{U}_1(s, 1 + 5\delta), \quad s \in [0, L]$$

$$\hat{u}(0, t) = \bar{U}_1(0, t), \quad \hat{u}_s(L, t) = 0, \quad t \in (1 - 3\delta, 1 + 5\delta).$$

Observe that we use a homogenous Neumann condition at the artificial outflow boundary $s = L$ and the parameters $L.\delta$ are chosen so that the boundary $t = 1 - 3\delta$ lies entirely within the inner circle.

We now specify the associated discrete problem on this patch. For the internal mesh points, where $0 < s_i < L$, $1 - 3\delta < t_j < 1 + 5\delta$,

$$-\varepsilon \mathscr{L}^{N,M}\hat{U} + \hat{a}(s_i, t_j)D_s^{\pm}\hat{U} + (s_i\hat{a}(s_i, t_j) - \varepsilon)D_t^{\pm}\hat{U} = \hat{f};$$

$$\text{where} \quad \mathscr{L}^{N,M}Y := \delta_{ss}^2 Y + 2s D_t^- D_s^- Y + (1 + s^2)\delta_{tt}^2 Y$$

and for the remaining mesh points

$$\hat{U}(s_i, t_j) = 1, \quad \text{if} \quad s_i^2 + (t_j - s_i^2/2)^2 \le 1;$$

$$\hat{U}(s_i, 1 + 5\delta) = \bar{U}_1(s_i, 1 + 5\delta), \quad 0 \le s_i \le L;$$

$$\hat{U}(0, t_j) = \bar{U}_1(0, t_j), \quad D_s^- \hat{U}(L, t_j) = 0, \quad 1 - 3\delta < t_j < 1 + 5\delta.$$

By requiring that the elements in the associated system matrix have a suitable sign pattern for an M-matrix structure, we can preserve inverse monotonicity of the finite difference operator by choosing

$$\delta = 0.025 \quad \text{and} \quad L := \sqrt{\min\{2\delta, \varepsilon^{2/3} \ln N_{\text{fine}}\}}. \tag{3.2}$$

We now form a correction to the initial numerical approximation $\bar{U}_1$ over a small section of the annulus A, which we denote as $\bar{U}_2$: For all $(x, y) \in A \cap \{x \le \delta\}$, define

$$\bar{U}_2 := \begin{cases} \bar{U}_1, & \text{if} \quad (x, y) \in (A \cap \{x \le \delta\}) \setminus (Q^+ \cup Q^-), \\ \hat{U}_I, & \text{if} \quad (x, y) \in Q^+ \cup Q^-, \end{cases}$$

where $\hat{U}_I$ is the bilinear interpolant of $\hat{U}$, using the (s, t) co-ordinate system. In Figs. 4 and 5, we display the approximate errors over this patched region. We see

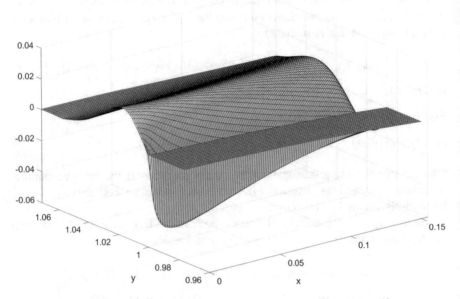

Fig. 4 Approximate error $\bar{U}^{128} - \bar{U}^{1024}$ over the parabolic mesh Ω_P^{128} for $\varepsilon = 2^{-10}$

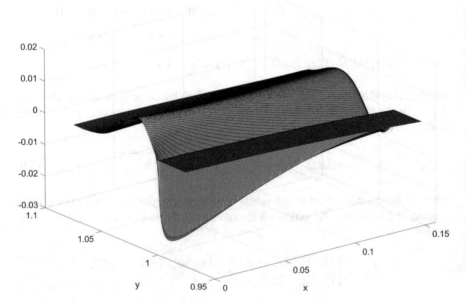

Fig. 5 Approximate error $\bar{U}^{256} - \bar{U}^{1024}$ over the parabolic mesh Ω_P^{256} for $\varepsilon = 2^{-10}$

a halving of the maximum pointwise error when we double the number of mesh points.

The introduction of this patch has allowed us extend the region of convergence on the annulus beyond $x \leq \varepsilon^{1/3}$ to $x \leq C\varepsilon^{1/3}(\ln N)^{1/2}$. This turns out to be a crucial improvement in the performance of the numerical algorithm.

4 Beyond the Characteristic Points

In the final phase of the numerical method, we generate an approximation for $x > L$. Now we use a rectangular co-ordinate system and solve the following problem: Find $\tilde{u}(x, y)$ such that

$$- \varepsilon \Delta \tilde{u} + \tilde{u}_x = 0, \quad (x, y) \in (L, R_2) \times (-R_2, R_2) \setminus \{(x, y) : x^2 + y^2 \leq 1\} \quad (4.1a)$$

$$\tilde{u}(x, y) = 1, \quad \text{if} \quad x^2 + y^2 \leq 1; \quad (4.1b)$$

$$\tilde{u}(x, -R_2) = \tilde{u}(x, R_2) = 0, \ x \in [L, R_2]; \quad \tilde{u}_x(R_2, y) = 0, \ y \in [-R_2, R_2]; \quad (4.1c)$$

$$\tilde{u}(L, y) = \bar{U}_2, R_2 > |y| > 1 - L^2. \quad (4.1d)$$

Observe that we have eliminated the artificial boundary layer (present in the problem (2.1) posed on the annulus) by applying a homogeneous Neumann condition at the outflow $x = R_2$. We computationally solve this problem (4.1) using standard upwinding on a rectangular Shishkin mesh Ω_C^N defined by:

Shishkin Mesh Ω_C^N *The mesh is uniform in the x-direction and in the vertical direction the interval $[-R_2, R_2]$ is discretised using a one-dimensional Shishkin mesh with*

$$\zeta_1 := \min\{\frac{1}{2}, \sqrt{\varepsilon} \ln N\} \quad \zeta_2 := \min\{\frac{R_2 - 1}{2}, \sqrt{\varepsilon} \ln N\}$$

so that $[-R_2, R_2]$ is subdivided into the intervals

$$[-R_2, -1 - \zeta_2], [-1 - \zeta_2, -1 + \zeta_1], [-1 + \zeta_1, 1 - \zeta_1], [1 - \zeta_1, 1 + \zeta_2], [1 + \zeta_2, R_2]$$

in the ratio $N/8 : N/4 : N/4 : N/4 : N/8$ subintervals.

The numerical solution generated over this mesh Ω_C^N will be denoted by U_R. Finally, the numerical solution across the entire domain is given by:

$$\bar{U} := \begin{cases} \bar{U}_2, & \text{if} \quad x \leq L, \\ \bar{U}_R, & \text{if} \quad x > L. \end{cases}$$

5   Computed Orders Over the Complete Domain

A sample of the composite mesh $\Omega_A^N \cup \Omega_P^N \cup \Omega_C^N$ is shown in Fig. 6. Although the inclusion of the patch in the vicinity of the characteristic points turns out to be a crucial ingredient in the algorithm, the grid over this patch is not visible in Fig. 6, given the dimensions of the patch.

The approximate global errors over the entire domain are presented in Figs. 7 and 8. They display a halving of the maximum global error as we double the number of mesh elements in each co-ordinate direction. These figures are, of course, only for one particular sample value of ε and N. The computed orders of convergence $\bar{p}_\varepsilon^N$ and the **parameter-uniform** orders of convergence $\bar{p}^N$ (defined in (2.3)) are presented in Table 2. It is important to point out that for any numerical method, in general, $\bar{p}^N \neq \min_\varepsilon \bar{p}_\varepsilon^N$, especially if the rates $\bar{p}_\varepsilon^N$ are not monotone in N [2]. Taking this into consideration, it is the parameter-uniform rates of convergence $\bar{p}^N$ which are of interest to us. The results in Table 2 indicate that the numerical solutions $\bar{U}$, computed over the composite domain, are converging uniformly for all values of the singular perturbation parameter within the range $\varepsilon \in [2^{-20}, 1]$.

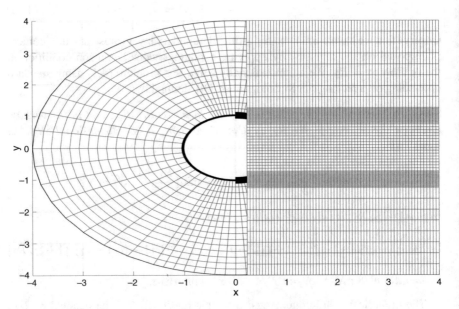

Fig. 6 Full composite mesh $\Omega_A^N \cup \Omega_P^N \cup \Omega_C^N$ for $N = 64$ and $\varepsilon = 2^{-10}$

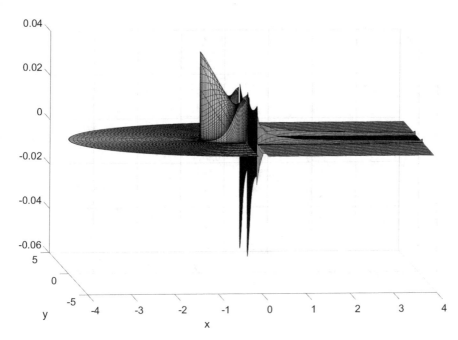

Fig. 7 Approximate error $\bar{U}^{128} - \bar{U}^{1024}$ for $\varepsilon = 2^{-10}$ over the composite mesh $\Omega_A^{128} \cup \Omega_P^{128} \cup \Omega_C^{128}$

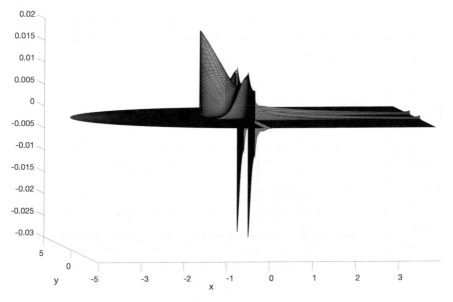

Fig. 8 Approximate error $\bar{U}^{256} - \bar{U}^{1024}$ for $\varepsilon = 2^{-10}$ over the composite mesh $\Omega_A^{256} \cup \Omega_P^{256} \cup \Omega_C^{256}$

Table 2 Computed double-mesh orders $\bar{p}_\epsilon^N$ and parameter-uniform orders $\bar{p}^N$ of global convergence for the composite meshes $\Omega_A^N \cup \Omega_P^N \cup \Omega_C^N$

ε	$\bar{p}_\epsilon^N$						
	8	16	32	64	128	256	512
1	1.0977	0.2922	0.6655	0.7427	0.8660	0.8180	0.8270
2^{-2}	0.6681	0.5705	0.3669	0.8542	0.7708	0.7660	0.7763
2^{-4}	0.3815	0.5800	0.6823	1.1740	0.5891	0.6868	0.7347
2^{-6}	0.4939	−0.1606	1.8700	−0.1396	0.2123	1.5108	0.3240
2^{-8}	0.2183	0.8220	1.6623	−0.6738	1.5720	−0.3670	1.1822
2^{-10}	0.8178	0.4077	0.8233	1.0415	0.6589	0.8509	0.9086
2^{-12}	0.7629	0.5676	0.7698	1.0034	1.0018	0.6660	0.7370
2^{-14}	0.5714	0.8034	0.5615	0.9343	0.8768	0.8613	1.0638
2^{-16}	0.4633	0.6751	0.6478	0.8441	0.9813	0.8658	1.0218
2^{-18}	0.4037	0.6131	0.5389	0.7511	0.9942	0.9435	1.0155
2^{-20}	0.3710	0.4480	0.3663	0.7591	1.0007	0.9806	1.0387
$\bar{p}^N$	0.3710	0.4480	0.3663	0.7591	1.0007	0.9806	1.0387

6   Conclusion

The numerical method gives globally accurate results for all values of ε between 2^{-20} and 1, which converge numerically uniformly with respect to ε within the specified finite domain. As noted in [9], as $x \to \infty$ the two interior layers eventually merge; the numerical method would need a modification if the right boundary $x = R_2$ were to be so extended.

References

1. Augustin, M., Caiazzo, A., Fiebach, A., Fuhrmann, J., John, V., Linke, A., Umla, R.: An assessment of discretizations for convection-dominated convection-diffusion equations. Comput. Methods Appl. Mech. Eng. **200**(47–48), 3395–3409 (2011)
2. Farrell, P.A., Hegarty, A.F.: On the determination of the order of uniform convergence. In: Proceedings of the 13th IMACS World Congress, Dublin, pp. 501–502 (1991)
3. Farrell, P.A., Hegarty, A.F., Miller, J.J.H., O'Riordan, E., Shishkin, G.I.: Robust Computational Techniques for Boundary Layers. Chapman and Hall/CRC Press, Boca Raton (2000)
4. Han, H., Huang, Z., Kellogg, R.B.: A tailored finite point method for a singular perturbation problem on an unbounded domain. J. Sci. Comput. **36**(2), 243–261 (2008)
5. Hegarty, A.F., O'Riordan, E.: Numerical results for singularly perturbed convection-diffusion problems on an annulus. In: Huang, Z., Stynes, M., Zhang, Z. (eds.) Proceedings of the International Conference on Boundary and Interior Layers - Computational and Asymptotic Methods, BAIL 2016, Beijing, August 2016. Lecture Notes in Computational Science and Engineering, vol. 120, pp. 101–112. Springer, New York (2017)
6. Hegarty, A.F. O'Riordan, E.: A parameter-uniform numerical method for a singularly perturbed convection-diffusion problem posed on an annulus. Comput. Math. Appl. **78**(10), 3329–3344 (2019)

7. Hemker, P.W.: A singularly perturbed model problem for numerical computation. J. Comp. Appl. Math. **76**, 277–285 (1996)
8. Il'in, A.M.: Matching of Asymptotic Expansions of Solutions of Boundary Value Problems. Mathematical Monographs, vol. 102. American Mathematical Society, Providence (1992)
9. Lagerstrom, P.A.: Matched Asymptotic Expansions: Ideas and Techniques. Applied Mathematical Sciences, vol. 76. Springer, New York (1988)

On Basic Iteration Schemes for Nonlinear AFC Discretizations

Abhinav Jha and Volker John

Abstract Algebraic flux correction (AFC) finite element discretizations of steady-state convection-diffusion-reaction equations lead to a nonlinear problem. This paper presents first steps of a systematic study of solvers for these problems. Two basic fixed point iterations and a formal Newton method are considered. It turns out that the fixed point iterations behave often quite differently. Using a sparse direct solver for the linear problems, one of them exploits the fact that only one matrix factorization is needed to become very efficient in the case of convergence. For the behavior of the formal Newton method, a clear picture is not yet obtained.

1 Introduction

A steady-state convection-diffusion-reaction equation is given by

$$-\varepsilon \Delta u + \boldsymbol{b} \cdot \nabla u + cu = f \quad \text{in } \Omega, \tag{1}$$

where $\Omega \subset \mathbb{R}^d$, $d \in \{2, 3\}$, is a bounded domain, $\varepsilon > 0$ is the diffusion coefficient, $\boldsymbol{b}$ is the convection field, c describes a reaction, and f is the source term. Problem (1) has to be equipped with boundary conditions on $\partial \Omega$.

The work of A. Jha has been supported by the Berlin Mathematical School (BMS).

A. Jha
Freie Universität Berlin, Department of Mathematics and Computer Science, Berlin, Germany
e-mail: jha@wias-berlin.de

V. John (✉)
Weierstrass Institute for Applied Analysis and Stochastics (WIAS), Berlin, Germany

Freie Universität Berlin, Department of Mathematics and Computer Science, Berlin, Germany
e-mail: john@wias-berlin.de

© Springer Nature Switzerland AG 2020
G. R. Barrenechea, J. Mackenzie (eds.), *Boundary and Interior Layers,*
Computational and Asymptotic Methods BAIL 2018, Lecture Notes in
Computational Science and Engineering 135,
https://doi.org/10.1007/978-3-030-41800-7_7

In applications, the convection-dominated regime $\varepsilon \ll \|b\|_{L^\infty(\Omega)}$ is of interest since the transport of the quantity u (temperature, concentration) by the convection field (velocity) is typically much stronger than the transport by molecular diffusion. In this regime, the solution of (1) possesses usually layers whose width is much smaller than the affordable mesh width.

It is well known that the discretization of (1) in the convection-dominated regime requires stabilized schemes. Two obviously desirable requirements are:

- the numerical solution should be accurate, in particular, it should exhibit sharp layers,
- the numerical solution must not have spurious oscillations.

The second requirement is particularly important in applications. Mathematically, it is formulated in the form of the discrete maximum principle (DMP). In the numerical analysis, the DMP is proved, e.g., with the sufficient condition that the discretization leads to a system with an M-matrix. However, it is known [17, Chap. 4.4] that a linear discretization of (1) in the limit case $\varepsilon = 0$ leading to an M-matrix cannot have a local discretization error of second order, thus it is only of low accuracy. A similar result for $\varepsilon > 0$ is not known, but experience shows that higher order discretizations lead to numerical solutions with spurious oscillations.

This situation led to the development of many nonlinear discretizations, where the nonlinearity arises from parameters that depend on the numerical solution. However, most of the proposed nonlinear schemes do not satisfy the DMP, see [11]. A notable exception is the class of so-called algebraic flux correction (AFC) schemes, proposed the first time for equations of type (1) in the context of finite element methods in [14]. In AFC schemes, one has to compute a so-called limiter, which depends on the finite element solution. The DMP for the Kuzmin limiter from [14] was proved in [3] and for another limiter, the so-called BJK limiter, in [4].

With the nonlinearity, a new issue arises:

- the numerical solution has to be computed efficiently.

So far, there is no discretization for convection-diffusion-reaction equations (1) that satisfies all three requirements. The construction of such a discretization is formulated as an important open problem in [13].

This paper studies numerical methods for the solution of the nonlinear problems arising in AFC schemes. There seems to be so far no systematic investigation of this topic in the literature. A few brief studies can be found in [1, 5], which state more or less that the solution of these problems might be problematic. The goal of this paper consists in performing first steps of a systematic study. Two fixed point iterations, which can be derived in a straightforward way, and a formal Newton method are included. Simulations were performed on academic problems in two dimensions. These studies should serve to obtain some insight into the properties of these schemes and, based on that, to develop ideas for improving them or for constructing new schemes.

2 AFC Schemes

This section provides a short presentation of AFC schemes.

Let $A\underline{u} = \underline{f}$, $A = (a_{ij})_{i,j=1}^n \in \mathbb{R}^{n \times n}$, $\underline{u}, \underline{f} \in \mathbb{R}^n$, be a linear system of equations from a conforming Galerkin discretization of (1). Ordering the unknowns such that the $(n - m)$, $m < n$, Dirichlet values are at the end of $\underline{u}$, this system can be written in the form

$$\sum_{j=1}^n a_{ij} u_j = f_i, \quad i = 1, \ldots, m,$$

$$u_i = u_i^b, \quad i = m+1, \ldots, n. \tag{2}$$

Defining the symmetric artificial diffusion matrix $D = (d_{ij})_{i,j=1}^n$ by

$$d_{ij} = d_{ji} = -\max\{a_{ij}, 0, a_{ji}\} \quad \text{for } i \neq j, \quad d_{ii} = -\sum_{i \neq j} d_{ij}, \tag{3}$$

leads to a system that is equivalent to (2)

$$(\hat{A}\underline{u})_i = f_i + (D\underline{u})_i, \quad i = 1, \ldots, n, \tag{4}$$

where $\hat{A} = A + D$. Since the row sums of the matrix D vanish, there is a representation

$$(D\underline{u})_i = \sum_{i \neq j} f_{ij}, \quad i = 1, \ldots, n,$$

with so-called fluxes $f_{ij} = d_{ij}(u_j - u_i) = -f_{ji}$ for all $i, j = 1, \ldots, n$.

AFC schemes limit those anti-diffusive fluxes f_{ij} that cause spurious oscillations. To this end, system (4) is modified to

$$(\hat{A}\underline{u})_i = f_i + \sum_{j \neq i} \alpha_{ij} f_{ij}, \quad i = 1, \ldots, n, \tag{5}$$

with solution-dependent coefficients $\alpha_{ij} = \alpha_{ij}(\underline{u}) \in [0, 1]$. It is important, for proving the existence of a solution of (5), see [3], and for the conservativity of the method, compare [15], that $\alpha_{ij} = \alpha_{ji}$, $i, j = 1, \ldots, n$. Rewriting (5) yields the following nonlinear system of equations

$$\sum_{j=1}^n a_{ij} u_j + \sum_{j=1}^n (1 - \alpha_{ij}) d_{ij}(u_j - u_i) = f_i, \quad i = 1, \ldots, m,$$

$$u_i = u_i^b, \quad i = m+1, \ldots, n. \tag{6}$$

In the literature, one finds several proposals of limiters for computing α_{ij}. Two of them are briefly presented here. More details, e.g., concerning some issues of the implementation, can be found in [3, 4].

The Kuzmin limiter. This limiter was proposed in [14]. Defining

$$P_i^+ = \sum_{\substack{j=1 \\ a_{ji} \leq a_{ij}}}^n f_{ij}^+, \quad P_i^- = \sum_{\substack{j=1 \\ a_{ji} \leq a_{ij}}}^n f_{ij}^-, \quad Q_i^+ = -\sum_{j=1}^n f_{ij}^-, \quad Q_i^- = -\sum_{j=1}^n f_{ij}^+, \quad (7)$$

$i = 1, \ldots, n$, where $f_{ij}^+ = \max\{0, f_{ij}\}$ and $f_{ij}^- = \min\{0, f_{ij}\}$, one computes

$$R_i^+ = \min\left\{1, \frac{Q_i^+}{P_i^+}\right\}, \quad R_i^- = \min\left\{1, \frac{Q_i^-}{P_i^-}\right\}, \quad i = 1, \ldots, m. \quad (8)$$

If P_i^+ or P_i^- is zero, one sets $R_i^+ = 1$ or $R_i^- = 1$, respectively. Also at Dirichlet nodes, one sets

$$R_i^+ = 1, \quad R_i^- = 1, \quad i = m + 1, \ldots, n. \quad (9)$$

Finally, for any $i, j \in \{1, \ldots, n\}$ such that $a_{ji} \leq a_{ij}$, the limiter is given by

$$\alpha_{ij} = \begin{cases} R_i^+ & \text{if } f_{ij} > 0 \\ 1 & \text{if } f_{ij} = 0 , \quad \alpha_{ji} = \alpha_{ij}. \\ R_i^- & \text{if } f_{ij} < 0 \end{cases} \quad (10)$$

The Kuzmin limiter can be applied to P_1 and Q_1 finite elements.

The BJK limiter. This limiter was developed in [4] for P_1 finite elements. It was proved that the corresponding AFC method is linearity preserving on arbitrary simplicial grids. First, one sets for $i = 1, \ldots, n$

$$u_i^{\max} = \max_{j \in S_i \cup \{i\}} u_j, \quad u_i^{\min} = \min_{j \in S_i \cup \{i\}} u_j, \quad q_i = \gamma_i \sum_{j \in S_i} d_{ij}, \quad (11)$$

where the index set S_i satisfies

$$\{j \in \{1, \ldots, n\} \setminus \{i\} : a_{ij} \neq 0 \text{ or } a_{ji} > 0\} \subset S_i \subset \{1, \ldots, n\},$$

and γ_i is a positive constant that has to be chosen sufficiently large. Now, one defines for $i = 1, \ldots, m$

$$P_i^+ = \sum_{j \in S_i} f_{ij}^+, \quad P_i^- = \sum_{j \in S_i} f_{ij}^-, \quad Q_i^+ = q_i(u_i - u_i^{\max}), \quad Q_i^- = q_i(u_i - u_i^{\min}),$$

$$(12)$$

and computes

$$R_i^+ = \min\left\{1, \frac{Q_i^+}{P_i^+}\right\}, \quad R_i^- = \min\left\{1, \frac{Q_i^-}{P_i^-}\right\}, \quad i = 1, \ldots, m.$$

If P_i^+ or P_i^- vanishes, one sets $R_i^+ = 1$ or $R_i^- = 1$, respectively. The final steps consist in setting (9) for the Dirichlet nodes, in computing

$$\bar{\alpha}_{ij} = \begin{cases} R_i^+ & \text{if } f_{ij} > 0 \\ 1 & \text{if } f_{ij} = 0 , \quad i = 1, \ldots, m, \; j = 1, \ldots, n, \\ R_i^- & \text{if } f_{ij} < 0 \end{cases} \tag{13}$$

and in setting

$$\alpha_{ij} = \min\{\bar{\alpha}_{ij}, \bar{\alpha}_{ji}\}, \quad i, j = 1, \ldots, m, \tag{14}$$

$$\alpha_{ij} = \bar{\alpha}_{ij}, \quad i = 1, \ldots, m, \; j = m+1, \ldots, n. \tag{15}$$

3 Methods for Solving the Nonlinear Problem

The methods considered here were already outlined in [5, Sec. 5].

Given an approximation $\underline{u}^{(v)}$, $v \geq 0$. Then, the next iterate is computed by

$$\underline{u}^{(v+1)} = \underline{u}^{(v)} + \omega^{(v)} \left(\hat{F}\left(\underline{u}^{(v)}\right) - \underline{u}^{(v)} \right), \tag{16}$$

where $\omega^{(v)}$ is a damping parameter and $\hat{F}$ is a map that is determined by the iterative method. The damping parameter is computed with an adaptive strategy that is described in detail in [12].

Consider the nonlinear problem (6) in the form $F(\underline{u}) = \underline{0}$ with

$$F_i(\underline{u}) = \sum_{j=1}^n a_{ij} u_j + \sum_{j=1}^n (1 - \alpha_{ij}(\underline{u})) d_{ij}(u_j - u_i) - f_i = 0, \quad i = 1, \ldots, m,$$

$$F_i(\underline{u}) = u_i - u_i^b = 0, \quad i = m+1, \ldots, n.$$

Then, the damped iteration (16) can be written as

$$\underline{u}^{(v+1)} = \underline{u}^{(v)} + \omega^{(v)} \left(B^{-1} \left[B\underline{u}^{(v)} - F\left(\underline{u}^{(v)}\right) \right] - \underline{u}^{(v)} \right), \tag{17}$$

with a non-singular matrix $B \in \mathbb{R}^{n \times n}$. A vector $\underline{u}$ is a solution of the nonlinear problem (6) if and only if it is a fixed point of (17).

A straightforward idea consists in considering linear problems where the currently available approximation of the limiter is used to assemble the matrix:

$$\sum_{j=1}^{n} a_{ij} u_j^{(v+1)} + \sum_{j=1}^{n} \left(1 - \alpha_{ij}^{(v)}\right) d_{ij} \left(u_j^{(v+1)} - u_i^{(v+1)}\right) = f_i, \quad i = 1, \ldots, m,$$

$$u_i^{(v+1)} = u_i^b, \quad i = m+1, \ldots, n,$$

$$(18)$$

with $\alpha_{ij}^{(v)} = \alpha_{ij}\left(u^{(v)}\right)$. In this iteration, the matrix B from (17) is given by

$$B\left(\underline{u}^{(v)}\right)_{ij} = \begin{cases} a_{ij} + d_{ij} - \alpha_{ij}^{(v)} d_{ij} & \text{if } i \neq j, \\ a_{ii} + d_{ii} + \sum_{j=1, j \neq i}^{n} \alpha_{ij}^{(v)} d_{ij} & \text{if } i = j, \end{cases}$$

for $i = 1, \ldots, m$, $j = 1, \ldots, n$. The last $n - m$ rows have just the diagonal entry 1.

Using that the row sums of the matrix D vanish, one can derive a second fixed point iteration where the limiter appears on the right-hand side, see [5] for details,

$$\sum_{j=1}^{n} (a_{ij} + d_{ij}) u_j^{(v+1)} = f_i + \sum_{j=1}^{n} \alpha_{ij}^{(v)} f_{ij}^{(v)}, \quad i = 1, \ldots, m,$$

$$u_i^{(v+1)} = u_i^b, \quad i = m+1, \ldots, n. \tag{19}$$

In (19), the matrix in iteration (17) is given by $B = A + D$, i.e., one has the same matrix in each step. Thus, using a (sparse) direct solver, a matrix factorization has to be performed only once.

Also, a formal Newton method can be derived. This method is formal because the limiters are not differentiable. The formal Jacobian is the matrix B of the scheme (17) and it is given by, compare [5],

$$B\left(\underline{u}^{(v)}\right)_{ij}$$

$$= \begin{cases} a_{ij} + d_{ij} - \alpha_{ij}^{(v)} d_{ij} - \sum_{k=1}^{n} \dfrac{\partial \alpha_{ik}^{(v)}}{\partial u_j} d_{ik} \left(u_k^{(v)} - u_i^{(v)}\right) & \text{if } i \neq j, \\ a_{ii} + d_{ii} + \sum_{\substack{j=1 \\ j \neq i}}^{n} \alpha_{ij}^{(v)} d_{ij} - \sum_{k=1}^{n} \dfrac{\partial \alpha_{ik}^{(v)}}{\partial u_i} d_{ik} \left(u_k^{(v)} - u_i^{(v)}\right) & \text{if } i = j, \end{cases} \tag{20}$$

for $i = 1, \ldots, m$, $j = 1, \ldots, n$. Again, The last $n - m$ rows have only the diagonal entry 1.

In the formal Newton method studied here, the non-smooth situations of the limiters are treated as follows. Non-smoothness is given by the maxima and minima in both limiters. In the definition of f_{ij}^+, f_{ij}^-, R_i^+, and R_i^-, one argument of the maximum or minimum is constant. Thus, there is a one-sided derivative that vanishes. In our approach, the derivative that appears in the formal Jacobian is set to be zero in these situations. Consider first the Kuzmin limiter and $a_{ki} \leq a_{ik}$. Then, the entry of the Jacobian is set to be zero if $(f_{ik} > 0) \wedge R_i^+ = 1$, $f_{ik} = 0$, or $(f_{ik} < 0) \wedge R_i^- = 1$. For the BJK limiter, this step is performed if $(f_{ik} > 0) \wedge R_i^+ = 1$, $f_{ik} = 0$ or $(f_{ik} < 0) \wedge R_i^- = 1$. In all other cases, the derivative $\partial \alpha_{ik}^{(\nu)} / \partial u_j$ can be computed. For brevity, details are omitted here.

4 Numerical Studies

All examples are defined in the unit square $\Omega = (0, 1)^2$. Various meshes were used in the simulations, see Fig. 1 for the coarsest level. A standard red refinement was performed. The described iterative methods for solving the nonlinear AFC problems were studied with respect to the number of used steps (iterations + rejections, a rejected step is of the same order of costs as an accepted step) and the used computing time. The following abbreviations will be used for the methods:

- *fixed point rhs*: fixed point iteration with changing right-hand side (19),
- *fixed point matrix*: fixed point iteration with changing matrix (18),
- *formal Newton*: formal Newton's method (20).

All schemes were started with the damping parameter $\omega = 1$. The linear systems of equations were solved with the sparse direct solver UMFPACK [7]. Two stopping criteria were applied. Either, the iteration was stopped if the Euclidean norm of the residual vector was below $\sqrt{\#\text{dof}}\, 10^{-10}$, where #dof is the number of degrees of freedom (including Dirichlet nodes). Or, the iteration terminated if the number of accepted steps reached 25,000. In this case, the iteration did not converge. For simplicity of presentation, we do not distinguish between simulations that diverged, giving *nan* or *inf*, and simulations which did not converge. Both cases are indicated with markers at about 25,000 iteration steps. All simulations were performed with

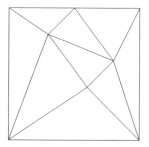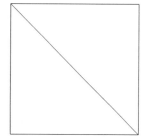

Fig. 1 Grid 1, 2 and 3, level 0

the code PARMOON [8, 18] at compute servers HP BL460c Gen9 2xXeon, Fourteen-Core 2600 MHz.

4.1 Example with a Smooth Solution

In this example, the prescribed solution is

$$u(x) = 100x^2(1-x)y(1-2y)(1-y),$$

the convection field is $b = (3,2)^T$, and the reaction coefficient is $c = 1$. Homogeneous Dirichlet boundary conditions are applied on the whole boundary. Results will be presented for two values of the diffusion coefficient: the moderately small value $\varepsilon = 10^{-3}$ and the much smaller value $\varepsilon = 10^{-6}$. This example serves for obtaining first impressions on the behavior of the iterative schemes. Simulations were performed on Grid 1 and Grid 2 from Fig. 1. Note that Grid 2 is not a Delaunay triangulation. For the initial iterate, all values were set to be zero.

In a first study, only the fixed point iterations *fixed point rhs* and *fixed point matrix* were considered. For $\varepsilon = 10^{-3}$, the number of iteration steps is presented in Fig. 2. One can already observe that the behavior of the methods is somewhat different for the different limiters. For the Kuzmin limiter, the method *fixed point rhs* had no difficulties to solve the nonlinear problems and the number of iterations decreased with refinement of the grids. A similar behavior can be observed for *fixed point matrix*, often with a similar number of iterations. For the BJK limiter, in contrast, the method *fixed point matrix* needed consistently much fewer iterations than *fixed point rhs*, apart of the coarsest uniform grid. Altogether, the nonlinear problems in the case of a moderately small value of the diffusion could be solved without real difficulties. We had similar observations for other examples. For this reason, no further results for moderately small diffusion coefficients will be presented.

Results for $\varepsilon = 10^{-6}$ are shown in Figs. 3 and 4. Figure 3 presents the reduction of the error $\|\nabla(u - u^h)\|_{L^2}$. On the uniform grid, the order of error convergence is

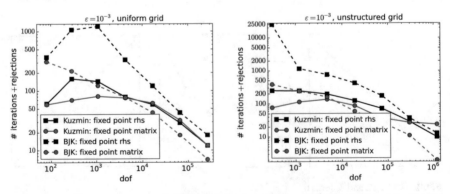

Fig. 2 Example 4.1. Number of iterations and rejections for $\varepsilon = 10^{-3}$, left: Grid 1, right: Grid 2

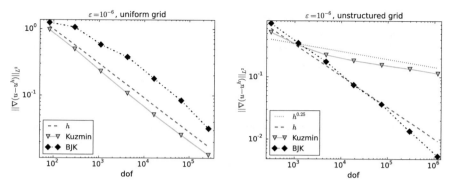

Fig. 3 Example 4.1. Errors of the computed solutions, left: Grid 1, right: Grid 2

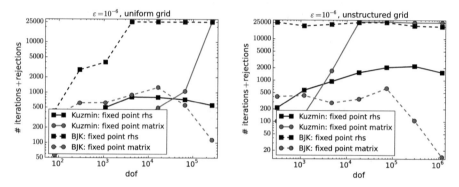

Fig. 4 Example 4.1. Number of iterations and rejections for $\varepsilon = 10^{-6}$, left: Grid 1, right: Grid 2

similar for both limiters, with the solution of the Kuzmin limiter being somewhat more accurate. For the unstructured grid, it can be observed that the BJK limiter worked well on this grid with an order of convergence of about 1. In contrast, the application of the Kuzmin limiter led to a clear reduction of this order. The behavior of the iterative methods is presented in Fig. 4. Now, there are fundamental differences considering both limiters. For the Kuzmin limiter, *fixed point rhs* worked satisfactorily, all problems were solved within the prescribed maximal number of iterations. But even on the uniform grid, *fixed point matrix* failed to converge on fine grids. In case of the BJK limiter, *fixed point rhs* did not converge on many grids, but *fixed point matrix* performed usually quite well.

Since the application of the Kuzmin limiter on the unstructured grid led to quite inaccurate numerical solutions, this limiter should not be used on this grid. This combination will not be considered in the further studies.

Next, the *formal Newton* method will be included in the studies. It is well known that Newton-type methods possess generally a smaller domain of convergence than simpler fixed point iterations. We could observe this behavior also here: applying *formal Newton* from the first step of the iteration led usually to unsatisfactory results concerning the number of steps. For brevity, those results are not presented here.

The first approach for involving the *formal Newton* method was quite simple. In the first part of the iteration, a fixed point method was applied until the Euclidean norm of the residual vector was below a switching tolerance tol_{sw}. Then, *formal Newton* was performed without any possibility of switching back. The current damping parameter ω was used in the first step of the *formal Newton* method. For the first part, we applied *fixed point rhs* as well as *fixed point matrix*. From the results obtained with these methods, Fig. 4, it can be expected that *fixed point rhs* is a better choice for the Kuzmin limiter and *fixed point matrix* for the BJK limiter. In fact, the numerical results confirmed these expectations. Thus, for brevity, only the corresponding results are presented in Figs. 5 and 6.

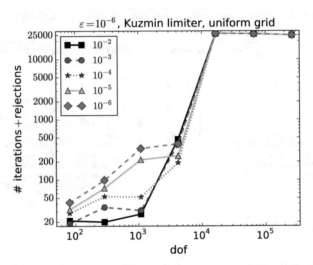

Fig. 5 Example 4.1. Number of iterations and rejections for $\varepsilon = 10^{-6}$, Kuzmin limiter and *formal Newton* method with *fixed point rhs* in the first part, different values for the parameter tol_{sw}, Grid 1

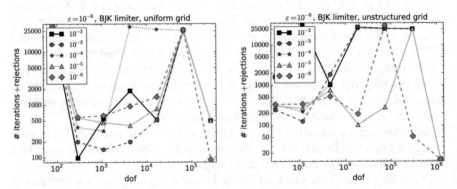

Fig. 6 Example 4.1. Number of iterations and rejections for $\varepsilon = 10^{-6}$, BJK limiter and *formal Newton* method with *fixed point matrix* in the first part, different values for the parameter tol_{sw}, left: Grid 1, right: Grid 2

For the Kuzmin limiter, Fig. 5, it can be seen that *formal Newton* worked well only on coarse grids. On finer grids, it did not converge even for small switching tolerances tol_{sw}. The observations for the BJK limiter are different. On some levels, *formal Newton* worked well, at least for sufficiently small tol_{sw}, but on other levels, this method failed to converge.

Examining the non-convergent simulations more closely, we found that often the Euclidean norm of the residual increased within a few steps after having switched to the *formal Newton* method, sometimes it increased considerably. A straightforward idea to mitigate this behavior consists in switching back to the fixed point iteration that was used in the first part after the norm of the residual exceeds a certain limit. This approach was implemented in the form that the back switch to the method from the first part took place always if the Euclidean norm of the residual became larger than $100 \cdot tol_{sw}$. While switching between the methods, the current damping parameter ω was not changed. However, the behavior of the *formal Newton* method generally did not improve. The only exception is presented in Fig. 7, where it can be seen that the choice $tol_{sw} = 10^{-5}$ led to a convergent method for the BJK limiter on all levels of the unstructured grid.

The last investigation for this example studies computing times. On Grid 1, the methods *fixed point rhs* for the Kuzmin limiter and *fixed point matrix* applied to the BJK limiter converged without difficulties, compare Fig. 4. The times for calculating the limiters were very similar. Thus, differences in computing times are mainly due to differences of the needed time for examining the iterations. All setups were simulated five times, the slowest and the fastest computing times were removed, and the averages of the other three times are presented in Fig. 8. One can observe that *fixed point rhs* is generally about one order of magnitude faster than *fixed point matrix*, although the number of iterations is similar for many levels and *fixed point matrix* needed even considerably fewer iterations on the finest grid, see Fig. 4. Hence, the possibility to use in *fixed point rhs* just one factorization of the matrix

Fig. 7 Example 4.1. Number of iterations and rejections for $\varepsilon = 10^{-6}$, BJK limiter and *formal Newton* method with *fixed point matrix* in the first part and switching back to *fixed point matrix* if the norm of the residual became too large, different values for the parameter tol_{sw}, Grid 2

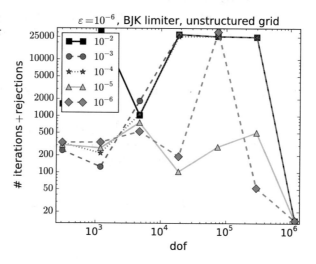

Fig. 8 Example 4.1.
Simulation times for
$\varepsilon = 10^{-6}$, Grid 1

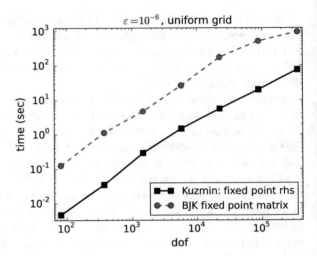

for the complete iteration has a strong positive impact on the computing times for this method.

In further studies, we could observe that one step with the *formal Newton* method is even more expensive than one step with *fixed point matrix*, because of the time needed for computing the entries of the formal Jacobian. For brevity, computing times for *formal Newton* are not presented here.

Already for an example with a smooth solution, there were only few of the considered methods that converged in the convection-dominated case on every refinement level. On the uniform grid, for the Kuzmin limiter only *fixed point rhs* worked well and for the BJK limiter only *fixed point matrix*. There were two satisfactory performing approaches for the BJK limiter on the unstructured grid: *fixed point matrix* and *formal Newton* with $tol_{sw} = 10^{-5}$, where *fixed point matrix* was used as starting method and it was switched back to *fixed point matrix* if the norm of the residual became too large. With respect to computing times, *fixed point rhs*, in the case of convergence, outperformed all other methods.

4.2 Example with Interior and Boundary Layers

This example, proposed in [10], is a standard academic example for numerical studies of steady-state convection-diffusion equations. It is given in $\Omega = (0, 1)^2$ with $b = (\cos(-\pi/3), \sin(-\pi/3))$, $c = f = 0$ and the Dirichlet boundary condition

$$u = \begin{cases} 1 & (y = 1 \wedge x > 0) \text{ or } (x = 0 \wedge y > 0.7), \\ 0 & \text{else.} \end{cases}$$

Fig. 9 Example 4.2. Solution
(computed with the BJK
limiter, Grid 3, level 9)

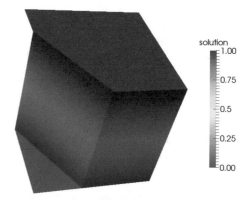

Again, the strongly convection-dominated case $\varepsilon = 10^{-6}$ is considered. Then, the solution exhibits an internal layer in the direction of the convection starting from the jump of the boundary condition at the left boundary and two exponential layers at the right and the lower boundary, see Fig. 9.

In this example, a study of the impact on choosing the initial iterate in different ways will be presented. For the initial iterate, we considered the following options:

- setting all non-Dirichlet degrees of freedom to zero (zero),
- using the solution of the upwind finite element method from [16] (upwind),
- using the solution of the SUPG method from [6, 9] (SUPG),
- using the solution of the Galerkin method (Galerkin).

Starting with the zero initial iterate is a usual approach if no information about the expected solution is available. With the upwind method as initial iterate, the positions of the layers are known from the beginning, but the layers are strongly smeared. The positions of the layers are also known with the SUPG method, the layers are sharp, but there are considerable spurious oscillations in the vicinity of the layers. The incorporation of the Galerkin finite element method in this study is just for completeness.

First, again the behavior of the fixed point iterations was studied, see Fig. 10, left picture. All simulations presented in this figure were started with the SUPG solution as initial iterate. In this example, *fixed point rhs* converged for both limiters on all levels, whereas *fixed point matrix* did not converge for both limiters on fine levels. For the Kuzmin limiter, the method *fixed point rhs* needed considerably fewer iterations than for the BJK limiter. Representative results for the *formal Newton* method, with *fixed point rhs* as scheme that was used if the norm of the residual was too large and $\text{tol}_{sw} = 10^{-5}$, are displayed in Fig. 10, right picture. On coarser levels,

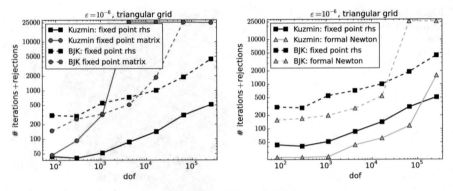

Fig. 10 Example 4.2. Number of iterations and rejections, Grid 3

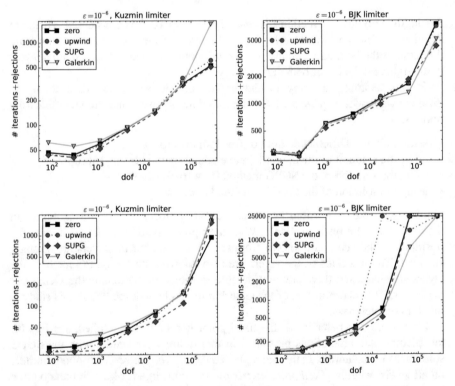

Fig. 11 Example 4.2. Number of iterations and rejections depending on the initial iterate, top: *fixed point rhs*, bottom: *formal Newton*, Grid 3

this approach needed less iterations than *fixed point rhs*, but on finer levels, it even failed in two cases.

The dependency of the number of iterations and rejections on the initial iterate is illustrated in Fig. 11. Generally, there are only minor differences between the four initial iterates. Often, using the SUPG solution proved to be a good choice.

5 Summary and Outlook

This paper presented first steps of a systematic study of schemes for solving the nonlinear problems arising in AFC finite element discretizations of steady-state convection-diffusion-reaction equations. Two basic fixed point iterations and a formal Newton method were included in these studies. The studies were performed for two limiters.

Consider only the results for the strongly convection-dominated situations:

- It could be observed that both fixed point iterations behaved rather differently, which we did not expect before the studies. Whereas *fixed point rhs* always converged for the Kuzmin limiter and in Example 4.2 also for the BJK limiter, *fixed point matrix* often failed to converge on fine grids.
- For the *formal Newton* method, there is no clear picture. Its behavior depended on the choice of tol_{sw}, sometimes it needed considerably fewer iterations than the fixed point methods, however, rather often it did not converge.
- For all methods, the choice of the initial iterate did generally not possess a big impact on the number of iterations. Usually, using the SUPG solution was an appropriate choice.
- From the point of view of efficiency, *fixed point rhs* exploited the fact that a sparse direct solver was used and this method requires only one matrix factorization for the whole iteration. In case of convergence, it was by far the most efficient approach.

The findings collected for the basic schemes will serve in our future work as basis for the development of schemes that behave hopefully better. We want to pursue the approaches of constructing more sophisticated transitions between the schemes, of appropriate combinations of schemes, of utilizing regularizations in Newton-type methods, and of using better damping strategies. In examples where it can be applied, a projection step to a space of admissible functions, as proposed in [2], should be utilized. Furthermore, more complex examples in two and three dimensions will be studied.

References

1. Augustin, M., Caiazzo, A., Fiebach, A., Fuhrmann, J., John, V., Linke, A., Umla, R.: An assessment of discretizations for convection-dominated convection-diffusion equations. Comput. Methods Appl. Mech. Eng. **200**(47–48), 3395–3409 (2011)
2. Badia, S., Bonilla, J.: Monotonicity-preserving finite element schemes based on differentiable nonlinear stabilization. Comput. Methods Appl. Mech. Eng. **313**, 133–158 (2017)
3. Barrenechea, G.R., John, V., Knobloch, P.: Analysis of algebraic flux correction schemes. SIAM J. Numer. Anal. **54**(4), 2427–2451 (2016)
4. Barrenechea, G.R., John, V., Knobloch, P.: An algebraic flux correction scheme satisfying the discrete maximum principle and linearity preservation on general meshes. Math. Models Methods Appl. Sci. **27**(3), 525–548 (2017)

5. Barrenechea, G.R., John, V., Knobloch, P., Rankin, R.: A unified analysis of algebraic flux correction schemes for convection–diffusion equations. SeMA J. **75**(4), 655–685 (2018)
6. Brooks, A.N., Hughes, T.J.R.: Streamline upwind/Petrov-Galerkin formulations for convection dominated flows with particular emphasis on the incompressible Navier-Stokes equations. Comput. Methods Appl. Mech. Eng. **32**(1–3), 199–259 (1982). FENOMECH '81, Part I, Stuttgart (1981)
7. Davis, T.A.: Algorithm 832: UMFPACK V4.3—an unsymmetric-pattern multifrontal method. ACM Trans. Math. Softw. **30**(2), 196–199 (2004)
8. Ganesan, S., John, V., Matthies, G., Meesala, R., Abdus, S., Wilbrandt, U.: An object oriented parallel finite element scheme for computing PDEs: design and implementation. In: IEEE 23rd International Conference on High Performance Computing Workshops (HiPCW) Hyderabad, pp. 106–115. IEEE, New York (2016)
9. Hughes, T.J.R., Brooks, A.: A multidimensional upwind scheme with no crosswind diffusion. In: Finite Element Methods for Convection Dominated Flows Papers, Winter Annual Meeting. AMD, vol. 34, pp. 19–35. American Society of Mechanical Engineers, New York (1979)
10. Hughes, T.J.R., Mallet, M., Mizukami, A.: A new finite element formulation for computational fluid dynamics. II. Beyond SUPG. Comput. Methods Appl. Mech. Eng. **54**(3), 341–355 (1986)
11. John, V., Knobloch, P.: On spurious oscillations at layers diminishing (SOLD) methods for convection-diffusion equations. I. A review. Comput. Methods Appl. Mech. Eng. **196**(17–20), 2197–2215 (2007)
12. John, V., Knobloch, P.: On spurious oscillations at layers diminishing (SOLD) methods for convection-diffusion equations. II. Analysis for P_1 and Q_1 finite elements. Comput. Methods Appl. Mech. Eng. **197**(21–24), 1997–2014 (2008)
13. John, V., Knobloch, P., Novo, J.: Finite elements for scalar convection-dominated equations and incompressible flow problems: a never ending story? Comput. Visual. Sci. **19**, 47–63 (2018)
14. Kuzmin, D.: Algebraic flux correction for finite element discretizations of coupled systems. In: Papadrakakis, M., Oñate, E., Schrefler, B. (eds.) Proceedings of the International Conference on Computational Methods for Coupled Problems in Science and Engineering, pp. 1–5. CIMNE, Barcelona (2007)
15. Kuzmin, D., Möller, M.: Algebraic flux correction. I. Scalar conservation laws. In: Flux-Corrected Transport. Scientific Computation, pp. 155–206. Springer, Berlin (2005)
16. Tabata, M.: A finite element approximation corresponding to the upwind finite differencing. Mem. Numer. Math. **1**(4), 47–63 (1977)
17. Wesseling, P.: Principles of Computational Fluid Dynamics. Springer Series in Computational Mathematics, vol. 29. Springer, Berlin (2001)
18. Wilbrandt, U., Bartsch, C., Ahmed, N., Alia, N., Anker, F., Blank, L., Caiazzo, A., Ganesan, S., Giere, S., Matthies, G., Meesala, R., Shamim, A., Venkatesan, J., John, V.: ParMooN—a modernized program package based on mapped finite elements. Comput. Math. Appl. **74**(1), 74–88 (2017)

A Linearly Implicit Splitting Method for Solving Time Dependent Semilinear Reaction-Diffusion Systems

Carmelo Clavero and Juan Carlos Jorge

Abstract In this paper we deal with the efficient resolution of a coupled system of two one dimensional, time dependent, semilinear parabolic singularly perturbed partial differential equations of reaction-diffusion type, with distinct diffusion parameters which may have different orders of magnitude. The numerical method is based on a linearized version of the fractional implicit Euler method, which avoids the use of iterative methods, and a splitting by components to discretize in time; so, only tridiagonal linear systems are involved in the time integration process. Consequently, the computational cost of the proposed method is lower than classical schemes used for the same type of problems. The solution of this singularly perturbed problem features layers, what are resolved on an appropriate piecewise uniform mesh of Shishkin type. We show that the method is uniformly convergent of first order in time and of almost second order in space. Numerical results are presented to corroborate the theoretical results.

1 Introduction

In this paper we consider a type of singularly perturbed parabolic initial and boundary value problem given by

C. Clavero
Department of Applied Mathematics and IUMA, University of Zaragoza, Zaragoza, Spain
e-mail: clavero@unizar.es

J. C. Jorge (✉)
Department of Computer Science, Mathematics & Statistics, ISC, Public University of Navarra, Pamplona, Spain
e-mail: jcjorge@unavarra.es

© Springer Nature Switzerland AG 2020
G. R. Barrenechea, J. Mackenzie (eds.), *Boundary and Interior Layers, Computational and Asymptotic Methods BAIL 2018*, Lecture Notes in Computational Science and Engineering 135,
https://doi.org/10.1007/978-3-030-41800-7_8

$$\begin{cases} L_\varepsilon \mathbf{u} \equiv \dfrac{\partial \mathbf{u}}{\partial t}(x,t) + \mathscr{L}_{x,\varepsilon}\mathbf{u}(x,t) + \mathbf{A}(x,t,\mathbf{u}) = \mathbf{0}, \ (x,t) \in Q \equiv \Omega \times (0,T], \\ \mathbf{u}(0,t) = \mathbf{g_0}(t), \ \mathbf{u}(1,t) = \mathbf{g_1}(t), \ \forall t \in [0,T], \ \mathbf{u}(x,0) = \boldsymbol{\varphi}(x), \ \forall x \in \bar{\Omega}, \end{cases}$$

$$(1)$$

where $\Omega = (0,1)$, the spatial differential operator $\mathscr{L}_{x,\varepsilon}$ is defined by

$$\mathscr{L}_{x,\varepsilon} \equiv -\mathscr{D}_\varepsilon \frac{\partial^2}{\partial x^2}, \tag{2}$$

$\mathbf{u} = (u_1, u_2)^T$, $\boldsymbol{\varphi} = (\varphi_1, \varphi_2)^T$, $\mathbf{g_0} = (g_{1,0}, g_{2,0})^T$, $\mathbf{g_1} = (g_{1,1}, g_{2,1})^T$, $\mathscr{D}_\varepsilon = \text{diag}$ $(\varepsilon_1, \varepsilon_2)$, and the reaction term is given by $\mathbf{A}(x,t,\mathbf{u}) = (a_1(x,t,\mathbf{u}), a_2(x,t,\mathbf{u}))^T$. We assume that the vector diffusion parameter $\boldsymbol{\varepsilon} = (\varepsilon_1, \varepsilon_2)^T$ satisfies $0 < \varepsilon_1 \le \varepsilon_2 \le 1$, and that both ε_i, $i = 1, 2$ can be very small and they can have different orders of magnitude; we also assume that the components $a_i(x,t,\mathbf{u})$, $i = 1, 2$ of the reaction matrix are sufficiently smooth functions and that sufficient compatibility conditions among all of these data hold, in order that the exact solution $\mathbf{u} \in C^{4,2}(\bar{Q})$, (see [11, 12] for a detailed discussion); more details are given posteriorly in (3).

In many previous works, see [2, 6, 8, 10, 13] and references therein, the linear case of problem (1) is analyzed; in this case the differential equation is given by

$$\frac{\partial \mathbf{u}}{\partial t}(x,t) + \left(-\mathscr{D}_\varepsilon \frac{\partial^2}{\partial x^2} + \mathscr{A} \right) \mathbf{u}(x,t) = \mathbf{f}(x,t),$$

where $\mathscr{A} = (a_{ij}(x,t))$, $i, j = 1, 2$ and $\mathbf{f}(x,t) = (f_1(x,t), f_2(x,t))^T$. In those works, the combination of classical schemes to discretize in space and time variables, together with the use of a priori or equidistributed special meshes in space, gave numerical methods which are uniformly convergent with respect to the diffusion parameter $\boldsymbol{\varepsilon}$. Nevertheless, in general, the computational cost of the numerical methods used to solve singularly coupled systems is high due to the coupling of the components of the discrete solution. To reduce this computational cost, in [2, 3] additive schemes were used to solve parabolic linear systems of reaction-diffusion type for one and two dimensional linear problems respectively, and in [4] a decomposition technique, named splitting by components, was used for parabolic one dimensional linear systems. Both techniques allow the decoupling of the components of the system to calculate the discrete solution.

Extending the main ideas of [4], in this paper we design a numerical algorithm to solve the semilinear problem (1) which is also uniformly convergent with respect to the diffusion parameters and its computational cost is similar to the proposed one in [4] for the linear case. Note that for nonlinear problems, the computational cost increases if an iterative method is used to solve them. Then, we construct our method by combining the idea of the splitting by components combined with a local linearization of the nonlinear reaction term; in this way, the numerical solution is obtained by solving only tridiagonal linear systems, without requiring the use of any iterative method. In this work we present some mathematical details only for the case of systems with two equations; nevertheless, the technique can be applied to systems with an arbitrary number of equations. Moreover, our proposal becomes more and more advantageous as long as the number of equations which compose the system increases. Up to our knowledge, paper [1] is the first one where problem (1) is considered; in that work, a nonlinear finite difference scheme is defined and a monotone iterative method, which constructs sequences of ordered upper and lower solutions, is used to construct a numerical solution; moreover, the method is uniformly convergent and it has first order in time and almost first order in space.

The paper is structured as follows. In Sect. 2, we describe the asymptotic behavior of the exact solution with respect to the diffusion parameters and we give appropriate estimates for its derivatives. In Sect. 3, we construct the spatial semidiscretization for (1) and we prove that if the discretization is defined on a piecewise uniform mesh of Shishkin type then it is uniformly convergent of almost second order. In Sect. 4, we give our proposal for the numerical integration in time; the algorithm combines the fractional implicit Euler method, a splitting by components of the discrete diffusion-reaction operator and a local linearization of the reaction term. We prove that this discretization is uniformly and unconditionally convergent of first order in time. Moreover, we prove that the fully discrete method, which follows of the combination of the time and space discretizations, is uniformly convergent of almost second order in space and of first order in time. Finally, in Sect. 5, we show the numerical results obtained for a test problem, which corroborates the uniform convergence of the algorithm.

Henceforth, $\| \cdot \|$ denotes the infinity norm in $\mathbf{R}^2$, C denotes a generic positive constant independent of the diffusion parameters ε_1, ε_2, and also of the discretization parameters N and M and $\|\mathbf{f}\|_G \equiv \max\{\|f_1\|_G, \|f_2\|_G\}$, where $\|f\|_G$ is the maximum norm of f on the closed set G.

2   Asymptotic Behavior of the Exact Solution

In this section we show the asymptotic behavior of the solution $\mathbf{u}$ of (1) with respect to the diffusion parameters $\boldsymbol{\varepsilon}$, and we give some estimates for its derivatives, which are useful below for the analysis of the uniform convergence of the numerical method. To do that, we assume that the coefficients of the reaction matrix satisfy

$$
\begin{aligned}
&\frac{\partial a_i}{\partial u_i}(x, t, \mathbf{v}) \geq \beta > 0, \quad \frac{\partial a_i}{\partial u_{i'}}(x, t, \mathbf{v}) \leq 0, \ i \neq i', \ i, i' = 1, 2, \\
&\min_{\mathbf{v} \in \mathbf{R}^2} \left(\frac{\partial a_i}{\partial u_1}(x, t, \mathbf{v}) + \frac{\partial a_i}{\partial u_2}(x, t, \mathbf{v}) \right) \geq \alpha > 0, \ i = 1, 2.
\end{aligned}
\tag{3}
$$

Under assumptions (3), problem (1) has a unique solution (see Theorem 3.1, Chap. 8 in [12]). Moreover, following [1], by using the mean-value theorem, we have

$$a_i(x, y, \mathbf{u}) = a_i(x, t, \mathbf{0}) + \frac{\partial a_i}{\partial u_1}(x, t, \mathbf{v})u_1 + \frac{\partial a_i}{\partial u_2}(x, t, \mathbf{v})u_2, \quad i = 1, 2, \tag{4}$$

and therefore the solution of problem (1) can be studied as the solution of a linear variant of it. Using the same argument, the following inverse positivity result can be deduced.

Theorem 1 *If* $-\mathbf{A}(x, t, \mathbf{0})$, $\mathbf{g}_0(t)$, $\mathbf{g}_1(t)$, $\boldsymbol{\varphi}(x)$ *have only non-negative components, then* $\mathbf{u}(x, t)$ *has only non-negative components.*

Proof See [5].

Using a similar reasoning as in [9] we obtain the following estimates for the time derivatives.

Lemma 1 *The solution of problem (1) satisfies*

$$\left\| \frac{\partial^k \mathbf{u}(x, t)}{\partial t^k} \right\| \leq C, \quad (x, t) \in \bar{Q}, \ k = 0, 1, 2. \tag{5}$$

On the other hand, in [1] the estimates

$$\left| \frac{\partial u_1(x, t)}{\partial x} \right| \leq C \left(1 + \varepsilon_1^{-1/2} B_{\varepsilon_1}(x) + \varepsilon_2^{-1/2} B_{\varepsilon_2}(x) \right), \tag{6}$$

$$\left| \frac{\partial u_2(x, t)}{\partial x} \right| \leq C \left(1 + \varepsilon_2^{-1/2} B_{\varepsilon_2}(x) \right),$$

were proven for first derivatives respect to spatial variable x; following [1], using (4) and the technique of [9] we deduce

$$\left| \frac{\partial^2 u_1(x, t)}{\partial x^2} \right| \leq C \left(1 + \varepsilon_1^{-1} B_{\varepsilon_1}(x) + \varepsilon_2^{-1} B_{\varepsilon_2}(x) \right), \quad \left| \frac{\partial^2 u_2(x, t)}{\partial x^2} \right| \leq C \left(1 + \varepsilon_2^{-1} B_{\varepsilon_2}(x) \right),$$

$$\left| \frac{\partial^k u_1(x, t)}{\partial x^k} \right| \leq C \left(1 + \varepsilon_1^{-k/2} B_{\varepsilon_1}(x) + \varepsilon_2^{-k/2} B_{\varepsilon_2}(x) \right), \quad k = 3, 4,$$

$$\left| \frac{\partial^k u_2(x, t)}{\partial x^k} \right| \leq C \left(1 + \varepsilon_2^{-1} (\varepsilon_1^{(2-k)/2} B_{\varepsilon_1}(x) + \varepsilon_2^{(2-k)/2} B_{\varepsilon_2}(x)) \right), \quad k = 3, 4,$$

for any $(x, t) \in \bar{Q}$, with

$$B_\gamma(x) = e^{-x\sqrt{\alpha/\gamma}} + e^{-(1-x)\sqrt{\alpha/\gamma}}, \tag{7}$$

where γ is a generic positive constant and α is the parameter defined in (3).

3   Spatial Semidiscretization

To construct our fully discrete scheme, which will solve numerically the continuous problem (1), firstly we discretize such problem only with respect to the spatial variable x. From the asymptotic behavior of the exact solution, described in previous section, it follows that, in general, two overlapping parabolic boundary layers can appear at both end points of the spatial domain. Then, we construct piecewise uniform meshes of Shishkin type, which concentrate the grid points in the boundary layer regions. To do that, taking multiple of 8 positive integers denoted N, we define the transition parameters

$$\sigma_{\varepsilon_2} = \min\left\{1/4, 2\sqrt{\varepsilon_2} \ln N\right\}, \quad \sigma_{\varepsilon_1} = \min\left\{\sigma_{\varepsilon_2}/2, 2\sqrt{\varepsilon_1} \ln N\right\}.$$

Then, the grid points of the mesh, $\overline{\Omega}_N \equiv \{0 = x_0 < x_1 < \ldots < x_N = 1\}$, are given by

$$x_j = \begin{cases} jh_{\varepsilon_1}, & j = 0, \ldots, N/8, \\ x_{N/8} + (j - N/8)h_{\varepsilon_2}, & j = N/8 + 1, \ldots, N/4, \\ x_{N/4} + (j - N/4)H, & j = N/4 + 1, \ldots, 3N/4, \\ x_{3N/4} + (j - 3N/4)h_{\varepsilon_2}, & j = 3N/4 + 1, \ldots, 7N/8, \\ x_{7N/8} + (j - 7N/8)h_{\varepsilon_1}, & j = 7N/8 + 1, \ldots, N, \end{cases} \tag{8}$$

where $h_{\varepsilon_1} = 8\sigma_{\varepsilon_1}/N$, $h_{\varepsilon_2} = 8(\sigma_{\varepsilon_2} - \sigma_{\varepsilon_1})/N$, $H = 2(1 - 2\sigma_{\varepsilon_2})/N$. Below, we denote by $h_i = x_i - x_{i-1}$, $i = 1, \ldots, N$, and $\overline{h}_i = (h_i + h_{i+1})/2$, $i = 1, \ldots, N-1$.

On these special piecewise uniform meshes, we approximate the exact solution at the grid points, $\mathbf{u}(x_i, t) \equiv (u_1(x_i, t), u_2(x_i, t))^T$, with $x_i \in \overline{\Omega}_N$ and $t \in [0, T]$, by the semidiscrete functions $\mathbf{U}_{N,i}(t) = (U_{N,i,1}(t), U_{N,i,2}(t))^T \in \mathcal{R}^2$, $i = 0, 1, \ldots, N$. These functions are the solutions of the following family of initial value problems

$$\begin{cases} (L_{\varepsilon,N}\mathbf{U}_N(t))_i \equiv \dfrac{d}{dt}\mathbf{U}_{N,i}(t) + (\mathscr{L}_{N,\varepsilon}\mathbf{U}_N(t))_i + \mathbf{A}(x_i, t, \mathbf{U}_{N,i}(t)) = \mathbf{0}, \; i = 1, \ldots, N-1, \\ \mathbf{U}_{N,0}(t) = \mathbf{g}_0(t), \; \mathbf{U}_{N,N}(t) = \mathbf{g}_1(t), \\ \mathbf{U}_N(0) = (\boldsymbol{\varphi}(x_0), \ldots, \boldsymbol{\varphi}(x_N)), \end{cases} \tag{9}$$

being

$$\mathbf{U}_N(t) \equiv (\mathbf{U}_{N,0}(t), \mathbf{U}_{N,1}(t), \ldots, \mathbf{U}_{N,N}(t)),$$

$$\mathscr{L}_{N,\varepsilon}\mathbf{U}_N(t) \equiv (\mathscr{L}_{N,\varepsilon,1}\mathbf{U}_N(t), \mathscr{L}_{N,\varepsilon,2}\mathbf{U}_N(t))^T,$$

with

$$(\mathscr{L}_{N,\varepsilon,1}\mathbf{U}_N(t))_i \equiv -\frac{\varepsilon_1}{\hbar_i}\left(\frac{U_{N,i+1,1}-U_{N,i,1}}{h_{i+1}} - \frac{U_{N,i,1}-U_{N,i-1,1}}{h_i}\right),$$
$$i = 1, \ldots N-1,$$

and

$$(\mathscr{L}_{N,\varepsilon,2}\mathbf{U}_N(t))_i \equiv -\frac{\varepsilon_2}{\hbar_i}\left(\frac{U_{N,i+1,2}-U_{N,i,2}}{h_{i+1}} - \frac{U_{N,i,2}-U_{N,i-1,2}}{h_i}\right),$$
$$i = 1, \ldots N-1.$$

Similar to [4], for this spatial discretization we obtain the following result, which is the discrete analogue of Theorem 1.

Theorem 2 *Assuming that all of the data* $(-\mathbf{A}(x_i, t, \mathbf{0}), \mathbf{g_0}(t), \mathbf{g_1}(t), \boldsymbol{\varphi}(x_i))$, $i = 0, \ldots, N$, *of problem (9) have only non-negative values in their components, its solution* $\mathbf{U}_N(t)$ *has only non-negative components.*

Using Theorem 2 and suitable discrete barrier functions on a linearized rewriting of (9), it follows that its solution satisfies

$$\|\mathbf{U}_N(t)\|_{\overline{\Omega}_N} \leq$$
$$\max\{\|\mathbf{g_0}(t)\|_{[0,T]}, \|\mathbf{g_1}(t)\|_{[0,T]}, \|[\boldsymbol{\varphi}(x)]_N\|_{\overline{\Omega}_N}, \frac{\|[-\mathbf{A}(x,t,\mathbf{0})]_N\|_{\overline{\Omega}_N \times [0,T]}}{\alpha}\}, \forall\, t \in [0, T], \tag{10}$$

where $[.]_N$ denotes the restriction of a function defined on $\overline{\Omega}$ to $\overline{\Omega}_N$. Therefore, problem (9) is well-posed independently of ε and N. Also, the result (10) can be viewed as a uniform stability property of the semidiscretization process.

The local error at any time $t \in [0, T]$, at the grid point $x_i \in \overline{\Omega}_N$, $i = 1, \ldots N-1$, is given by

$$v_{i,N}(t) \equiv L_\varepsilon \mathbf{u}(x_i, t) - (L_{\varepsilon,N}[\mathbf{u}(x,t)]_N)_i. \tag{11}$$

Taking into account that the contribution of the time derivatives to the local error is zero, using appropriate Taylor expansions and the estimates (7) and the same technique as in [9] we prove that

$$\|v_N\|_{\overline{\Omega}_N} \leq C(N^{-1}\ln N)^2. \tag{12}$$

Combining this result of uniform consistency and the uniform stability, the following uniform convergence result for the spatial semidiscretization is deduced.

Theorem 3 *The global error associated to the spatial discretization (9) on the Shishkin mesh (8) satisfies*

$$\|[\mathbf{u}(x,t)]_N - \mathbf{U}_N(t)\|_{\overline{\Omega}_N} \leq C(N^{-1}\ln N)^2, \quad \forall\, t \in [0, T]. \tag{13}$$

Proof See [5].

4   The Fully Discrete Scheme: Uniform Convergence

The second step to construct our numerical algorithm consists of integrating in time, numerically, the family of initial value problems (9) introduced in previous section. For simplicity, we consider a uniform mesh $\overline{w}_M = \{t_m = m\tau, \ m = 0, 1, \ldots, M\}$, with $\tau = T/M$. Let us denote by $\mathbf{U}^m = (U_1^m, U_2^m)^T$ the approximations to $\mathbf{u}(t_m) = (u_1(t_m), u_2(t_m))^T$ on the grid points of $\overline{\Omega}_N$ at each time level $t_m, \ m = 0, 1, \ldots, M$. In order to get an efficient integration process we have chosen a linearized variant of the fractional implicit Euler method. Then, the fully discrete scheme can be written as

$$
\begin{cases}
\mathbf{U}_N^0 = \mathbf{U}_N(0), \\[4pt]
\text{For } m = 0, 1, \ldots, M - 1, \\[4pt]
\text{First half step} \\
\quad
\begin{cases}
U_{N,0,1}^{m+1/2} = g_{1,0}(t_{m+1}), \\[4pt]
\dfrac{U_{N,i,1}^{m+1/2} - U_{N,i,1}^m}{\tau} + (\mathscr{L}_{N,\varepsilon,1}\mathbf{U}_N^{m+1/2})_i + a_1\left(x_i, t_{m+1}, \mathbf{U}_{N,i}^m\right) + \\[6pt]
\qquad\qquad \dfrac{\partial a_1}{\partial u_1}\left(x_i, t_{m+1}, \mathbf{U}_{N,i}^m\right)\left(U_{N,i,1}^{m+1/2} - U_{N,i,1}^m\right) = 0, \ i = 1, \ldots, N-1, \\[6pt]
U_{N,N,1}^{m+1/2} = g_{2,0}(t_{m+1}),
\end{cases} \\
U_{N,i,2}^{m+1/2} = U_{N,i,2}^m, \ i = 0, \ldots, N, \\[4pt]
\text{Second half step} \\
U_{N,i,1}^{m+1} = U_{N,i,1}^{m+1/2}, \ i = 0, \ldots, N, \\[4pt]
\quad
\begin{cases}
U_{N,0,2}^{m+1} = g_{1,1}(t_{m+1}), \\[4pt]
\dfrac{U_{N,i,2}^{m+1} - U_{N,i,2}^{m+1/2}}{\tau} + (\mathscr{L}_{N,\varepsilon,2}\mathbf{U}_N^{m+1})_i + a_2\left(x_i, t_{m+1}, \mathbf{U}_{N,i}^{m+1/2}\right) + \\[6pt]
\qquad\qquad \dfrac{\partial a_2}{\partial u_2}\left(x_i, t_{m+1}, \mathbf{U}_{N,i}^{m+1/2}\right)\left(U_{N,i,2}^{m+1} - U_{N,i,2}^{m+1/2}\right) = 0, \ i = 1, \ldots, N-1, \\[6pt]
U_{N,N,2}^{m+1} = g_{2,1}(t_{m+1}).
\end{cases}
\end{cases}
$$

$$\tag{14}$$

Note that this algorithm has many advantages from a numerical point of view. The first one is that the linearization which we have carried out avoids having to solve nonlinear systems at each time level; on the other hand, the splitting by components decouples the approximation of both components, and therefore only one tridiagonal linear system must be solved at each half step. The combination of these facts makes that the computational cost of the algorithm is considerably smaller than the associated one to classical implicit schemes.

Next, we state the main theoretical results which permit to prove that this method is unconditionally and uniformly convergent of first order in time.

Theorem 4 *The resolution of the half steps defined in (14) involves systems of the form*

$$
A_{N,j} U_N^{m+j/2} = b_{N,j}, \ j = 1, 2,
$$

whose matrices are tridiagonal, inverse positive and satisfy

$$\|(A_{N,j})^{-1}\|_{\infty} \leq \frac{1}{1+\beta\tau}, \quad j = 1, 2. \tag{15}$$

This result plays a main role to obtain the uniform and unconditional stability, as well as the uniform and unconditional consistency of first order for method given by (14). Such properties are stated in the following two theorems.

Theorem 5 (Uniform Stability) *Assuming the following Lipschitz type restrictions on the reaction terms*

$$c_1 \geq \frac{\partial a_i}{\partial u_i}(x, t, \mathbf{u}), \quad i = 1, 2,$$

$$\|\mathbf{A}(x, t, \mathbf{u}) - \mathbf{A}(x, t, \tilde{\mathbf{u}})\| \leq c_2 \|\mathbf{u} - \tilde{\mathbf{u}}\|, \tag{16}$$

$$\left| \frac{\partial a_i}{\partial u_i}(x, t, \mathbf{u})u_i - \frac{\partial a_i}{\partial u_i}(x, t, \tilde{\mathbf{u}})\tilde{u}_i \right| \leq c_3 |u_i - \tilde{u}_i|, \quad i = 1, 2,$$

two solutions of (14), obtained with different initial conditions $\mathbf{U}_N^0$ and $\tilde{\mathbf{U}}_N^0$, satisfy

$$\|\mathbf{U}_N^{m+1} - \tilde{\mathbf{U}}_N^{m+1}\|_{\Omega_N} \leq \frac{1+c\tau}{1+\beta\tau}\|\mathbf{U}_N^m - \tilde{\mathbf{U}}_N^m\|_{\Omega_N}, \tag{17}$$

where c depends only of c_1, c_2 and c_3.

To study the consistency of the discretization, we introduce in a standard way the concept of the local error at time t_m, denoted by e_N^m, as the difference $\mathbf{U}_N(t_m) - \widehat{\mathbf{U}}_N^m$, being $\widehat{\mathbf{U}}_N^m$ the solution of

$$
\begin{cases}
\widehat{U}_{N,0,1}^{m-1/2} = g_{1,0}(t_m), \\
\dfrac{\widehat{U}_{N,i,1}^{m-1/2} - U_{N,i,1}(t_{m-1})}{\tau} + (\mathscr{L}_{N,\varepsilon,1}\widehat{\mathbf{U}}_N^{m-1/2})_i + a_1\left(x_i, t_m, \mathbf{U}_{N,i}(t_{m-1})\right) + \\
\qquad \dfrac{\partial a_1}{\partial u_1}\left(x_i, t_m, \mathbf{U}_{N,i}(t_{m-1})\right)\left(\widehat{U}_{N,i,1}^{m-1/2} - U_{N,i,1}(t_{m-1})\right) = 0, \\
\qquad\qquad i = 1, \dots, N-1, \\
\widehat{U}_{N,N,1}^{m-1/2} = g_{2,0}(t_m),
\end{cases}
$$

$$\widehat{U}_{N,i,2}^{m-1/2} = U_{N,i,2}(t_m), \quad i = 0, \dots, N,$$

$$\widehat{U}_{N,i,1}^m = \widehat{U}_{N,i,1}^{m-1/2}, \quad i = 0, \dots, N,$$

$$
\begin{cases}
\widehat{U}_{N,0,2}^m = g_{1,1}(t_m), \\
\dfrac{\widehat{U}_{N,i,2}^m - \widehat{U}_{N,i,2}^{m-1/2}}{\tau} + (\mathscr{L}_{N,\varepsilon,2}\widehat{\mathbf{U}}_N^m)_i + a_2\left(x_i, t_m, \widehat{\mathbf{U}}_{N,i}^{m-1/2}\right) + \\
\qquad \dfrac{\partial a_2}{\partial u_2}\left(x_i, t_m, \widehat{\mathbf{U}}_{N,i}^{m-1/2}\right)\left(\widehat{U}_{N,i,2}^m - \widehat{U}_{N,i,2}^{m-1/2}\right) = 0, \\
\qquad\qquad i = 1, \dots, N-1, \\
\widehat{U}_{N,N,2}^m = g_{2,1}(t_m).
\end{cases}
$$

$$\tag{18}$$

Then, the following consistency result follows.

Theorem 6 *Under the previous assumptions for the data of (1), it holds*

$$\|e_N^m\|_{\Omega_N} \leq CM^{-2}, \quad \forall \tau \in (0, \tau_0], \ m = 1, 2, \ldots, M. \tag{19}$$

Proof It is analogue to the proof of Theorem 7 in [5].

To conclude the analysis of the time integration process we introduce the global error in time, at time t_m as

$$E_N^m \equiv \mathbf{U}_N(t_m) - \mathbf{U}_N^m$$

and combining the last two results the following uniform and unconditional first order convergence result is deduced.

Theorem 7 *Under the previous assumptions for the data of (1), the global error of the time discretization holds*

$$\|E_N^m\|_{\Omega_N} \leq CM^{-1}. \ m = 1, \ldots M. \tag{20}$$

Finally, combining the results (3) and (20) we are ready to state the main result for our numerical algorithm.

Theorem 8 *Under the previous assumptions for (1), the global errors for (14), satisfy that*

$$\max_{0 \leq m \leq M, \ 0 \leq i \leq N} \|\mathbf{U}_{N,i}^m - \mathbf{u}(x_i, t_m)\| \leq C\left((N^{-1}\ln N)^2 + M^{-1}\right), \tag{21}$$

where, as we mentioned before, C is a positive constant independent of the diffusion parameters $\varepsilon_1, \varepsilon_2$ and the discretization parameters N and M. Then, the fully discrete scheme is a uniformly and unconditionally convergent method of first order in time and of almost second order in space.

5 Numerical Results

In this section we show the numerical results obtained with the algorithm proposed here to solve a test problem of type (1). The data of the example are given by

$$a_1(x, t, \mathbf{u}) = 2u_1 + t(u_1 - \sin(u_1)) - u_2 - te^{3t}\sin(\pi x),$$

$$a_2(x, t, \mathbf{u}) = -u_1 + 2u_2 + (1 - e^{-t})\frac{u_2}{1 + u_2^2}, \tag{22}$$

$$\mathbf{g}_0(t) = \mathbf{g}_1(t) = (8t + e^{-8t}, 16t + e^{-16t})^T, \ \boldsymbol{\varphi}(x) = (1, 1)^T.$$

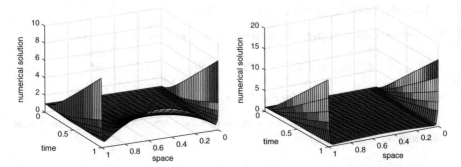

Fig. 1 Components 1 (left) and 2 (right) of problem (22) for $\varepsilon_1 = 10^{-6}$, $\varepsilon_2 = 10^{-4}$ with $N = M = 32$

Figure 1 displays the numerical approximation for both components, for specific small values of ε_1 and ε_2, showing the boundary layers at $x = 0$ and $x = 1$.

As the exact solution is unknown, to approximate the norm of errors

$$\|\mathbf{U}_{N,i}^m - \mathbf{u}(x_i, t_m)\|, \ i = 0, 1, \ldots, N, \ m = 0, 1, \ldots, M,$$

we use a variant of the double-mesh principle (see [7] for instance), i.e, we calculate

$$d_{\varepsilon,j}^{N,M} = \max_{0 \le m \le M} \max_{0 \le i \le N} \|U_{N,i,j}^m - \widehat{U}_{2N,2i,j}^{2m}\|, \ d_j^{N,M} = \max_{\varepsilon} d_{\varepsilon,j}^{N,M}, \quad j = 1, 2, \tag{23}$$

where $\{\widehat{U}_{2N}^m\}$ is the numerical solution on a finer mesh $\{(\hat{x}_i, \hat{t}_m)\}$ that is composed of the mesh points of the coarse mesh joint to their midpoints, i.e.,

$$\hat{x}_{2i} = x_i, \ i = 0, \ldots, N, \quad \hat{x}_{2i+1} = (x_i + x_{i+1})/2, \ i = 0, \ldots, N-1,$$
$$\hat{t}_{2m} = t_m, \ m = 0, \ldots, M, \quad \hat{t}_{2m+1} = (t_m + t_{m+1})/2, \ m = 0, \ldots, M-1. \tag{24}$$

From the double-mesh differences computed in (23), we obtain the corresponding orders of convergence by

$$p_{\varepsilon,j}^{N,M} = \log(d_{\varepsilon,j}^{N,M}/d_{\varepsilon,j}^{2N,2M})/\log 2, \ p_j^{uni} = \log(d_j^{N,M}/d_j^{2N,2M})/\log 2, \quad j = 1, 2. \tag{25}$$

Table 1 shows maximum errors and their corresponding orders of convergence for some values of ε_2 when ε_1 is in the set $R = \{\varepsilon_1; \varepsilon_1 = \varepsilon_2, 2^{-2}\varepsilon_2, \ldots, 2^{-32}\}$, and some values of the discretization parameter N and $M = N/4$. Concretely, each cell of this table has four numbers which, from above to below, correspond to $d_{\varepsilon,1}^{N,M}, p_{\varepsilon,1}^{N,M}, d_{\varepsilon,2}^{N,M}$ and $p_{\varepsilon,2}^{N,M}$. From them, the uniformly convergent behavior of the method can be observed; concretely, for large values of ε_2 clearly we see

Table 1 Maximum errors and orders of convergence for problem (22)

ε_2	$N=64$ $M=16$	$N=128$ $M=32$	$N=256$ $M=64$	$N=512$ $M=128$	$N=1024$ $M=256$
2^{-6}	1.3444E$-$1	7.0148E$-$2	3.5733E$-$2	1.8022E$-$2	9.0486E$-$3
	0.9385	0.9732	0.9875	0.9940	
	4.6886E$-$2	2.3334E$-$2	1.1718E$-$2	5.8721E$-$3	2.9393E$-$3
	1.0067	0.9937	0.9968	0.9984	
2^{-8}	1.3416E$-$1	6.9993E$-$2	3.5650E$-$2	1.7980E$-$2	9.0278E$-$3
	0.9387	0.9733	0.9875	0.9939	
	7.7328E$-$2	2.8120E$-$2	1.2624E$-$2	6.0502E$-$3	2.9821E$-$3
	1.4594	1.1554	1.0611	1.0207	
2^{-10}	1.8148E$-$1	7.5122E$-$2	3.5567E$-$2	1.7918E$-$2	8.9931E$-$3
	1.2725	1.0787	0.9891	0.9946	
	3.0621E$-$1	8.4278E$-$2	2.1304E$-$2	7.2289E$-$3	3.2007E$-$3
	1.8613	1.9840	1.5593	1.1754	
2^{-12}	2.5950E$-$1	1.1579E$-$1	5.0747E$-$2	2.1799E$-$2	9.4647E$-$3
	1.1643	1.1901	1.2191	1.2036	
	6.2623E$-$1	2.5739E$-$1	9.0294E$-$2	2.8867E$-$2	8.7490E$-$3
	1.2827	1.5113	1.6452	1.7222	
2^{-14}	2.6025E$-$1	1.1596E$-$1	5.0842E$-$2	2.1856E$-$2	9.4969E$-$3
	1.1663	1.1895	1.2180	1.2025	
	6.2650E$-$1	2.5757E$-$1	9.0395E$-$2	2.8921E$-$2	8.7784E$-$3
	1.2824	1.5106	1.6441	1.7201	
$\ldots$	$\ldots$	$\ldots$	$\ldots$	$\ldots$	$\ldots$
$\ldots$	$\ldots$	$\ldots$	$\ldots$	$\ldots$	$\ldots$
2^{-24}	1.7776E$-$1	8.0687E$-$2	3.2542E$-$2	1.4798E$-$2	7.1548E$-$3
	1.1395	1.3100	1.1370	1.0484	
	5.2191E$-$1	1.9279E$-$1	6.0360E$-$2	1.7090E$-$2	4.4647E$-$3
	1.4368	1.6754	1.8204	1.9365	
$d_1^{N,M}$	2.6096E$-$1	1.1625E$-$1	5.0926E$-$2	2.1905E$-$2	9.5253E$-$3
p_1^{uni}	1.1666	1.1908	1.2171	1.2014	
$d_2^{N,M}$	6.2675E$-$1	2.5773E$-$1	9.0490E$-$2	2.8972E$-$2	8.8061E$-$3
p_2^{uni}	1.2820	1.5101	1.6431	1.7181	

first order of convergence for both components; for smaller values of ε_2, for the first component we again observe first order and for the second component almost second appears.

To highlight the uniformly convergent behavior in space, we have made a second table (see Table 2), where we diminish the influence of the errors in time by taking $M = 4N$. From it, we observe more clearly the almost second order of uniform convergence in space of the algorithm, according to the theoretical results.

Table 2 Maximum errors and orders of convergence for problem (22)

ε_2	N = 64 M = 256	N = 128 M = 512	N = 256 M = 1024	N = 512 M = 2048	N = 1024 M = 4096
2^{-6}	1.1008E−1	5.0851E−2	1.6506E−2	5.0973E−3	1.4731E−3
	1.1141	1.6233	1.6952	1.7909	
	1.7266E−2	4.2097E−3	1.1968E−3	4.6202E−4	2.0285E−4
	2.0362	1.8145	1.3732	1.1875	
2^{-8}	1.1328E−1	5.2820E−2	1.7146E−2	5.2405E−3	1.5328E−3
	1.1008	1.6232	1.7101	1.7735	
	7.7744E−2	2.0060E−2	5.0136E−3	1.2147E−3	2.9565E−4
	1.9544	2.0004	2.0452	2.0386	
2^{-10}	1.4628E−1	6.4476E−2	2.0290E−2	6.1259E−3	1.8047E−3
	1.1819	1.6680	1.7278	1.7632	
	3.0408E−1	8.4140E−2	2.1498E−2	5.3480E−3	1.3012E−3
	1.8536	1.9686	2.0071	2.0392	
2^{-12}	1.9452E−1	7.6378E−2	2.4677E−2	8.8429E−3	3.0575E−3
	1.3487	1.6300	1.4806	1.5322	
	6.2165E−1	2.5651E−1	9.0218E−2	2.8941E−2	8.8326E−3
	1.2771	1.5075	1.6403	1.7122	
2^{-14}	1.9473E−1	7.6467E−2	2.4695E−2	8.8516E−3	3.0616E−3
	1.3486	1.6306	1.4802	1.5317	
	6.2164E−1	2.5652E−1	9.0230E−2	2.8946E−2	8.8346E−3
	1.2770	1.5074	1.6402	1.7121	
...	...	...	...	...	...
...	...	...	...	...	...
2^{-24}	1.5729E−1	6.4520E−2	2.3492E−2	7.4140E−3	2.1920E−3
	1.2856	1.4575	1.6639	1.7580	
	5.2469E−1	1.9429E−1	6.1474E−2	1.7819E−2	4.9137E−3
	1.4332	1.6602	1.7866	1.8585	
$d_1^{N,M}$	1.9493E−1	7.6552E−2	2.4711E−2	8.8593E−3	3.0652E−3
p_1^{uni}	1.3484	1.6313	1.4799	1.5312	
$d_2^{N,M}$	6.2165E−1	2.5654E−1	9.0242E−2	2.8951E−2	8.8365E−3
p_2^{uni}	1.2769	1.5073	1.6402	1.7121	

6 Conclusions

In this work we design, analyze and test a new numerical method for solving semilinear, time dependent, singularly perturbed diffusion reaction systems. The method is the result of combining a standard central difference scheme on a special mesh of Shishkin type, to discretize in space, and a linearized version of the fractional implicit Euler method which permits to split by components the linear systems involved in the time integration process. It is shown that the method is uniformly convergent of almost second order in space and of first order in time; besides, the computational cost of this numerical algorithm is very low because only two small tridiagonal systems per time step must be solved to advance in time.

References

1. Boglaev, I.: Uniformly convergent monotone iterates for nonlinear parabolic reaction-diffusion systems. Lect. Notes Comput. Sci. Eng. **120**, 35–48 (2017)
2. Clavero, C., Gracia, J.L.: Uniformly convergent additive finite difference schemes for singularly perturbed parabolic reaction-diffusion system. Comput. Math. Appl. **67**, 655–670 (2014)
3. Clavero, C., Gracia, J.L.: Uniformly convergent additive schemes for 2D singularly perturbed parabolic systems of reaction-diffusion type. Numer. Algorithms (2018). https://doi.org/10.1007/s11075-018-0518-y
4. Clavero, C., Jorge, J.C.: Solving efficiently one dimensional parabolic singularly perturbed reaction-diffusion systems: a splitting by components. J. Comp. Appl. Math. **344**, 1–14 (2018)
5. Clavero, C., Jorge, J.C.: An efficient and uniformly convergent scheme for one dimensional parabolic singularly perturbed semilinear systems of reaction-diffusion type. Numer. Algorithms (2019). https://doi.org/10.1007/s11075-019-00850-3
6. Das, P., Natesan, S.: Optimal error estimate using mesh equidistribution technique for singularly perturbed system of reaction-diffusion boundary-value problems. Appl. Math. Comp. **249**, 265–277 (2014)
7. Farrell, P.A., Hegarty, A.F., Miller, J.J.H., O'Riordan, E., Shishkin, G.I.: Robust Computational Techniques for Boundary Layers. Applied Mathematics, vol. 16. Chapman and Hall/CRC, Boca Raton (2000)
8. Franklin, V., Paramasivam, M., Miller, J.J.H., Valarmathi, S.: Second order parameter-uniform convergence for a finite difference method for a singularly perturbed linear parabolic system. Int. J. Numer. Anal. Model. **10**, 178–202 (2013)
9. Gracia, J.L., Lisbona, F.: A uniformly convergent scheme for a system of reaction–diffusion equations. J. Comp. Appl. Math. **206**, 1–16 (2007)
10. Gracia, J.L., Lisbona, F., O'Riordan, E.: A coupled system of singularly perturbed parabolic reaction-diffusion equations. Adv. Comput. Math. **32**, 43–61 (2010)
11. Ladyzhenskaya, O.A., Solonnikov, V.A., Uraltseva, N.N.: Linear and Quasilinear Equations of Parabolic Type. Translations of Mathematical Monographs, vol. 23. American Mathematical Society, Providence, RI (1967)
12. Pao, C.V.: Nonlinear Parabolic and Elliptic Equations. Plenum Press, New York (1992)
13. Shishkina, L.P., Shishkin, G.I.: Robust numerical method for a system of singularly perturbed parabolic reaction-diffusion equations on a rectangle. Math. Model. Anal. **13**, 251–261 (2008)

Improved Energy-Norm A Posteriori Error Estimates for Singularly Perturbed Reaction-Diffusion Problems on Anisotropic Meshes

Natalia Kopteva

Abstract In the recent article (Kopteva, Numer Math 137:607–642, 2017) the author obtained residual-type a posteriori error estimates in the energy norm for singularly perturbed semilinear reaction-diffusion equations on unstructured anisotropic triangulations. The error constants in these estimates are independent of the diameters and the aspect ratios of mesh elements and of the small perturbation parameter. The purpose of this note is to improve the weights in the jump residual part of the estimator. This is attained by using a novel sharper version of the scaled trace theorem for anisotropic elements, in which the hat basis functions are involved as weights.

1 Introduction

Consider finite element approximations to singularly perturbed semilinear reaction-diffusion equations of the form

$$Lu := -\varepsilon^2 \Delta u + f(x, y; u) = 0 \quad \text{for } (x, y) \in \Omega, \qquad u = 0 \quad \text{on } \partial\Omega, \tag{1.1}$$

posed in a, possibly non-Lipschitz, polygonal domain $\Omega \subset \mathbb{R}^2$. Here $0 < \varepsilon \leqslant 1$. We also assume that f is continuous on $\Omega \times \mathbb{R}$ and satisfies $f(\cdot; s) \in L_\infty(\Omega)$ for all $s \in \mathbb{R}$, and the one sided Lipschitz condition $f(x, y; v) - f(x, y; w) \geqslant C_f[v - w]$ whenever $v \geqslant w$, with some constant $C_f \geqslant 0$. Then there is a unique solution

N. Kopteva (✉)
University of Limerick, Limerick, Ireland
e-mail: natalia.kopteva@ul.ie

© Springer Nature Switzerland AG 2020
G. R. Barrenechea, J. Mackenzie (eds.), *Boundary and Interior Layers,
Computational and Asymptotic Methods BAIL 2018*, Lecture Notes in
Computational Science and Engineering 135,
https://doi.org/10.1007/978-3-030-41800-7_9

143

$u \in W_\ell^2(\Omega) \subseteq W_q^1 \subset C(\bar{\Omega})$ for some $\ell > 1$ and $q > 2$ [3, Lemma 1]. We additionally assume that $C_f + \varepsilon^2 \geq 1$ (as (1.1) can always be reduced to this case by a division by $C_f + \varepsilon^2$).

For this problem, the recent articles [6, 7] gave residual-type a posteriori error estimates on unstructured anisotropic meshes. In particular, in [7] the error was estimated in the energy norm $\|\!|\!| \cdot |\!|\!|_{\varepsilon\,;\Omega}$, which is an appropriately scaled $W_2^1(\Omega)$ norm naturally associated with our problem, defined for any $\mathscr{D} \subseteq \Omega$ by $\|\!|\!| v |\!|\!|_{\varepsilon\,;\mathscr{D}} := \left\{ \varepsilon^2 \|\nabla v\|_{2\,;\mathscr{D}}^2 + \|v\|_{2\,;\mathscr{D}}^2 \right\}^{1/2}$. Linear finite elements were used to discretize (1.1) with a piecewise-linear finite element space $S_h \subset H_0^1(\Omega) \cap C(\bar{\Omega})$ relative to a triangulation $\mathscr{T}$, and the the computed solution $u_h \in S_h$ satisfying

$$\varepsilon^2 \langle \nabla u_h, \nabla v_h \rangle + \langle f_h^I, v_h \rangle = 0 \quad \forall\, v_h \in S_h, \qquad f_h(\cdot) := f(\cdot\,; u_h). \tag{1.2}$$

Here $\langle \cdot, \cdot \rangle$ denotes the $L_2(\Omega)$ inner product, and f_h^I is the standard piecewise-linear Lagrange interpolant of f_h.

To give a flavour of the results in [7], assuming that all mesh elements are anisotropic, one estimator reduces to

$$\|\!|\!| u_h - u |\!|\!|_{\varepsilon\,;\Omega} \leqslant C \left\{ \sum_{z \in \mathscr{N}} \min\{|\omega_z|, \lambda_z\} \left\| \varepsilon [\![\nabla u_h]\!] \right\|_{\infty\,;\gamma_z}^2 \right.$$

$$\left. + \sum_{z \in \mathscr{N}} \left\| \min\{1, H_z \varepsilon^{-1}\} f_h^I \right\|_{2\,;\omega_z}^2 + \left\| f_h - f_h^I \right\|_{2\,;\Omega}^2 \right\}^{1/2},$$

$$\tag{1.3}$$

where C is independent of the diameters and the aspect ratios of elements in $\mathscr{T}$, and of ε. Here $\mathscr{N}$ is the set of nodes in $\mathscr{T}$, $[\![\nabla u_h]\!]$ is the standard jump in the normal derivative of u_h across an element edge, ω_z is the patch of elements surrounding any $z \in \mathscr{N}$, γ_z is the set of edges in the interior of ω_z, $H_z = \mathrm{diam}(\omega_z)$, and $h_z \simeq H_z^{-1} |\omega_z|$.

A version of (1.3) obtained in [7] involves a somewhat surprising weight $\lambda_z = \varepsilon H_z^2 h_z^{-1}$ at the jump residual terms. The main purpose of this note is to improve the jump residual part of the latter estimator and establish its sharper version with a more natural $\lambda_z = \varepsilon H_z$. This will be attained by employing a novel sharper version of the scaled trace theorem, in which the hat basis functions are involved as weights (see Remark 3.1). As the improvement that we present here applies to the jump residual terms only, we restrict our analysis to these terms.

Note that the new shaper version of the jump residual part of the estimator works not only for (1.3) (see Theorem 4.3 below), but can be also combined with a shaper bound for the interior residual terms given by [7, Theorem 6.2]. The latter is more intricate and was obtained under some additional assumptions on the mesh, so we shall not give it here. Comparing it to (1.3), roughly speaking, the weight $\min\{1, H_z\varepsilon^{-1}\}$ is replaced by a sharper $\min\{1, h_z\varepsilon^{-1}\}$ with a few additional terms included.

Note also that a similar improved jump residual part of the estimator is also obtained in [8, (1.2)] using an entirely different (and more complicated in the context of residual-type estimation) approach for a version of (1.2) (with a special anisotropic quadrature used for the reaction term).

Our interest in locally anisotropic meshes is due to that they offer an efficient way of computing reliable numerical approximations of layer solutions. (In the context of (1.1) with $\varepsilon \ll 1$, see, e.g., [4, 9, 14] and references therein.) But such anisotropic meshes are frequently constructed a priori or by heuristic methods, while the majority of available a posteriori error estimators assume shape regularity of the mesh [1]. In the case of shape-regular triangulations, residual-type a posteriori error estimates for equations of type (1.1) were proved in [16] in the energy norm, and more recently in [3] in the maximum norm. The case of anisotropic meshes having a tensor-product structure was addressed in [15] for the Laplace equation and in [2, 5] for problems of type (1.1), with the error estimators given, respectively, in the H^1 norm and the maximum norm. For unstructured anisotropic meshes, a posteriori error estimates can be found in [10, 12] for the Laplace equation in the H^1 norm, and in [11, 12] for a linear constant-coefficient version of (1.1) in the energy norm.

Note that the error constants in the estimators of [10–12] involve the so-called matching functions; the latter depend on the unknown error and take moderate values only when the grid is either isotropic, or, being anisotropic, is aligned correctly to the solution, while, in general, they may be as large as mesh aspect ratios. The presence of such matching functions in the estimator is clearly undesirable. It is entirely avoided in the more recent papers [6–8], as well as here.

The paper is organized as follows. In Sects. 2 and 3, we respectively describe our triangulation assumptions and give a novel shaper version of the scaled trace theorem for anisotropic elements. In Sect. 4, we derive the main result of the paper, a new shaper jump residual part of the estimator. A simplified version of this analysis is given in Sect. 4.1 for partially structured anisotropic meshes, while more general anisotropic meshes are addressed in Sect. 4.2.

Notation We write $a \simeq b$ when $a \lesssim b$ and $a \gtrsim b$, and $a \lesssim b$ when $a \leqslant Cb$ with a generic constant C depending on Ω and f, but not on either ε or the diameters and the aspect ratios of elements in $\mathcal{T}$. Also, for $\mathcal{D} \subset \bar{\Omega}$, $1 \leqslant p \leqslant \infty$, and $k \geqslant 0$, let $\| \cdot \|_{p;\mathcal{D}} = \| \cdot \|_{L_p(\mathcal{D})}$ and $| \cdot |_{k,p;\mathcal{D}} = | \cdot |_{W_p^k(\mathcal{D})}$, where $| \cdot |_{W_p^k(\mathcal{D})}$ is the standard Sobolev seminorm, and $\operatorname{osc}(v; \mathcal{D}) = \sup_{\mathcal{D}} v - \inf_{\mathcal{D}} v$ for $v \in L_\infty(\mathcal{D})$.

2 Basic Triangulation Assumptions

We shall use $z = (x_z, y_z)$, S and T to respectively denote particular mesh nodes, edges and elements, while $\mathcal{N}$, $\mathcal{S}$ and $\mathcal{T}$ will respectively denote their sets. For each $T \in \mathcal{T}$, let H_T be the maximum edge length and $h_T := 2H_T^{-1}|T|$ be the minimum height in T. For each $z \in \mathcal{N}$, let ω_z be the patch of elements surrounding any $z \in \mathcal{N}$, $\mathcal{S}_z$ the set of edges originating at z, and

$$H_z := \operatorname{diam}(\omega_z), \quad h_z := H_z^{-1}|\omega_z|, \quad \gamma_z := \mathcal{S}_z \setminus \partial\Omega, \quad \mathring{\gamma}_z := \{S \subset \gamma_z : |S| \lesssim h_z\}. \tag{2.1}$$

Throughout the paper we make the following triangulation assumptions.

- *Maximum Angle condition.* Let the maximum interior angle in any triangle $T \in \mathcal{T}$ be uniformly bounded by some positive $\alpha_0 < \pi$.
- *Local Element Orientation condition.* For any $z \in \mathcal{N}$, there is a rectangle $R_z \supset \omega_z$ such that $|R_z| \simeq |\omega_z|$. Furthermore, if $z \in \mathcal{N} \cap \partial\Omega$ is not a corner of Ω, then R_z has a side parallel to the segment $\mathcal{S}_z \cap \partial\Omega$.
- Also, let the number of triangles containing any node be uniformly bounded.

Note that the above conditions are automatically satisfied by shape-regular triangulations.

Additionally, we restrict our analysis to the following two node types defined using a fixed small constant c_0 (to distinguish between anisotropic and isotropic elements), with the notation $a \ll b$ for $a < c_0 b$.

(1) *Anisotropic Nodes*, the set of which is denoted by $\mathcal{N}_{\text{ani}}$, are such that

$$h_z \ll H_z, \qquad h_T \simeq h_z \text{ and } H_T \simeq H_z \quad \forall T \subset \omega_z. \tag{2.2}$$

Note that the above implies that $\mathcal{S}_z$ contains at most two edges of length $\lesssim h_z$ (see also Fig. 3, left).

(2) *Regular Nodes*, the set of which is denoted by $\mathcal{N}_{\text{reg}}$, are those surrounded by shape-regular mesh elements.

Note that most of our analysis applies to more general node types that were considered in [6, 7]; see Remarks 3.2 and 4.3 for details.

3 Sharper Scaled Trace Theorem for Anisotropic Elements

Our task is to get an improved bound for the jump residual terms (see I in (4.2) below). The key to this will be to employ the following sharper version of the scaled trace theorem for anisotropic elements, which is the main result of this section.

Lemma 3.1 *For any node* $z \in \mathcal{N} = \mathcal{N}_{\text{ani}} \cup \mathcal{N}_{\text{reg}}$, *any function* $v \in W_1^1(\omega_z)$, *and any edge* $S \subset \gamma_z$, *one has*

$$\|v\phi_z\|_{1;S} \lesssim \|\nabla v\|_{1;\omega_z} + \|v\|_{1;\omega_z} \begin{cases} H_z^{-1} & \text{if } S \subset \mathring{\gamma}_z, \\ h_z^{-1} & \text{if } S \subset \gamma_z \backslash \mathring{\gamma}_z, \end{cases} \tag{3.1}$$

$$|S|^{-1}\|v\phi_z\|_{1;S}^2 \lesssim \|v\|_{2;\omega_z}\|\nabla v\|_{2;\omega_z} + \|v\|_{2;\omega_z}^2 \begin{cases} H_z^{-1} & \text{if } S \subset \mathring{\gamma}_z, \\ h_z^{-1} & \text{if } S \subset \gamma_z \backslash \mathring{\gamma}_z, \end{cases} \tag{3.2}$$

where ϕ_z *is the hat basis function associated with* z.

Remark 3.1 Similar versions of the scaled trace theorem for anisotropic elements were obtained in [6, Lemma 3.1] and [7, §3]. Lemma 3.1 is an improvement in the sense that in the case of long edges (i.e. $S \subset \gamma_z \backslash \mathring{\gamma}_z$), the weights at $\|\nabla v\|_{p;\omega_z}$ are sharper. To be more precise, the version of (3.1) in [6, Lemma 3.1] has the weight $H_z/h_z \gg 1$ at $\|\nabla v\|_{1;\omega_z}$, while the version of (3.2) given by [7, Corollory 3.2] also involves the weight $H_z/h_z \gg 1$ at $\|\nabla v\|_{2;\omega_z}$. Importantly, for the shaper bounds of Lemma 3.1 to hold true, one needs to estimate $\|v\phi_z\|_{p;S}$ rather than $\|v\|_{p;S}$ bounded in [6, 7]. Note that this improvement is crucial for getting an improved weight in the jump residual part of our estimator.

Remark 3.2 An inspection of the proof of Lemma 3.1 shows that this lemma remains valid for the more general node types introduced in [7, §2].

To prove Lemma 3.1, we shall employ the following auxiliary result.

Lemma 3.2 *For any sufficiently smooth function* $v \geq 0$ *on a triangle* T *with vertices* z, z' *and* z'' *and their respective opposite edges* S, S' *and* S'', *one has*

$$\sin \angle(S', S'') \|v\phi_z\|_{1;S'} \lesssim \|\nabla v\|_{1;T} + |S''|^{-1}\|v\|_{1;T}, \tag{3.3a}$$

$$|S'|^{-1}\|v\phi_z\|_{1;S'} \lesssim |S''|^{-1}\|v\phi_z\|_{1;S''} + |S||T|^{-1}\|\nabla v\|_{1;T}. \tag{3.3b}$$

Proof For (3.3a), let $\boldsymbol{\mu}''$ be the unit vector along S'' directed from z' to z so that $\nabla\phi_z \cdot \boldsymbol{\mu}'' = |S''|^{-1}$. Note that $\nabla \cdot (v\phi_z\boldsymbol{\mu}'') = \nabla(v\phi_z) \cdot \boldsymbol{\mu}''$, so the divergence theorem yields

$$\int_{\partial T} (v\phi_z\boldsymbol{\mu}'') \cdot \boldsymbol{v} = \int_T \nabla(v\phi_z) \cdot \boldsymbol{\mu}'' = \int_T (\phi_z \nabla v \cdot \boldsymbol{\mu}'' + |S''|^{-1}v).$$

Here, to evaluate the integral $\int_{\partial T}$, note that $\boldsymbol{\mu}'' \cdot \boldsymbol{v} = 0$ on S'' and $\phi_z = 0$ on S, while $\boldsymbol{\mu}'' \cdot \boldsymbol{v} = \sin \angle(S', S'')$ on S', so $\int_{\partial T}(v\phi_z\boldsymbol{\mu}'') \cdot \boldsymbol{v} = \sin \angle(S', S'') \int_{S'} v\phi_z$. The desired bound (3.3a) follows.

To get (3.3b), we modify the proof of [7, Lemma 7.1]. Set $w = v\phi_z$ and also $\mathscr{A}_S w := |S|^{-1}\int_S w$ for any edge S. Now, with the ζ-axis having the inward normal

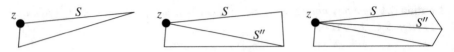

Fig. 1 Illustration to the proof of (3.1) in Lemma 3.1: case (i) (left); case (ii) with a single application of (3.3b) (centre); case (ii) with a double application of (3.3b) (right)

direction to S, and $\hbar := 2|T||S|^{-1}$, one gets $\mathscr{A}_{S'} w - \mathscr{A}_{S''} w = \hbar^{-1} \int_0^{\hbar} \left(w|_{S'} - w|_{S''} \right) d\zeta$. This yields (3.3b) as ϕ_z does not change in the direction normal to ζ. $\square$

Proof of Lemma 3.1. First, note that (3.2) follows from (3.1) as $|S|^{-1} \|v\phi_z\|_{1;S}^2 \leqslant \|v^2\phi_z^2\|_{1;S} \leqslant \|v^2\phi_z\|_{1;S}$, while $\nabla v^2 = v\nabla v$. With regard to (3.1), it suffices to prove it for the case $v \geqslant 0$, as if v changes sign on ω_z, apply (3.1) with v replaced by $v_\tau := \sqrt{v^2 + \tau^2} \geqslant 0$, where τ is a small positive constant (while $|\nabla v_\tau| \leqslant |\nabla v|$), and then let $\tau \to 0^+$ so that $v_\tau \to |v|$.

Thus it remains to show (3.1) for $v \geqslant 0$. When $S \subset \mathring{\gamma}_z$, this bound follows from a similar bound on $\|v\|_{1;S}$ in [6, Lemma 3.1]. Now consider $S \subset \gamma_z \backslash \mathring{\gamma}_z$. Then S is a long edge shared by two anisotropic triangles. Consider two cases; see Fig. 1. Case (i): If in at least one of these triangles, T, the angle at z is $\gtrsim 1$, then an application of (3.3a) yields $\|v\phi_z\|_{1;S} \lesssim \|\nabla v\|_{1;T} + h_z^{-1}\|v\|_{1;T}$, and (3.1) follows. Case (ii): Otherwise, in any triangle T sharing the edge S, the other edge S'' originating at z is also of length $\simeq H_z$, while the edge opposite to z is of length $\simeq h_z$. Then an application of (3.3b) yields $H_z^{-1}\|v\phi_z\|_{1;S} \lesssim H_z^{-1}\|v\phi_z\|_{1;S''} + H_z^{-1}\|\nabla v\|_{1;T}$ or, equivalently, $\|v\phi_z\|_{1;S} \lesssim \|v\phi_z\|_{1;S''} + \|\nabla v\|_{1;\omega_z}$. Thus, a possibly repeated application of (3.3b) reduces this case to case (i); see Fig. 1. $\square$

4 A Posteriori Error Bounds for Jump Residual Terms

Assuming $\|u_h - u\|_{\varepsilon;\Omega} > 0$, let

$$G := \frac{u_h - u}{\|u_h - u\|_{\varepsilon;\Omega}} \quad \Rightarrow \quad \|G\|_{\varepsilon;\Omega} = 1, \qquad g := G - G_h, \qquad (4.1)$$

where $G_h \in S_h$ is some interpolant of G. Now, a relatively standard calculation yields the following error representation [7, §4]

$$\|u_h - u\|_{\varepsilon;\Omega} \lesssim \sum_{z \in \mathcal{N}} \varepsilon^2 \int_{\gamma_z} (g - \bar{g}_z)\phi_z [\![\nabla u_h]\!] \cdot \nu + \sum_{z \in \mathcal{N}} \int_{\omega_z} f_h^I (g - \bar{g}_z)\phi_z + |\langle f_h - f_h^I, G\rangle|$$

$$=: I + II + \mathscr{E}_{\text{quad}}, \qquad (4.2)$$

which holds for any $G_h \in S_h$ and any set of real numbers $\{\bar{g}_z\}_{z\in\mathcal{N}}$ such that $\bar{g}_z = 0$ whenever $z \in \partial\Omega$. (To be precise, $\bar{g}_z$ will be specified later as a certain average of

$g = G - G_h$ near z.) Here ϕ_z denotes the standard hat basis function corresponding to $z \in \mathcal{N}$.

In the following proofs it will be convenient to use, with $p = 1, 2$, the scaled $W_p^1(\mathcal{D})$ norm defined by

$$\interleave v \interleave_{p;\mathcal{D}} := \|\nabla v\|_{p;\mathcal{D}} + (\operatorname{diam}\mathcal{D})^{-1}\|v\|_{p;\mathcal{D}} \Rightarrow \interleave v \interleave_{p;\omega_z} = \|\nabla v\|_{p;\omega_z} + H_z^{-1}\|v\|_{p;\omega_z}.$$

4.1 Jump Residual for a Partially Structured Anisotropic Mesh

To illustrate our approach in a simpler setting, we first present a version of the analysis for a simpler, partially structured, anisotropic mesh in a square domain $\Omega = (0, 1)^2$. So, throughout this section, we make the following triangulation assumptions.

A1. Let $\{x_i\}_{i=0}^n$ be an arbitrary mesh on the interval $(0, 1)$ in the x direction. Then, let each $T \in \mathcal{T}$, for some i,

  (i) have the shortest edge on the line $x = x_i$;
  (ii) have a vertex on the line $x = x_{i+1}$ or $x = x_{i-1}$ (see Fig. 2, left).

A2. Let $\mathcal{N} = \mathcal{N}_{\text{ani}}$, i.e. each mesh node z satisfies (2.2).
A3. *Quasi-non-obtuse anisotropic elements.* Let the maximum angle in any triangle be bounded by $\frac{\pi}{2} + \alpha_1 \frac{h_T}{H_T}$ for some positive constant α_1.

These conditions essentially imply that all mesh elements are anisotropic and aligned in the x-direction. They also imply that if $x_z = x_i$, then

$$\omega_z \subseteq \omega_z^* := (x_{i-1}, x_{i+1}) \times (y_z^-, y_z^+), \quad y_z^+ - y_z^- \simeq h_z, \quad \operatorname{diam}\omega_z^* \simeq H_z, \quad (4.3)$$

where (y_z^-, y_z^+) is the range of y within ω_z, while $x_{-1} := x_0$ and $x_{n+1} := x_n$.

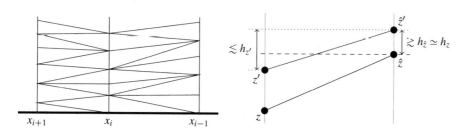

Fig. 2 Partially structured anisotropic mesh (left); illustration for Remark 4.1 (right): for any fixed edge $z\hat{z}$ and any edge $z'\hat{z}'$ intercepting the dashed horizontal line via $\hat{z}$, the figure shows that $h_z \lesssim h_{z'}$, so there is a uniformly bounded number of edges of type $z'\hat{z}'$, so $\omega_z^* \subset \omega_z^{(J)}$ with $J \lesssim 1$

Remark 4.1 The above conditions (in particular A3) imply that there is $J \lesssim 1$ such that $\omega_z^* \subset \omega_z^{(J)}$ for all $z \in \mathcal{N}$, where $\omega_z^{(0)} := \omega_z$, and $\omega_z^{(j+1)}$ denotes the patch of elements in/touching $\omega_z^{(j)}$. This conclusion is illustrated on Fig. 2 (right). (Note that $J = 1$ for any non-obtuse triangulation, i.e. for the case $\alpha_1 = 0$ in A3.)

Following [6, 7], the choice of $\bar{g}_z$ in (4.2) is related to the orientation of anisotropic elements, and is crucial in our analysis. Let $\bar{g}_z = 0$ for $z \in \partial\Omega$, and, otherwise, for $x_z = x_i$ with some $1 \leqslant i \leqslant n - 1$, let

$$\int_{x_{i-1}}^{x_{i+1}} (g(x, y_z) - \bar{g}_z)\, \varphi_i(x)\, dx = 0. \tag{4.4}$$

Here we use the standard one-dimensional hat function $\varphi_i(x)$ associated with the mesh $\{x_i\}$ (i.e. it has support on (x_{i-1}, x_{i+1}), equals 1 at $x = x_i$, and is linear on (x_{i-1}, x_i) and (x_i, x_{i+1})).

Theorem 4.1 *Let $g = G - G_h$ with G from (4.1) and any $G_h \in S_h$, while*

$$\Theta := \varepsilon^2 \|\nabla g\|_{2;\Omega}^2 + \sum_{z \in \mathcal{N}} (1 + \varepsilon^2 H_z^{-2}) \|g\|_{2;\omega_z}^2 . \tag{4.5}$$

Then $\|u_h - u\|_{\varepsilon;\Omega} \lesssim I + II + \mathscr{E}_{\text{quad}}$, where the right-hand side terms are specified in (4.2), and, under conditions A1–A3,

$$|I| \lesssim \left\{ \Theta \sum_{z \in \mathcal{N}} \left[\min\{|\omega_z|, \varepsilon h_z\} \left(\varepsilon \mathring{J}_z\right)^2 + \min\{|\omega_z|, \varepsilon H_z\} \left(\varepsilon J_z\right)^2 \right] \right\}^{1/2}, \tag{4.6}$$

where $\mathring{J}_z := \|[\![\nabla u_h]\!]\|_{\infty;\mathring{\gamma}_z}$ and $J_z := \|[\![\nabla u_h]\!]\|_{\infty;\gamma_z \setminus \mathring{\gamma}_z}$.

Corollary 4.2 *Under conditions A1–A3, one has (1.3) with $\lambda_z = \varepsilon H_z$.*

Proof To get the desired result, combine (4.6) with the bound [7, (5.8)] on II, the straightforward bound $|\mathscr{E}_{\text{quad}}| \leqslant \|f_h - f_h^I\|_{2;\Omega}$, and $\Theta \lesssim \|G\|_{\varepsilon;\Omega} = 1$ (the latter is given by [7, Theorem 7.4] under more general conditions than A1–A3). □

Proof of Theorem 4.1. Split I of (4.2) as $I = \sum_{z \in \mathcal{N}} (\mathring{I}_z + I_z)$, where

$$\mathring{I}_z := \varepsilon^2 \int_{\mathring{\gamma}_z} (g - \bar{g}_z)\phi_z [\![\nabla u_h]\!] \cdot \boldsymbol{v}, \qquad I_z := \varepsilon^2 \int_{\gamma_z \setminus \mathring{\gamma}_z} (g - \bar{g}_z)\phi_z [\![\nabla u_h]\!] \cdot \boldsymbol{v}. \tag{4.7}$$

First, consider $\bar{g}_z$, the definition of which (4.4) implies that $H_z|\bar{g}_z| \lesssim \|g\varphi_i\|_{1;\bar{S}_z}$, where $\bar{S}_z$ is the segment joining the points (x_{i-1}, y_z) and (x_{i+1}, y_z), so $|\bar{S}_z| \simeq H_z$. Versions of (3.1) and (3.2) then respectively yield

$$H_z|\bar{g}_z| \lesssim \|\nabla g\|_{1;\omega_z^*} + h_z^{-1}\|g\|_{1;\omega_z^*}, \quad H_z|\bar{g}_z|^2 \lesssim \|g\|_{2;\omega_z^*}(\|\nabla g\|_{2;\omega_z^*} + h_z^{-1}\|g\|_{2;\omega_z^*}). \tag{4.8}$$

These two bounds will be used when estimating both $\mathring{I}_z$ and I_z.

We now proceed to estimating $\mathring{I}_z$. Note that (3.1) implies that $\|(g - \bar{g}_z)\phi_z\|_{1;\mathring{\gamma}_z} \lesssim \||g\||_{1;\omega_z^*} \lesssim |\omega_z|^{1/2}\||g\||_{2;\omega_z^*}$, where we also used $\|\bar{g}_z\phi_z\|_{1;\mathring{\gamma}_z} \simeq h_z|\bar{g}_z|$ combined with the first bound from (4.8). Similarly, $\|(g - \bar{g}_z)\phi_z\|_{1;\mathring{\gamma}_z}^2 \lesssim h_z\|g\|_{2;\omega_z^*}\||g\||_{2;\omega_z^*}$, where we employed (3.2) and the second bound from (4.8). Now, from the definition of $\mathring{I}_z$ in (4.7) combined with the two bounds on $\|(g - \bar{g}_z)\phi_z\|_{1;\mathring{\gamma}_z}$, one concludes that

$$|\mathring{I}_z| \lesssim \mathring{\theta}_z^{1/2} \mathring{\lambda}_z^{1/2} (\varepsilon \mathring{J}_z), \qquad \mathring{\theta}_z := \frac{\varepsilon^2 \min\left\{|\omega_z|\||g\||_{2;\omega_z^*}^2, \; h_z\|g\|_{2;\omega_z^*}\||g\||_{2;\omega_z^*}\right\}}{\mathring{\lambda}_z}.$$

Set $\mathring{\lambda}_z := \min\{|\omega_z|, \varepsilon h_z\}$. Then, to get the bound of type (4.6) for $\sum_{z \in \mathcal{N}} \mathring{I}_z$, it remains to show that $\sum_{z \in \mathcal{N}} \mathring{\theta}_z \lesssim \Theta$. For the latter, in view of

$$\min\{aa', bb'\}/\min\{a', b'\} \leq a + b \qquad \forall\, a, a', b, b' > 0, \tag{4.9}$$

one gets $\mathring{\theta}_z \lesssim \varepsilon^2\||g\||_{2;\omega_z^*}^2 + \varepsilon\|g\|_{2;\omega_z^*}\||g\||_{2;\omega_z^*}$, which leads to $\sum_{z \in \mathcal{N}} \mathring{\theta}_z \lesssim \Theta$, also using Remark 4.1.

For I_z, first, recall the bound $|I_z| \lesssim \varepsilon\||g\||_{1;\omega_z^*}(\varepsilon J_z)$ from [7, (5.12)], which implies $|I_z| \lesssim \varepsilon|\omega_z|^{1/2}\||g\||_{2;\omega_z^*}(\varepsilon J_z)$. An alternative bound on I_z follows from $\|(g - \bar{g}_z)\phi_z\|_{1;\gamma_z\backslash\mathring{\gamma}_z}^2 \lesssim H_z\|g\|_{2;\omega_z^*}(\|\nabla g\|_{2;\omega_z^*} + h_z^{-1}\|g\|_{2;\omega_z^*})$, where the latter is obtained by an application of (3.2) for g, while the second bound from (4.8) is employed for $\bar{g}_z$. Combining the two bounds on I_z, we arrive at

$$|I_z| \lesssim \theta_z^{1/2} \lambda_z^{1/2} (\varepsilon J_z),$$

$$\theta_z := \frac{\varepsilon^2 \min\left\{|\omega_z|\||g\||_{2;\omega_z^*}^2, \; H_z\|g\|_{2;\omega_z^*}(\|\nabla g\|_{2;\omega_z^*} + h_z^{-1}\|g\|_{2;\omega_z^*})\right\}}{\lambda_z}.$$

$$\tag{4.10}$$

Here set $\lambda_z := \min\{|\omega_z|, \varepsilon H_z(1 + \varepsilon h_z^{-1})\}$. Now, again using (4.9), one gets

$$\theta_z \lesssim \varepsilon^2\||g\||_{2;\omega_z^*}^2 + \varepsilon\|g\|_{2;\omega_z^*}(\|\nabla g\|_{2;\omega_z^*} + h_z^{-1}\|g\|_{2;\omega_z^*})/(1 + \varepsilon h_z^{-1}), \tag{4.11}$$

and hence $\sum_{z \in \mathcal{N}} \theta_z \lesssim \Theta$. Finally, to get the bound of type (4.6) for $\sum_{z \in \mathcal{N}} I_z$, it remains to note that $\lambda_z = \min\{|\omega_z|, \varepsilon H_z[1 + \varepsilon h_z^{-1}]\} \simeq \min\{|\omega_z|, \varepsilon H_z\}$. $\qquad\square$

Remark 4.2 While the definition (4.4) for $\bar{g}_z$ is quite different from a standard choice (see, e.g., [13, Lecture 5]), its role may not be immediately obvious in the proof of Theorem 4.1. To clarify this, note that it is crucial for the bound $|I_z| \lesssim \varepsilon\||g\||_{1;\omega_z^*}(\varepsilon J_z)$ quoted from [7, (5.12)]. To be more precise, the latter bound is obtained in [7] using the representation

$$I_z = I_z' + I_z'' + I_z''' := \varepsilon^2 \int_{\gamma_z \backslash \mathring{\gamma}_z} (g - \bar{g}_z) \phi_z [\![\partial_x u_h]\!] \, \boldsymbol{\nu}_x$$

$$+ \varepsilon^2 \int_{\gamma_z \backslash \mathring{\gamma}_z} [g - g(x, y_z)] \phi_z [\![\partial_y u_h]\!] \, \boldsymbol{\nu}_y$$

$$+ \varepsilon^2 \int_{\gamma_z \backslash \mathring{\gamma}_z} [g(x, y_z) - \bar{g}_z] \phi_z [\![\partial_y u_h]\!] \, \boldsymbol{\nu}_y \, ,$$

where $[\![w]\!]$, for any w, is understood as the jump in w across any edge in γ_z evaluated in the anticlockwise direction about z. Importantly, here $I_z''' = 0$ due to our choice of $\bar{g}_z$ (as well as due to the partial structure of our mesh; in a more general case, the estimation of I_z''' is more intricate).

4.2 Jump Residual for General Anisotropic Meshes

Theorem 4.3 *Suppose that $\mathcal{N} = \mathcal{N}_{\mathrm{ani}} \cup \mathcal{N}_{\mathrm{reg}}$ and all corners of Ω are in $\mathcal{N}_{\mathrm{reg}}$. Let $g = G - G_h$ with G from (4.1) and any $G_h \in S_h$, while Θ is defined by (4.5). Then $\| u_h - u \|_{\varepsilon; \Omega} \lesssim I + II + \mathscr{E}_{\mathrm{quad}}$, where the right-hand side terms are specified in (4.2), and*

$$|I| \lesssim \left\{ \Theta \sum_{z \in \mathcal{N}} \min\{|\omega_z|, \, \varepsilon H_z\} \left\| \varepsilon [\![\nabla u_h]\!] \right\|_{\infty; \gamma_z}^2 \right\}^{1/2}. \tag{4.12}$$

Corollary 4.4 *Under the conditions of Theorem 4.3, one has (1.3) with $\lambda_z = \varepsilon H_z$.*

Proof To get the desired result, combine (4.12) with the bound [7, (6.2)] on II, the straightforward bound $|\mathscr{E}_{\mathrm{quad}}| \leqslant \| f_h - f_h^I \|_{2;\Omega}$, and $\Theta \lesssim \| G \|_{\varepsilon;\Omega} = 1$ (the latter follows from [7, Theorem 7.4] as $\mathcal{N} = \mathcal{N}_{\mathrm{ani}} \cup \mathcal{N}_{\mathrm{reg}}$). □

Remark 4.3 In view of Remark 3.2, an inspection of the proof of Theorem 4.3 shows that this theorem remains valid for the more general node types introduced in [7, §2], and furthermore, can be combined with a shaper bound for the interior residual terms given by [7, Theorem 6.2].

Proof of Theorem 4.3. Split I of (4.2) as $I = \sum_{z \in \mathcal{N}} I_z$, where I_z is defined as in (4.7), only with $\gamma_z \backslash \mathring{\gamma}_z$ replaced by γ_z. It suffices to show that for some edge subset $\mathscr{S}^* \subset \mathscr{S}$ with some quantities $\mathscr{I}_{S;z}$ associated with any $S \in \mathscr{S}_z \cap \mathscr{S}^*$ (to be specified below), one has

$$\sum_{z \in \mathcal{N}} \sum_{S \in \mathcal{S}_z \cap \mathcal{S}^*} \mathcal{I}_{S;z} = 0, \tag{4.13a}$$

$$|I_z + \sum_{S \in \mathcal{S}_z \cap \mathcal{S}^*} \mathcal{I}_{S;z}| \lesssim \varepsilon \| g \|_{1 \, ; \omega_z} \, \| \varepsilon [\![\nabla u_h]\!] \|_{\infty \, ; \gamma_z}$$

$$\lesssim \varepsilon |\omega_z|^{1/2} \| g \|_{2 \, ; \omega_z} \, \| \varepsilon [\![\nabla u_h]\!] \|_{\infty \, ; \gamma_z}, \tag{4.13b}$$

$$|I_z| + \sum_{S \in \mathcal{S}_z \cap \mathcal{S}^*} |\mathcal{I}_{S;z}| \lesssim \varepsilon \left\{ H_z \| g \|_{2 \, ; \omega_z} (\| \nabla g \|_{2 \, ; \omega_z} + h_z^{-1} \| g \|_{2 \, ; \omega_z}) \right\}^{1/2} \| \varepsilon [\![\nabla u_h]\!] \|_{\infty \, ; \gamma_z}. \tag{4.13c}$$

Indeed, (4.13a) implies that $I = \sum_{z \in \mathcal{N}} I_z = \sum_{z \in \mathcal{N}} (I_z + \sum_{S \in \mathcal{S}_z \cap \mathcal{S}^*} \mathcal{I}_{S;z})$, while (4.13b), (4.13c) yield

$$|I_z + \sum_{S \in \mathcal{S}_z \cap \mathcal{S}^*} \mathcal{I}_{S;z}| \lesssim \theta_z^{1/2} \lambda_z^{1/2} \| \varepsilon [\![\nabla u_h]\!] \|_{\infty \, ; \gamma_z},$$

$$\theta_z := \frac{\varepsilon^2 \min \left\{ |\omega_z| \| g \|_{2 \, ; \omega_z}^2, \, H_z \| g \|_{2 \, ; \omega_z} (\| \nabla g \|_{2 \, ; \omega_z} + h_z^{-1} \| g \|_{2 \, ; \omega_z}) \right\}}{\lambda_z}. \tag{4.14}$$

Here set $\lambda_z := \min\{|\omega_z|, \, \varepsilon H_z(1 + \varepsilon h_z^{-1})\}$. Then (4.14) becomes a version of (4.10) with ω_z^* replaced by ω_z, so proceeding as in the proof of Theorem 4.1 (i.e. again employing (4.9)), one gets a version (4.11) with ω_z^* replaced by ω_z, which leads to $\sum_{z \in \mathcal{N}} \theta_z \lesssim \Theta$. Now, to get the desired bound (4.12), it remains to note that $\lambda_z = \min\{|\omega_z|, \, \varepsilon H_z(1 + \varepsilon h_z^{-1})\} \simeq \min\{|\omega_z|, \, \varepsilon H_z\}$.

So, to complete the proof, we need to establish (4.13). Relations (4.13a) and (4.13b) immediately follow from [8, (6.10), (6.11a), (6.11b)] for a certain choice of $\{\bar{g}_z\}_{z \in \mathcal{N}}$, the edge subset $\mathcal{S}^* \subset \mathcal{S}$ and the quantities $\mathcal{I}_{S;z}$ associated with any $S \in \mathcal{S}_z \cap \mathcal{S}^*$. We need to recall their definitions to prove the remaining required bound (4.13c) (which is a sharper version of [8, (6.11c)]).

First, we recall the definition of $\{\bar{g}_z\}_{z \in \mathcal{N}}$. In view of the Local Element Orientation condition (see Sect. 2), for each fixed $z \in \mathcal{N}$, introduce the following local notation. Let the local cartesian coordinates (ξ, η) be such that $z = (0, 0)$, and the unit vector i_ξ in the ξ direction lies along the longest edge $\hat{S}_z \in \mathcal{S}_z$ (see Fig. 3 (left)). For $z \in \mathcal{N}_{ani} \cap \partial \Omega$ (hence z is not a corner of Ω), let i_ξ be either parallel or orthogonal to $\partial \Omega$ at z (depending on whether ω_z is, roughly speaking, parallel or orthogonal to $\partial \Omega$).

Next, split $\mathcal{S}_z = \mathring{\mathcal{S}}_z \cup \mathcal{S}_z^+ \cup \mathcal{S}_z^-$, where $\mathring{\mathcal{S}}_z = \{S \subset \mathcal{S}_z : |S| \lesssim h_z\}$ (so $\gamma_z = \mathring{\mathcal{S}}_z \setminus \partial \Omega$). Here we also use $\mathcal{S}_z^\pm := \{S \subset \mathcal{S}_z \setminus \mathring{\mathcal{S}}_z : S_\xi \subset \mathbb{R}_\pm\}$, where $S_\xi = \text{proj}_\xi(S)$ denotes the projection of S onto the ξ-axis. Now, let (ξ_z^-, ξ_z^+) be the maximal interval such that $(\xi_z^-, 0) \subset S_\xi$ for all $S \in \mathcal{S}_z^-$ and $(0, \xi_z^+) \subset S_\xi$ for all

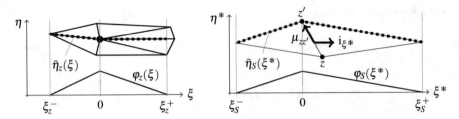

Fig. 3 Local notation associated with a node $z \in \mathcal{N}_{\mathrm{ani}}$ (left), and an edge $S \in \mathcal{S}^*$ with endpoints z and z' (right).

$S \in \mathcal{S}_z^+$. Also, let $\varphi_z(\xi)$ be the standard piecewise-linear hat-function with support on (ξ_z^-, ξ_z^+) and equal to 1 at $\xi = 0$. Note that if $\mathcal{S}_z^- = \emptyset$ (and $\mathcal{S}_z^+ = \emptyset$), then we set $\xi_z^- = 0$ (and $\xi_z^+ = 0$) and do not use φ_z for $\xi < 0$ (and $\xi > 0$).

Next, for $\xi \in [\xi_z^-, \xi_z^+]$ define a continuous function $\bar{\eta}_z(\xi)$ as follows: (i) $\bar{\eta}_z(\xi)$ is linear on $[\xi_z^-, 0]$ and $[0, \xi_z^+]$; (ii) $\bar{\eta}_z(0) = 0$; (iii) $(\xi, \bar{\eta}_z(\xi)) \in \omega_z$ for all $\xi \in (\xi_z^-, \xi_z^+)$. (For example, one may choose $\bar{\eta}_z(\xi)$ so that $\{(\xi, \bar{\eta}_z(\xi)) : \xi \in (\xi_z^-, 0)\}$ lies on any edge in $\mathcal{S}_z^-$, while $\{(\xi, \bar{\eta}_z(\xi)) : \xi \in (0, \xi_z^+)\}$ lies on any edge in $\mathcal{S}_z^+$; see Fig. 3 (left).)

We are now prepared to specify $\bar{g}_z$. Let $\bar{g}_z := 0$ if $z \in \partial\Omega$ or $z \in \mathcal{N}_{\mathrm{reg}}$ (as for the latter, $\xi_z^- = \xi_z^+ = 0$), and, otherwise, let

$$\int_{\xi_z^-}^{\xi_z^+} \left[g(\xi, \bar{\eta}_z(\xi)) - \bar{g}_z \right] \varphi_z(\xi)\, d\xi = 0. \tag{4.15}$$

Also, let $\bar{S}_z^- := \{(\xi, \bar{\eta}_z(\xi)) : \xi \in (\xi_z^-, 0)\}$ and $\bar{S}_z^+ := \{(\xi, \bar{\eta}_z(\xi)) : \xi \in (0, \xi_z^+)\}$, i.e. $\bar{S}_z^\pm$ is the segment joining $(0, 0)$ and $(\xi_z^\pm, \bar{\eta}_z(\xi_z^\pm))$.

We can now proceed to getting a bound of type (4.13c) for $|I_z|$. First, consider $\bar{g}_z$, the definition of which (4.15) implies that $H_z |\bar{g}_z| \lesssim \|g \varphi_z\|_{1; \bar{S}_z^- \cup \bar{S}_z^-}$, where $|\bar{S}_z^- \cup \bar{S}_z^-| \simeq H_z$. Using (3.1) and (3.2) then yields a version of (4.8), only with ω_z^* replaced by ω_z (as now $\bar{S}_z^- \cup \bar{S}_z^+ \subset \omega_z$). Next, we get $\|(g - \bar{g}_z)\phi_z\|_{1; \gamma_z}^2 \lesssim H_z \|g\|_{2;\omega_z} (\|\nabla g\|_{2;\omega_z} + h_z^{-1} \|g\|_{2;\omega_z})$, which is obtained by an application of (3.2) for g, while the second bound from (4.8) is employed for $\bar{g}_z$. Combining this with the definition of I_z immediately yields a bound of type (4.13c) for $|I_z|$.

To establish a bound of type (4.13c) for $|\mathscr{I}_{S;z}|$, we now recall the definitions of the edge subset $\mathscr{S}^* \subset \mathscr{S}$ and the quantities $\mathscr{I}_{S;z}$ associated with any $S \in \mathscr{S}_z \cap \mathscr{S}^*$ from [7]. Let $\mathscr{S}^* := \cup_{z \in \mathcal{N}_{\mathrm{ani}} \setminus \partial\Omega} \mathscr{S}_z$, and for any $S \in \mathscr{S}^*$ with endpoints z and z', define

$$\mathscr{I}_{S;z} := \varepsilon^2 \alpha_S\, \boldsymbol{\mu}_{zz'} \cdot \mathbf{i}_{\xi^*}\, J_S, \qquad \alpha_S := \int_0^{\xi_S^+} [g(\xi^*, \bar{\eta}_S(\xi^*)) - \bar{g}_S]\, \varphi_S(\xi^*)\, d\xi^*. \tag{4.16}$$

Here J_S is the standard signed version of $|\llbracket \nabla u_h \rrbracket|$ on S, $\boldsymbol{\mu}_{zz'}$ is the unit vector directed from z to z', and i_{ξ^*} is the unit vector along the ξ^*-axis. The local cartesian coordinates (ξ^*, η^*) are associated with S and coincide with the local coordinates (ξ, η) associated with either $z \in \mathcal{N}_{\mathrm{ani}} \setminus \partial\Omega$ or $z' \in \mathcal{N}_{\mathrm{ani}} \setminus \partial\Omega$ (at least one of them is always in $\mathcal{N}_{\mathrm{ani}} \setminus \partial\Omega$). The above α_S is defined by a version $\int_{\xi_S^-}^{\xi_S^+} [g(\xi^*, \bar\eta_S(\xi^*)) - \bar g_S] \varphi_S(\xi^*) \, d\xi^* = 0$ of (4.15). The one-dimensional hat function $\varphi_S(\xi^*)$ is associated with the interval (ξ_S^-, ξ_S^+); the latter is the projection of $\omega_z \cap \omega_{z'}$ (which includes at most two triangles) onto the ξ^*-axis. The piecewise-linear function $\bar\eta_S(\xi^*)$ is defined similarly to $\bar\eta_z(\xi)$ under the restriction that any point $(\xi^*, \bar\eta(\xi^*)) \in \omega_z \cap \omega_{z'}$ (see Fig. 3(right)).

Under this definition, a bound of type (4.13c) for $|\mathscr{I}_{S;z}|$ is established similarly to a similar bound for $|I_z|$. (Note also that $\boldsymbol{\mu}_{zz'} + \boldsymbol{\mu}_{z'z} = 0$ in (4.16), so $\mathscr{I}_{S;z} + \mathscr{I}_{S;z'} = 0$, which implies (4.13a).) This completes the proof of (4.13), and hence of (4.12). $\qquad \square$

References

1. Ainsworth, M., Oden, J.T.: A Posteriori Error Estimation in Finite Element Analysis. Wiley, New York (2000)
2. Chadha, N.M., Kopteva, N.: Maximum norm a posteriori error estimate for a 3d singularly perturbed semilinear reaction-diffusion problem. Adv. Comput. Math. **35**, 33–55 (2011)
3. Demlow, A., Kopteva, N.: Maximum-norm a posteriori error estimates for singularly perturbed elliptic reaction-diffusion problems. Numer. Math. **133**, 707–742 (2016)
4. Kopteva, N.: Maximum norm error analysis of a 2d singularly perturbed semilinear reaction-diffusion problem. Math. Comp. **76**, 631–646 (2007)
5. Kopteva, N.: Maximum norm a posteriori error estimate for a 2d singularly perturbed reaction-diffusion problem. SIAM J. Numer. Anal. **46**, 1602–1618 (2008)
6. Kopteva, N.: Maximum-norm a posteriori error estimates for singularly perturbed reaction-diffusion problems on anisotropic meshes. SIAM J. Numer. Anal. **53**, 2519–2544 (2015)
7. Kopteva, N.: Energy-norm a posteriori error estimates for singularly perturbed reaction-diffusion problems on anisotropic meshes. Numer. Math. **137**, 607–642 (2017)
8. Kopteva, N.: Fully computable a posteriori error estimator using anisotropic flux equilibration on anisotropic meshes. Submitted for publication (2017). arXiv:1704.04404
9. Kopteva, N., O'Riordan, E.: Shishkin meshes in the numerical solution of singularly perturbed differential equations. Int. J. Numer. Anal. Model. **7**, 393–415 (2010)
10. Kunert, G.: An a posteriori residual error estimator for the finite element method on anisotropic tetrahedral meshes. Numer. Math. **86**, 471–490 (2000)
11. Kunert, G.: Robust a posteriori error estimation for a singularly perturbed reaction-diffusion equation on anisotropic tetrahedral meshes. Adv. Comput. Math. **15**, 237–259 (2001)
12. Kunert, G., Verfürth, R.: Edge residuals dominate a posteriori error estimates for linear finite element methods on anisotropic triangular and tetrahedral meshes. Numer. Math. **86**, 283–303 (2000)
13. Nochetto, R.H.: Pointwise a posteriori error estimates for monotone semi-linear equations. Lecture Notes at 2006 CNA Summer School Probabilistic and Analytical Perspectives on Contemporary PDEs. http://www.math.cmu.edu/cna/Summer06/lecturenotes/nochetto/

14. Roos, H.-G., Stynes, M., Tobiska, T.: Robust Numerical Methods for Singularly Perturbed Differential Equations. Springer, Berlin (2008)
15. Siebert, K.G.: An a posteriori error estimator for anisotropic refinement. Numer. Math. **73**, 373–398 (1996)
16. Verfürth, R.: Robust a posteriori error estimators for a singularly perturbed reaction-diffusion equation. Numer. Math. **78**, 479–493 (1998)

Implicit LES with High-Order H(div)-Conforming FEM for Incompressible Navier-Stokes Flows

Gert Lube and Philipp W. Schroeder

Abstract Consider the transient incompressible Navier-Stokes flow at high Reynolds numbers. A high-order H(div)-conforming FEM with pointwise divergence-free discrete velocities is applied to implicit large-eddy-simulation in two limit cases: (1) decaying turbulence in periodic domains, (2) wall bounded channel flow.

1   H(div)-Conforming dGFEM for Navier-Stokes Problem

Consider a flow in a bounded polyhedron $\Omega \subset \mathbb{R}^d$, $d \leq 3$ with boundary $\partial \Omega = \Gamma_0 \cup \Gamma_{per}$ and outer unit normal $\mathbf{n} = (n_i)_{i=1}^d$. Set $Q_T := (0, T) \times \Omega$ and denote $\mathbf{f}$ as source term. We want to find velocity $\mathbf{u} : Q_T \to \mathbb{R}^d$ and pressure $p : Q_T \to \mathbb{R}$ s.t.

$$\partial_t \mathbf{u} - \nu \Delta \mathbf{u} + (\mathbf{u} \cdot \nabla)\mathbf{u} + \nabla p = \mathbf{f} \qquad \text{in } Q_T, \tag{1}$$

$$\nabla \cdot \mathbf{u} = 0 \qquad \text{in } Q_T, \tag{2}$$

$$\mathbf{u} = \mathbf{0} \qquad \text{on } (0, T) \times \Gamma_0, \tag{3}$$

$$\mathbf{u} = \mathbf{u}_0 \quad \text{on } \{0\} \times \Omega, \tag{4}$$

and periodic boundary conditions on Γ_{per}. Let $\mathbf{H} = [L^2(\Omega)]^d$ with inner product $(\cdot, \cdot)_{\mathbf{H}}$ and assume $\mathbf{u}_0 \in \mathbf{H}$, $\mathbf{f} \in L^2(0, T; \mathbf{H})$. The inner product in $L^2(\Omega)$ is $(\cdot, \cdot)_\Omega$.

G. Lube (✉) · P. W. Schroeder
Georg-August University Göttingen, Institut for Numerical and Applied Mathematics, Göttingen, Germany
e-mail: lube@math.uni-goettingen.de; p.schroeder@math.uni-goettingen.de

© Springer Nature Switzerland AG 2020
G. R. Barrenechea, J. Mackenzie (eds.), *Boundary and Interior Layers, Computational and Asymptotic Methods BAIL 2018*, Lecture Notes in Computational Science and Engineering 135,
https://doi.org/10.1007/978-3-030-41800-7_10

A variational formulation of the transient incompressible Navier-Stokes problem (1)–(4) is to find $(\mathbf{u}, p) \in \mathbf{X} \times Q \subseteq [H^1(\Omega)]^d \times L^2(\Omega)$ for $t \in (0, T)$ a.e. from

$$(\partial_t \mathbf{u}(t), \mathbf{v})_{\mathbf{H}} + \nu a(\mathbf{u}(t), \mathbf{v}) + c(\mathbf{u}(t), \mathbf{u}(t), \mathbf{v}) + b(p(t), \mathbf{v}) = (\mathbf{f}(t), \mathbf{v})_{\mathbf{H}}, \quad (5)$$

$$-b(q, \mathbf{u}(t)) = 0, \quad (6)$$

$$\mathbf{u}(0) = \mathbf{u}_0. \quad (7)$$

with bounded bilinear resp. trilinear forms

$$a(\mathbf{u}, \mathbf{v}) := \nu(\nabla \mathbf{u}, \nabla \mathbf{v})_\Omega, \quad b(q, \mathbf{v}) := -(q, \nabla \cdot \mathbf{v})_\Omega, \quad c(\mathbf{w}, \mathbf{u}, \mathbf{v}) := ((\mathbf{w} \cdot \nabla)\mathbf{u}, \mathbf{w})_\Omega. \quad (8)$$

Consider $\mathbf{H}(\text{div})$-conforming, discontinuous Galerkin methods (dGFEM) with

$$\mathbf{H}(\text{div}; \Omega) := \{\mathbf{w} \in \mathbf{H} : \nabla \cdot \mathbf{w} \in L^2(\Omega)\}, \quad (9)$$

$$\mathbf{H}_{\Gamma_0}(\text{div}; \Omega) := \{\mathbf{v} \in \mathbf{H}(\text{div}; \Omega) : \mathbf{v} \cdot \mathbf{n}|_{\Gamma_0} = 0\}. \quad (10)$$

Let $\mathcal{T}_h$ be a shape-regular decomposition of $\Omega \subset \mathbb{R}^d$. Moreover, denote $\mathcal{E}_h$ the set of (open) edges ($d = 2$) or faces ($d = 3$) in $\mathcal{T}_h$. $\mathcal{E}_h^B \subset \mathcal{E}_h$ is the set of all $E \in \mathcal{E}_h$ with $E \cap \Gamma_0 \neq \emptyset$ and $\mathcal{E}_h^I := \mathcal{E}_h \setminus \mathcal{E}_h^B$ the set of interior edges. Please note that edges/faces on Γ_{per} are considered as interior edges/faces. Consider adjacent elements $K, K' \in \mathcal{T}_h$ with $\partial K \cap \partial K' = E$ and unit normal vector $\boldsymbol{\mu}_E$. For a scalar function v in the broken Sobolev space $H^1(\Omega, \mathcal{T}_h)$ denote jump resp. average of v across E by

$$[|v|]_E := v|_{\partial K \cap E} - v|_{\partial K' \cap E}, \qquad \{\!\{v\}\!\}_E := (v|_{\partial K \cap E} + v|_{\partial K' \cap E})/2. \quad (11)$$

For $\mathbf{v} \in [H^1(\Omega, \mathcal{T}_h)]^d$, jump and average are understood component-wise.

Lemma 1 ([2]) *Let* $\mathbf{W}_h$ *be a space of vector-valued polynomials w.r.t.* $\mathcal{T}_h$*. Then* $\mathbf{W}_h \subset \mathbf{H}(div; \Omega)$ *if* $[|\mathbf{v}|]_E \cdot \boldsymbol{\mu}_E = 0$ *for all* $\mathbf{v} \in \mathbf{W}_h$ *and all* $E \in \mathcal{E}_h^I$*.*

Owing to Lemma 1 $[|\mathbf{v}|]_{t,E} = [|\mathbf{v} - (\mathbf{v} \cdot \boldsymbol{\mu}_E)\boldsymbol{\mu}_E|]_E$ is the tangential jump across $E \in \mathcal{E}_h$.

Example 1 Examples of $\mathbf{H}(\text{div})$-conforming FEM are given in [1]. On simplicial grids one can apply Raviart-Thomas (RT) or Brezzi-Douglas-Marini (BDM) spaces

$$RT_k = \{\mathbf{w}_h \in \mathbf{H}_{\Gamma_0}(\text{div}; \Omega) : \mathbf{w}_h|_K \in \mathbb{P}_k(K) \oplus x\mathbb{P}_k(K) \ \forall K \in \mathcal{T}_h\}, \ k \in \mathbb{N}_0 \quad (12)$$

$$BDM_k = \{\mathbf{w}_h \in \mathbf{H}_{\Gamma_0}(\text{div}; \Omega) : \mathbf{w}_h|_K \in \mathbb{P}_k(K) \ \forall K \in \mathcal{T}_h\}, \ k \in \mathbb{N}. \quad (13)$$

On quadrilateral meshes, local Raviart-Thomas (RT) elements of degree $k \in \mathbb{N}_0$ are $RT_k(K) = (\mathbb{P}_{k+1,k}(K), \mathbb{P}_{k,k+1}(K))^t, d = 2$. For $d = 3$, one has similarly $RT_k(K) = (\mathbb{P}_{k+1,k,k}(K), \mathbb{P}_{k,k+1,k}(K), \mathbb{P}_{k,k,k+1}(K))^t$. $\qquad \square$

Let $\mathbf{w}_h \in \mathbf{W}_h \subset \mathbf{H}(\mathrm{div}; \Omega)$ with $\mathbf{W}_h \in \{RT_k, BDM_k\}$. The spaces $\mathbf{W}_h \not\subset [H^1(\Omega)]^d$ are not $[H^1(\Omega)]^d$-stable, hence not directly applicable to the Navier-Stokes problem. As a remedy, we modify the diffusion bilinear form a using a symmetric interior penalty (SIP) dGFEM-approach with the broken gradient $\nabla_h \mathbf{v} := \nabla(\mathbf{v}|_K)$: For sufficiently smooth $\mathbf{u} \in [H^s(\Omega)]^d, s > \frac{3}{2}$, we define by adding two consistent terms

$$a_h(\mathbf{u}, \mathbf{w}_h) := \int_\Omega \nabla_h \mathbf{u} : \nabla_h \mathbf{w}_h \, dx + \sum_{E \in \mathscr{E}_h} \sigma h_E^{-1} \int_E |[\mathbf{u}]|_t |[\mathbf{w}]|_t \, ds \tag{14}$$

$$- \sum_{E \in \mathscr{E}_h} \int_E \left(\{\!\!\{ \nabla_h \mathbf{u} \cdot \mu_E \}\!\!\} [\mathbf{w}_h]_t + \{\!\!\{ \nabla_h \mathbf{w}_h \cdot \mu_E \}\!\!\} [\mathbf{u}]_t \right) ds \quad \forall \mathbf{w}_h \in \mathbf{W}_h$$

with $h_E := \mathrm{diam}(E)$ and parameter $\sigma > 0$ (to be chosen according to next lemma). Define the following discrete H^1-norms $\|\mathbf{w}\|_{1,h}$ and $\|\mathbf{w}\|_{1,h,*}$

$$\|\mathbf{w}\|_{1,h}^2 := \sum_{K \in \mathscr{T}_h} \|\nabla \mathbf{w}\|_{L^2(K)}^2 + \sum_{E \in \mathscr{E}_h} h_E^{-1} \|[\mathbf{w}]_\tau\|_{L^2(E)}^2, \tag{15}$$

$$\|\mathbf{w}\|_{1,h,*}^2 := \|\mathbf{w}\|_{1,h}^2 + \sum_{E \in \mathscr{E}_h} h_E \|\{\!\!\{ \nabla_h \mathbf{w} \cdot \mu_E \}\!\!\}\|_{L^2(E)}^2. \tag{16}$$

Lemma 2 ([2]) *There exists constant σ_0 (depending only on k and on shape regularity of $\mathscr{T}_h$) s.t. for $\sigma \geq \sigma_0$ one has:*

$$a_h(\mathbf{w}_h, \mathbf{w}_h) \geq \frac{1}{2} \|\mathbf{w}\|_{1,h}^2 \quad \forall \mathbf{w}_h \in \mathbf{W}_h, \tag{17}$$

$$a_h(\mathbf{v}, \mathbf{w}_h) \leq C \|\mathbf{v}\|_{1,h,*} \|\mathbf{w}_h\|_{1,h} \quad \forall \mathbf{w}_h \in \mathbf{W}_h \text{ and } \mathbf{v} \in [H^s(\Omega)]^d, \ s > \frac{3}{2}. \tag{18}$$

Lemma 3 ([1]) *RT- and BDM-spaces, together with appropriate discrete spaces Q_h*

$$\mathbf{W}_h = RT_k \text{ with } Q_h := \{q_h \in L^2(\Omega) : q_h|_K \in \mathbb{P}_k(K) \, \forall K \in \mathscr{T}_h\} \quad \text{and}$$

$$\mathbf{W}_h = BDM_k \text{ with } Q_h := \{q_h \in L^2(\Omega) : q_h|_K \in \mathbb{P}_{k-1}(K) \, \forall K \in \mathscr{T}_h\}$$

form inf-sup stable pairs w.r.t. discrete H^1-norm:

$$\exists \beta_h \geq \beta_0 > 0 \text{ s.t.} \quad \inf_{q_h \in Q_h \setminus \{0\}} \sup_{\mathbf{w}_h \in \mathbf{W}_h \setminus \{0\}} \frac{(\nabla \cdot \mathbf{w}_h, q_h)_\Omega}{\|\mathbf{w}_h\|_{1,h} \|q_h\|_{L^2(\Omega)}} \geq \beta_h. \tag{19}$$

By construction $\nabla \cdot \mathbf{W}_h = Q_h$, these spaces are globally pointwise divergence-free:

$$\{\mathbf{w}_h \in \mathbf{W}_h : (\nabla \cdot \mathbf{w}_h, q_h)_\Omega = 0 \, \forall q_h \in Q_h\} = \{\mathbf{w}_h \in \mathbf{W}_h : \nabla \cdot \mathbf{w}_h = 0\}. \tag{20}$$

For an exactly divergence-free field $\mathbf{b} \in [L^\infty(\Omega)]^d \cap \mathbf{H}(\mathrm{div}; \Omega)$ we modify the convective term c as in [2] by

$$c_h(\mathbf{b}; \mathbf{u}, \mathbf{v}) := \sum_{K \in \mathscr{T}_h} ((\mathbf{b} \cdot \nabla)\mathbf{u}, \mathbf{v})_K$$

$$- \sum_{E \in \mathscr{E}_h^i} ((\mathbf{b} \cdot \boldsymbol{\mu}_E)([\![\mathbf{u}]\!], \{\!\{\mathbf{v}\}\!\}))_E + \frac{1}{2} \sum_{E \in \mathscr{E}_h^i} (|\mathbf{b} \cdot \boldsymbol{\mu}_E|[\![\mathbf{u}]\!], [\![\mathbf{v}]\!]))_E. \quad (21)$$

The first right-hand side terms corresponds to the standard form of the convective term. The last two facet terms, the *upwind* discretization, are consistent perturbations of the standard form of the convective term for $\mathbf{u}, \mathbf{v} \in \mathbf{X}$. The impact of these terms is included in the jump semi-norm $|\mathbf{v}|_{\mathbf{b},\mathrm{upw}}$ defined via

$$|\mathbf{v}|_{\mathbf{b},\mathrm{upw}}^2 := \frac{1}{2} \sum_{E \in \mathscr{E}_h^i} |\mathbf{b} \cdot \boldsymbol{\mu}_E| \, \||[\![\mathbf{v}]\!]|\|_{L^2(E)}^2. \quad (22)$$

In case of exactly divergence-free fields $\mathbf{b}$, one has $c_h(\mathbf{b}; \mathbf{v}, \mathbf{v}) = |\mathbf{v}|_{\mathbf{b},\mathrm{upw}}^2$.

We consider now the $\mathbf{H}(\mathrm{div})$-conforming dGFEM for the transient Navier-Stokes problem (5)–(7) with $\mathbf{f} \in L^2(0, T; \mathbf{H})$. Combining the SIP-form of the diffusive term and the upwind-discretization of the convective term, one obtains:
Find $(\mathbf{u}_h, p_h) : (0, T) \rightarrow \mathbf{W}_h \times Q_h$ with $\mathbf{u}_h(0) = \mathbf{u}_{0,h}$ s.t. for all $(\mathbf{v}_h, q_h) \in \mathbf{W}_h \times Q_h$:

$$(\partial_t \mathbf{u}_h, \mathbf{v}_h)_{\mathbf{H}} + \nu a_h(\mathbf{u}_h, \mathbf{v}_h) + c_h(\mathbf{u}_h; \mathbf{u}_h, \mathbf{v}_h) + b(p_h, \mathbf{v}_h) = (\mathbf{f}, \mathbf{v}_h)_{\mathbf{H}}, \quad (23)$$

$$-b(q_h, \mathbf{u}_h) = 0. \quad (24)$$

All computations have been done using a hybridized variant of (23)–(24) implemented in the high-order sofware package NGSolve [9].

We will consider method (23)–(24) as tool for implicit large-eddy-simulation (ILES) in two limit cases: (i) decaying turbulence in periodic 2D and 3D domains (see Sect. 2) and (ii) wall bounded flow in a 3D-channel (see Sect. 3).

2 Decaying 2D- and 3D-Turbulent Flows

2.1 Stability and Error Analysis for Decaying Flows

Consider now decaying flows, i.e. we consider problem (23)–(24) with $\mathbf{f} \equiv \mathbf{0}$. Using the mesh-dependent expressions (15) and (22) and setting $\mathbf{v}_h = \mathbf{u}_h$ in the semidiscrete problem (23)–(24), one obtains with $\|\mathbf{v}\|_e^2 := a_h(\mathbf{v}, \mathbf{v})$ the balance

$$\frac{d}{dt} \left(\frac{1}{2} \|\mathbf{u}_h(t)\|_{L^2(\Omega)}^2 \right) + \nu \|\mathbf{u}_h(t)\|_e^2 + |\mathbf{u}_h(t)|_{\mathbf{u}_h,\mathrm{upw}}^2 = 0. \quad (25)$$

This implies existence of $(\mathbf{u}_h, p_h)$ and bounds for kinetic and dissipation energies:

$$\frac{1}{2}\|\mathbf{u}_h(t)\|^2_{L^2(\Omega)} \leq \frac{1}{2}\|\mathbf{u}_{0h}\|^2_{L^2(\Omega)} \exp(-\nu t/C_F^2), \quad (26)$$

$$\int_0^t \left(\frac{\nu}{2}\|\mathbf{u}_h(\tau)\|^2_e + |\mathbf{u}_h(\tau)|^2_{\mathbf{u}_h,\text{upw}}\right) d\tau \leq \frac{1}{2}\|\mathbf{u}_{0h}\|^2_{L^2(\Omega)}. \quad (27)$$

In case of smooth velocity with $\mathbf{u} \in L^1(0, T; [W^{1,\infty}(\Omega)]^d)$, we obtain the following pressure-robust and Re-semi-robust error estimate.

Theorem 1 ([10]) *Let* $\mathbf{u} \in L^2(0, T; \mathbf{H}^{\frac{3}{2}+\epsilon}(\Omega))$, $\epsilon > 0$, $\nabla \mathbf{u} \in L^1(0, T, [L^\infty(\Omega)]^d)$ *and* $\mathbf{u}_h(0) = \pi_S \mathbf{u}_0$ *with Stokes projector* $\pi_S \mathbf{u}$, *i.e.* $a_h(\mathbf{u} - \pi_S \mathbf{u}, \mathbf{v}_h) = 0 \; \forall \mathbf{v}_h \in W_h$; *then:*

$$\frac{1}{2}\|\mathbf{u}_h - \pi_S \mathbf{u}\|^2_{L^\infty(0,T;L^2(\Omega))} + \int_0^T \left[\frac{\nu}{2}\|\mathbf{u}_h - \pi_S \mathbf{u}\|^2_{1,h} + |\mathbf{u}_h - \pi_S|^2_{\mathbf{u}_h,\text{upw}}\right] d\tau$$

$$\leq e^{G_{\mathbf{u}}(T)} \int_0^T \left[\|\partial_t \eta\|^2_{L^2(\Omega)} + \|\mathbf{u}\|_{L^\infty(\Omega)}\|\nabla_h \eta\|^2_{L^2(\Omega)} + h^{-2}\|\nabla \mathbf{u}\|_{L^\infty(\Omega)}\|\eta\|^2_{L^2(\Omega)}\right] d\tau$$

with $\eta := \mathbf{u} - \pi_S \mathbf{u}$ *and Gronwall factor*

$$G_{\mathbf{u}}(T) := T + \|\mathbf{u}\|_{L^1(0,T;[L^\infty(\Omega)]^d)} + C\|\nabla \mathbf{u}\|_{L^1(0,T;[L^\infty(\Omega)]^d)}.$$

The vorticity equation for $\omega := \nabla \times \mathbf{u}$ describes the dynamics of decaying flows:

$$\partial_t \omega + \mathbf{u} \cdot \nabla \omega - \nu \Delta \omega = \omega \cdot \nabla \mathbf{u}, \qquad \nabla \cdot \mathbf{u} = 0. \quad (28)$$

The vortex stretching term $\omega \cdot \nabla \mathbf{u}$ vanishes for $d = 2$ which leads to a completely different behavior for $d = 2$ and $d = 3$.

2.2 Decaying 2D-Turbulent Flow

Consider the following 2D-turbulent flow problem with a unique solution of (5)–(7).

Example 2 2D-lattice flow
Consider on $\Omega = (-1, 1)^2$ the following solution of the steady Euler model $(\nu = 0)$

$$\mathbf{u}_0(x) = (-\Psi_{x_2}(x), \Psi_{x_1}(x))^t, \qquad \Psi(x) := \frac{1}{2\pi} \sin(2\pi x_1) \cos(2\pi x_2).$$

The initial vorticity $\omega_0 = \nabla \times \mathbf{u}_0$ is shown in Fig. 1 for $t = 0$. The Taylor cells $\mathbf{u}(t, x) = \mathbf{u}_0(x)e^{-4\pi^2 \nu t}$ are the (unique!) solution of the transient Navier-Stokes model. For this very smooth solution, a high-order FEM is preferable.

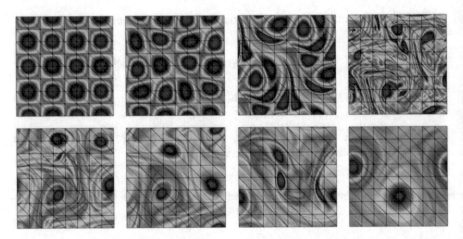

Fig. 1 Example 2: Snapshots of vorticity $\omega_h = \nabla \times \mathbf{u}_h$ of high-order FEM with $k = 8$, $N = 8$ with $t \in \{0, 22, 23, 26\}$ (see first row) and $t \in \{30, 35, 40, 50\}$ (see second row)

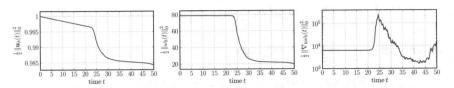

Fig. 2 Example 2: Temporal development of kinetic energy, enstrophy and palinstrophy

This is a generalized Beltrami flow, since $(\mathbf{u} \cdot \nabla)\mathbf{u} = -\nabla p$. Due to pressure-robustness, a linearization via dropping $(\mathbf{u} \cdot \nabla)\mathbf{u}$ preserves the *coherent* structures of the initial solution [4]. For order $k = 8$ and $h = \frac{1}{4}$, Fig. 1 shows snapshots of the discrete vorticity on the time interval $0 \le t \le T = 50$ for $\nu = 10^{-6}$. We observe a self-organization of vortical structures which deviates from the unique solution.

Consider now the behavior of the kinetic energy $\frac{1}{2}\|\mathbf{u}_h\|^2_{L^2(\Omega)}$, enstrophy $\frac{1}{2}\|\omega_h\|^2_{L^2(\Omega)}$ and palinstrophy $\frac{1}{2}\|\nabla_h \omega_h\|^2_{L^2(\Omega)}$ for $0 \le t \le 50$, see Fig. 2. Around $t = 22.0$ the solution deviates from coherent structures of the exact solution, also visible in the strong reduction of the amplitude of the kinetic energy. The exponential growth of the L^2- and H^1-errors of the velocity (according to Theorem 1) is shown in Fig. 3. The initial condition $\mathbf{u}_0$ of the planar lattice flow induces a flow structure which, due to its saddle point structure, is *"dynamically unstable so that small perturbations result in a very chaotic motion"* as stated in Majda & Bertozzi [6]. A convincing discussion of self-organization in 2D-flows is given by van Groesen [12].

Note that the preservation of the coherent structures (of the unique solution) can be extended in time by higher order k and/or h-refinement. Moreover, compared to standard mixed non-pressure-robust FEM, the application of pressure-robust FEM leads to much longer existence of such structures, see [4].   □

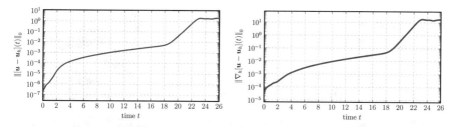

Fig. 3 Example 2: Error plots of high-order FEM for $\nu = 10^{-6}$, $k = 8$, $N = 8$, $\Delta t = 10^{-3}$

Remark 1

(i) A similar behavior of 2D-decaying turbulent flows is known for the 2D Kelvin-Helmholtz instability. We refer to careful numerical studies in [11].
(ii) The smallest scales depend on d. For $d = 3$, one has Kolmogorov-length $\lambda_{3D} \approx L Re^{-\frac{3}{4}}$ whereas for $d = 2$, the Kraichnan-length is $\lambda_{2D} \approx L Re^{-\frac{1}{2}}$. As conclusion, a direct numerical simulation (DNS) of 2D-flows at $Re \gg 1$ is much more realistic than in 3D, see [5].   □

2.3 Decaying 3D-Turbulent Flows

From the vorticity equation (28) we concluded a completely different behavior of high Re-number flows for $d = 3$ as compared to $d = 2$. The following example highlights the effect of vortex stretching term $(\omega \cdot \nabla)\mathbf{u}$.

Example 3 3D-lattice flow
Consider the exact solution of the transient incompressible Navier-Stokes problem

$$\mathbf{u}(t, x) = \mathbf{u}_0(x)e^{-4\pi^2\nu t}, \quad \mathbf{u}_0(x_1, x_2) = (-\Psi_{x_2}, \Psi_{x_1}, \sqrt{2}\Psi)^t (x_1, x_2)$$

in $\Omega = (0, 1)^3$ with stream function Ψ as in Example 2, with $\mathbf{f} = \mathbf{0}$ and $\frac{1}{\nu} = 2000$. This problem can be seen as 3D-extension of the 2D-lattice flow [6].

The snapshots of the solution in Fig. 4 show that until $t \approx 6$, the numerical method tries to preserve the 2D-behavior of the 2D-lattice flow. This can be seen from the "vortex tubes" (presented by the 5.0-isocontour of the so-called Q-criterion, colored with vorticity). Then the vortex stretching starts to deform the vortex tubes until $t = 7.5$. Later on, i.e. around $t = 10$, there starts the eddy-breakdown in the inertial range. Here we observe the transition to *homogeneous isotropic* turbulence. Finally, in Fig. 5, we consider the influence of the Reynolds number for $\frac{1}{\nu} \in \{2000, 4000, 100000\}$. We apply again the high-order H(div)-dGFEM (here with $k = 8$, $h = \frac{1}{8}$). In the first row, one observes the strongly

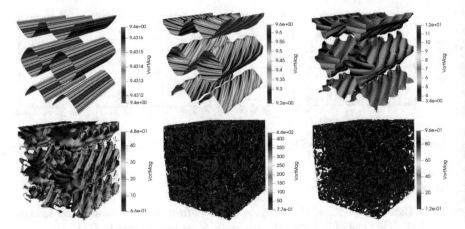

Fig. 4 Example 3: Transition to decaying homogeneous isotropic 3D-turbulence: 5.0 isocontour of Q-criterion, colored with vorticity at $t \in \{0.0, \ 6.5, \ 7.0, \ 7.5, \ 10.0, \ 20.0\}$

decaying kinetic energy and the effect of vortex stretching in the (scaled) dissipation rate in time.

The solution is still a Beltrami flow since $(\nabla \times \mathbf{u}) \times \mathbf{u} = \mathbf{0}$. Thus a linearization via $p \mapsto P := p + \frac{1}{2}|\mathbf{u}|^2$ would retain *coherent* structures as in 2D. This corresponds to the formal exact solution, see dashed lines. Solid lines correspond to the discrete solutions with $k = 8$ and $h = \frac{1}{8}$. The deviation of the discrete solution from the (formal) exact solution starts earlier for increasing Reynolds number. On the other hand, the deviation can be shifted to larger times if the FEM-order k is increased and/or an h-refinement is performed.

In the second row of Fig. 5, we consider the L^2- and H^1-errors for $\mathbf{u} - \mathbf{u}_h$. According to the estimate in Thm. 1, one observes the exponential behavior of both errors in time. This again indicates that, after a certain time, the discrete solution deviates from the (formal) exact solution.

Example 4   3D-Taylor-Green vortex at $Re = 1600$
A typical LES-benchmark is the 3D-Taylor-Green vortex problem at $Re = \frac{UL}{\nu} = 1600$ with $\mathbf{f} = \mathbf{0}$ and initial condition

$$\mathbf{u}_0(x) = U\left(\sin\frac{x_1}{L} \cos\frac{x_2}{L} \cos\frac{x_3}{L}, -\cos\frac{x_1}{L} \sin\frac{x_2}{L} \cos\frac{x_3}{L}, 0 \right)^t.$$

As in the previous example we observe the breakdown of large eddies into smaller and smaller eddies, see Fig. 6. This indicates that the typical behavior of *homogeneous isotropic turbulence* develops already for this relative small Reynolds number $Re = 1600$ where we set $U = L = 1$.

Consider now the temporal development of kinetic energy resp. the L^2-energy spectrum, see Fig. 7. For $\mathbf{f} = \mathbf{0}$, we found in Sect. 3 a weak exponential decay of kinetic energy according to (26). As reference solution serves the solution of a

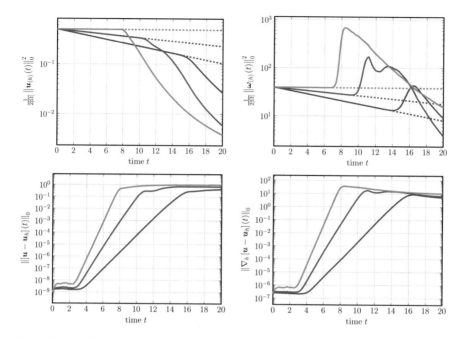

Fig. 5 Example 3. First row: t-dependent kinetic energy and enstrophy, Second row: t-dependence of errors in L^2 and H^1, Legend: $\frac{1}{\nu} = 2.0 \times 10^3$ (blue), $\frac{1}{\nu} = 4.0 \times 10^3$ (red), $\frac{1}{\nu} = 1.0 \times 10^5$ (green)

Fig. 6 Example 4. Behavior like decaying homogeneous isotropic 3D-turbulence: 0.1-isocontour of Q-criterion, coloured with velocity at $t \in \{0.0, \ 2.0, \ 4.0, \ 9.0\}$

spectral method with 512^3 grid points (ooo). For increasing values of FEM-order k and/or increasing spatial resolution (via refinement of $h = 1/N$), we observe grid convergence for the kinetic energy, see Fig. 7 (left).

In Fig. 7 (right) we plot the spectra of the kinetic energy at $t = 10$ for different values of k and $h = 1/N$. In particular, no pile-up of the spectra for large wave numbers k occurs. The Kolmogorov rate of $E(k) = \mathcal{O}(k^{-5/3})$ is not reached since $Re = 1600$ is too small but will be reached at larger values of Re.

Consider now the temporal development of the kinetic energy dissipation rate for which we obtained estimate (27). This quantity is much harder to approximate. For increasing values of FEM-order k and/or increasing resolution (via refinement

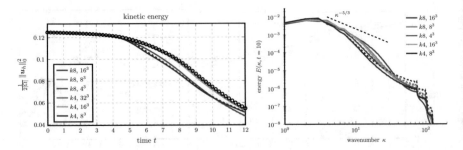

Fig. 7 Example 4. Left: Temporal development of kinetic energy for 3D-Taylor-Green vortex at $Re = 1600$ for different values of order k and $N = 1/h$; Right: Spectrum of kinetic energy at $t = 10$ for different values of k and $h = 1/N$

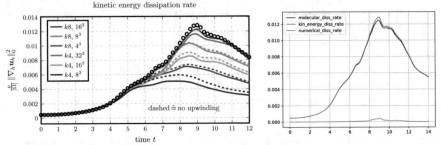

Fig. 8 Example 4. Left: Temporal development of energy dissipation rate for Taylor-Green vortex at $Re = 1600$ for different values of order k and $N = 1/h$; Right: Balance of dissipation rates

of $h = 1/N$), we observe nearly convergence for the energy dissipation rate. In particular, we find that upwind stabilization (see solid lines) decreases the energy dissipation rate on the coarser grids, see Fig. 8 (left).

Finally, consider the balance of dissipation rates according to

$$\frac{d}{dt}\left(\frac{1}{2}\|\mathbf{u}_h(t)\|^2_{L^2(\Omega)}\right) + va_h(\mathbf{u}_h(t), \mathbf{u}_h(t)) + |\mathbf{u}_h(t)|^2_{\mathbf{u}_h, \text{upw}} = 0.$$

The results are plotted for a relatively fine resolution with order $k = 8$ and $h = 1/N = 1/16$. This corresponds to 128^3 grid points. We observe a very good agreement between molecular dissipation rate $v\|\nabla_h\mathbf{u}_h\|^2_{L^2(\Omega)}$ and kinetic energy dissipation rate $\frac{d}{dt}\left(\frac{1}{2}\|\mathbf{u}_h(t)\|^2_{L^2(\Omega)}\right)$, since the numerical dissipation rate (stemming from SIP penalty and upwinding) reaches not more than 3% of the other rates around the peak rate of molecular dissipation rate, see Fig. 8 (right).

Please note that no explicit turbulence modeling has been applied. The price for such results is the H(div)-dGFEM simulation with around 9×10^6 unknowns.

3   Wall-Bounded Flow

For wall-bounded turbulent flows, one striking problem is the presence of strong boundary layers, e.g. at walls. Another problem is to apply a splitting $\mathbf{u}_{(h)} = \langle \mathbf{u}_{(h)} \rangle + \mathbf{u}'_{(h)}$ of the solution into an averaged velocity with some filter $\langle \cdot \rangle$ and fluctuations.

Example 5 (3D Channel Flow)

Figure 9 (left) presents a laminar channel flow with a uniquely defined deterministic solution. A snapshot of the turbulent channel flow at $Re_\tau = 180$ is shown on the right. The latter is slightly above the transition from laminar to turbulent flow. The chaotic solution of turbulent channel flow can be averaged in time and in x_1- and x_3-directions. One obtains, after a certain time of averaging, a relatively simple structure of the flow with $\langle u_1 \rangle = \langle u_1 \rangle (x_2)$.

Prandtl's boundary layer theory leads to the so-called *law of-the-wall*, visible in Fig. 10. The mean viscous stress at the wall, the *wall-shear stress*, is $\tau_W = \nu \partial_{x_2} \langle u_1 \rangle |_{x_2 = 0}$. An appropriate velocity resp. length-scale in the near-wall region are the friction velocity $U_\tau = \sqrt{\tau_W}$ resp. $\eta_\nu = \nu / \sqrt{\tau_W} = \nu / U_\tau$. The friction-based Reynolds number is defined as $Re_\tau = U_\tau H / \nu$ with channel half width H. The layer can be characterized via the non-dimensional distance from wall in wall units $x_2^+ = x_2 / \eta_\nu = U_\tau x_2 / \nu$. It is characterized by the viscous wall region $x_2^+ < 50$ with dominance of molecular viscosity, including the steep *viscous sublayer* at the wall with $x_2^+ < 5$, and by the outer layer with $x_2^+ > 50$.

The standard approach to resolve boundary layers is to use a (strongly) anisotropic mesh with refinement towards the wall(s). Very recent results with a L^2-based dGFEM-code by Fehn et al. [3] indicate that a strong anisotropic h-refinement can be relaxed to a (very) coarse h-mesh if higher-order FEM are applied. It turns out that for such (highly) under-resolved turbulent flows a "medium order" $(k = 4, \ldots, 8)$ is most efficient. Another point is that a purely numerical approach to stabilization is applied, i.e. no physical LES or VMS model is used.

Figure 10 shows results for the H(div)-dGFEM for the channel flow at $Re_\tau = 180$. It turns out that a method of order $k = 2$ is not sufficient, but $k = 3$ provides good results. Very coarse grids with $N = 4$ resp. $N = 8$ elements in each x_i-direction with slightly anisotropic refinement in x_2-direction towards the wall (indicated by vertical lines in Fig. 10) are used.

Fig. 9 3D channel flow: Laminar flow (left), Turbulent $Re_\tau = 180$-flow (right)

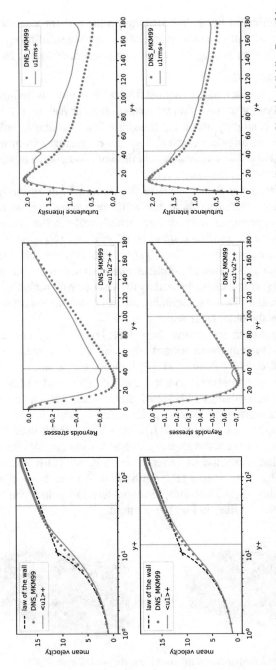

Fig. 10 3D Channel flow at $Re_\tau = 180$ with ILES: First row: $k = 3$, $N = 4$, Second row: $k = 3$, $N = 8$; Left: Mean profile U^+, Middle: Reynolds stress $\langle u_1' u_2' \rangle^+$, Right: rms turbulence intensity u_{RMS}^+

No explicit physical LES model is applied. In the ILES approach only numerical dissipation (basically from SIP and upwind) is used. The results for the averaged mean profile of U^+, the Reynolds stress component $\langle u_1' u_2' \rangle^+$ and the rms turbulence intensity values u_{RMS}^+, compared to the DNS-data by Moser et al. [7], are surprisingly good on this very coarse grids with 12^3 resp. 24^3 grid points.   □

Results in [3] indicate that such approach is also possible for larger values of Re_τ.

4   Outlook

The following features of H(div)-dGFEM are exploited in the numerical simulation of turbulent flows via implicit LES for incompressible Navier-Stokes flows:

- *Minimal stabilization:* Numerical dissipation may only result from the SIP term for the diffusive term a_h and upwind term in c_h.
- *Simple form of convective form:* There is no need to modify the convective term c_h since an exactly divergence-free FEM has a clean energy balance *a priori*.
- *Pressure robustness*: H(div)-conforming FEM have the relevant property that changing source term $\mathbf{f}$ to $\mathbf{f} + \nabla \psi$ changes the solution $(\mathbf{u}_h, p_h)$ to $(\mathbf{u}_h, p_h + \psi)$.
- *Re-semi-robust error estimates:* Right-hand-side terms of the error estimate, see Theorem 1, including the Gronwall-term do not explicitly depend on $1/\nu$.

We considered an ILES approach to simple turbulent flows with very resonable results. Turbulent flows in practice are clearly much more complex. A challenge is the flow around a high-lift airfoil, see Fig. 11, with complicated interplay of attached laminar and turbulent layers, separation, vortex structures etc. For a careful numerical study of such flows see [8]. A full DNS is still unfeasable. It would be of strong interest to develop new numerical concepts for such complex flows

Fig. 11 Complex flow around three-element high-lift airfoil

which clearly go beyond the limit cases (homogeneous isotropic turbulence and turbulent channel flows) under consideration. Nevertheless, the proposed ILES approach with high-order and pointwise divergence-free H(div)-dGFEM is a very promising approach. Another important point is that the flow in the previous example is governed by the *compressible* Navier-Stokes model. Many aspects of incompressible flows can be extended to the compressible case, e.g. the approach in boundary layer regions.

References

1. Boffi, D., Brezzi, F., Fortin, M.: Mixed Finite Element Methods and Applications. Springer, Berlin/Heidelberg (2013)
2. Di Pietro, D.A., Ern, A.: Mathematical Aspects of Discontinuous Galerkin Methods. Springer, Berlin (2012)
3. Fehn, N., Wall, W., Kronbichler, M.: Robust and efficient discontinuous Galerkin methods for under-resolved turbulent incompressible flows. J. Comput. Phys. **372**, 667–693 (2018)
4. Gauger, N.R., Linke, A., Schroeder, P.W.: On high-order pressure-robust space discretisations, their advantages for incompressible high Reynolds number generalised Beltrami flows and beyond. SMAI J. Comp. Math. **5**, 89–129 (2019)
5. John, V.: Finite Element Methods for Incompressible Flow Problems. Springer, Cham (2016)
6. Majda, A.J., Bertozzi, A.L.: Vorticity and Incompressible Flows. Cambridge Univ. Press, Cambridge (2002)
7. Moser, R.D., Kim, J., Mansour, N.N.: DNS of turbulent channel flows up to $Re_\tau = 590$. Phys. Fluids **11**, 943–945 (1999)
8. Reuß, S.: A grid-adaptive algebraic hybrid RANS/LES method, Ph.D. thesis, Göttingen (2015)
9. Schöberl, J.: C++11 implementation of finite elements in NGSolve. TU Vienna (2014)
10. Schroeder, P.W., et al.: Towards computable flows and robust estimates for inf-sup stable FEM applied to the time-dependent incompressible Navier-Stokes equations. SeMA J. (2018). https://doi.org/10.1007/s40324-018-0157-1
11. Schroeder, P.W., et al.: On reference solutions and the sensitivity of the 2D Kelvin-Helmholtz instability problem. Submitted to CAMWA. Comput. and Math. with Appl. **77**, 1010–1028 (2019)
12. Van Groesen, E.: Time-asymptotics and the self-organization hypothesis for 2D Navier-Stokes equation. Physica **1438A**, 312–330 (1988)

Numerical Methods for Singularly Perturbed Parabolic Problems with Incompatible Boundary-Initial Data in Two Space Dimensions

José Luis Gracia and Eugene O'Riordan

Abstract In this paper we consider a singularly perturbed parabolic problem of reaction-diffusion type, posed on a two dimensional domain in space. We focus our attention on the nature of the singularity in the solution, caused by an incompatibility between the initial and boundary conditions. We use an analytical/numerical approach to approximate the discontinuous solution. This consists of subtracting from the solution an explicit analytical function, which matches that incompatibility, and the remainder is approximated numerically using a standard finite difference operator on an appropriated fitted Shishkin mesh. The numerical results for several test examples indicate that the method proposed is globally and uniformly convergent in the maximum norm.

1   Introduction

Boundary and interior layers typically appear in singularly perturbed problems with smooth data. With sufficient *a priori* information about the location and width of all layers present in the solution, piecewise-uniform Shishkin meshes [7] can be constructed to allow parameter-uniform numerical methods [1] to be designed for these problems. We stress that these methods generate not only yield parameter-uniform nodal accuracy, but, more importantly, they also exhibit global pointwise accuracy using classical polynomial interpolation. However, if there is a discontinuity in the boundary/initial data [5, 6], then the nature of the singularity can mean that a fitted mesh may not exist for such a problem [8] that will yield global pointwise accuracy

J. L. Gracia
IUMA and Department of Applied Mathematics, University of Zaragoza, Zaragoza, Spain
e-mail: jlgracia@unizar.es

E. O'Riordan (✉)
School of Mathematical Sciences, Dublin City University, Dublin, Ireland
e-mail: eugene.oriordan@dcu.ie

© Springer Nature Switzerland AG 2020
G. R. Barrenechea, J. Mackenzie (eds.), *Boundary and Interior Layers,
Computational and Asymptotic Methods BAIL 2018*, Lecture Notes in
Computational Science and Engineering 135,
https://doi.org/10.1007/978-3-030-41800-7_11

[3]. Based on an idea in [2], an analytical/numerical approach was examined in [4], which generated parameter-uniform global approximations to singularly perturbed parabolic problems, with a discontinuity in the boundary/initial data. However, in [4], this method was restricted to problems in one space dimension. In this paper, we examine the numerical performance of the method developed in [4] (which combines ideas from [2] with appropriate Shishkin meshes) to ascertain if the approach can be easily extended to problems in two space dimensions; and if the extension of the method also retains parameter-uniform convergence.

Here we shall consider singularly perturbed reaction-diffusion problems of the form

$$Lu := \varepsilon(u_t - \Delta u) + b(x, y, t)u$$

$$= f(x, y, t), \text{ in } Q := \Omega \times (0, 1], \quad \Omega := (0, 1)^2, \tag{1a}$$

$$u(x, y, t) = 0, \quad (x, y) \in \partial\Omega, \ t > 0, \quad b(x, y, t) \geq \beta > 0, \text{ in } \bar{Q}; \tag{1b}$$

$$u(x, y, 0) = g(x, y), \quad (x, y) \in \Omega, \quad \text{and} \tag{1c}$$

$$g(1, y) = g(x, 0) = g(x, 1) = 0, \ 0 \leq x, y \leq 1, \ g(0, y) \neq 0, \ 0 < y < 1; \tag{1d}$$

where $0 < \varepsilon \leq 1$ is the singular perturbation parameter.

The functions b, f and g are assumed to be smooth, but the solution u has a singular behaviour due to the presence of the singular perturbation parameter ε and the incompatibility between the initial condition $g(x, y) \neq 0$ and the homogenous boundary condition along the face $x = 0$. The solution u of this problem is discontinuous along $x = 0$ and it typically exhibits boundary layers near the four faces $x = 0, x = 1, y = 0, y = 1$, initial layers near the plane $t = 0$ and initial-boundary layers near the edge $x = 0, t = 0, y \geq 0$.

2 Analytical/Numerical Approach

2.1 Analytical Decomposition

Similarly to the approach used for the one dimensional problem [4], the solution u of the two dimensional problem (1) is decomposed as follows

$$u = s + w,$$

where the singular function s is defined as

$$s(x, y, t) := g(x, y)e^{-b(0,0,0)t/\varepsilon}\text{erf}\left(\frac{x}{2\sqrt{t}}\right) \quad \text{and} \quad \text{erf}(z) := \frac{2}{\sqrt{\pi}}\int_{s=0}^{z} e^{-s^2}\, ds.$$

The remainder w is the solution of the problem

$$Lw = f(x, y, t) - (b(x, y, t) - b(0, 0, 0))s(x, y, t)$$

$$+ G(x, y, t)e^{-b(0,0,0)t/\varepsilon}, \quad (x, y, t) \in Q, \tag{2a}$$

$$w(x, y, t) = (u - s)(x, y, t) \equiv 0, \quad (x, y, t) \in \bar{Q} \setminus Q, \tag{2b}$$

$$G(x, y, t) := \left(\varepsilon \Delta g(x, y) \mathrm{erf}\left(\frac{x}{2\sqrt{t}}\right) + \frac{2\varepsilon g_x(x, y)}{\sqrt{\pi t}} e^{-\frac{x^2}{4t}} \right). \tag{2c}$$

Unlike the original problem (1), observe that problem (2) satisfies zero-order compatibility condition between the boundary and initial data. However, the right-hand side is a nonsmooth function. Similarly to the 1D problem, the error function (in the form $\mathrm{erf}(0.5x/\sqrt{t})$) is present in the right-hand side of (2a). In spite of this nonsmooth term, estimates of the derivatives of the solution of the 1D problem were proved in [4] and they were used to analyse the uniform convergence in the maximum norm of a standard finite difference scheme defined on a Shishkin mesh applied to the one-dimensional analogue of problem (2).

However, in the case of the two dimensional problem (2), there is another nonsmooth term in (2a) (denoted by G), which does not have a counterpart in the one dimensional case. In the special case where the initial condition g is a constant function, then $G \equiv 0$. Otherwise, the term G is in general unbounded near $t = 0$, if $g_x(0, y) \neq 0$ for some y. This can be seen, for example, by considering

$$\lim_{t \to 0, \, x=\sqrt{t}} \frac{2\varepsilon g_x(x, y)}{\sqrt{\pi t}} e^{-\frac{x^2}{4t}}. \tag{3}$$

If one assumes that

$$g_x(0, y) = 0, \tag{4}$$

then

$$|G(x, y, t)| \leq C\varepsilon \left(1 + \frac{x\|g_{xx}\|_{\bar{Q}}}{\sqrt{\pi t}} e^{-\frac{x^2}{4t}} \right) \leq C\varepsilon.$$

Furthermore, note that even if (4) is assumed and the new term G was bounded, it is still discontinuous over the closed domain (unless $\Delta g(0, 0) = 0$). In the numerical section we examine the influence of the constraint (4) on the orders of convergence of the numerical method, which is used to generate a numerical approximation to the remainder $w(x, y, t)$.

2.2 Numerical Scheme

The solution w of problem (2) exhibits initial and boundary layers [4]. So, we construct a tensor product of one dimensional piecewise uniform Shishkin meshes, condensing in these layer regions. The mesh is denoted by $\bar{Q}^{N,M} := \omega^N \times \omega_t^M$ with $\omega^N = \omega_x^N \times \omega_y^N$ and N, M are two positive integers. We denote the interior mesh by $Q^{N,M} := \bar{Q}^{N,M} \cap Q$.

In the time variable, the Shishkin mesh $\omega_t^M := \{t_j\}_{j=0}^M$ splits the time domain into two subintervals

$$[0, \tau] \cup [\tau, 1], \quad \tau := \min\left\{\frac{1}{2}, \frac{\varepsilon}{\beta} \ln M\right\}, \tag{5}$$

and the mesh points are distributed uniformly across these two subintervals in the ratio $M/2 : M/2$. For the space variables, the Shishkin mesh $\omega_x^N := \{x_i\}_{i=0}^N$ in the x-direction splits the interval $[0, 1]$ as follows:

$$[0, \sigma] \cup [\sigma, 1 - \sigma] \cup [1 - \sigma, 1], \quad \sigma := \min\left\{\frac{1}{4}, 2\sqrt{\frac{\varepsilon}{\beta}} \ln N\right\}. \tag{6a}$$

The N space mesh points are distributed in the ratio $N/4 : N/2 : N/4$ across these three subintervals. The mesh points in the y-direction are analogously defined. For the sake of simplicity, we use the same number of grid points in both spatial directions. The mesh steps in each co-ordinate direction are denoted by $h_i := x_i - x_{i-1}, h_j = y_j - y_{j-1}, k_m := t_m - t_{m-1}$.

We use a standard central finite difference and backward Euler methods on the Shishkin mesh $\bar{Q}^{N,M}$

$$L^{N,M} W_{i,j}^m := \varepsilon(D_t^- W_{i,j}^m - \delta_x^2 W_{i,j}^m - \delta_y^2 W_{i,j}^m) + b(x_i, y_j, t_m) W_{i,j}^m$$

$$= Lw(x_i, y_j, t_m), \quad \text{for } (x_i, y_j) \in Q^{N,M}, \quad t_m > 0, \tag{7}$$

with the finite difference operators defined by

$$D_t^- Y_{i,j}^m := \frac{Y_{i,j}^m - Y_{i,j}^{m-1}}{k_m}, \quad \delta_x^2 Y_{i,j}^m := \frac{2}{h_i + h_{i+1}} \left(\frac{Y_{i+1,j}^m - Y_{i,j}^m}{h_{i+1}} - \frac{Y_{i,j}^m - Y_{i-1,j}^m}{h_i}\right),$$

and the discrete operator δ_y^2 is defined similarly. The formulation of the discrete problem is completed with the values of the numerical solution at the boundary of the domain given by

$$W_{i,j}^m = 0 \text{ on } \partial \Omega^N \quad \text{and} \quad W_{i,j}^0 = g(x_i, y_j) \text{ on } \Omega^N.$$

Thus, $W_{i,j}^m$ is a nodal approximation to $w(x_i, y_j, t_m)$, which is the solution of problem (2). Simple trilinear interpolation is used to form a global approximation $\bar{W}^{N,M}$ to $w(x, y, t)$.

3   Numerical Experiments

We first compare the numerical results generated by applying the method to two test problems, which only differ in the initial condition. It is shown numerically that the condition $g_x(0, y) = 0$ has an influence on the convergence rates. In the last example the incompatibility between the initial and boundary conditions is located along the faces $x = 0$ and $y = 0$ of the domain.

In all three of the test examples, the exact solution is unknown and the orders of convergence are estimated using the two-mesh principle [1]. We denote the maximum two-mesh global differences by

$$D_\varepsilon^{N,M} := \|\bar{U}^{N,M} - \bar{U}^{2N,2M}\|_{Q^{N,M} \cup Q^{2N,2M}} = \|\bar{W}^{N,M} - \bar{W}^{2N,2M}\|_{Q^{N,M} \cup Q^{2N,2M}}$$

where $W^{N,M}$ and $W^{2N,2M}$ are the computed solutions with the standard scheme (7) on the Shishkin meshes $Q^{N,M}$ and $Q^{2N,2M}$, respectively and $\bar{W}^{N,M}$ and $\bar{W}^{2N,2M}$ denote the trilinear interpolation of the discrete solutions $W^{N,M}$ and $W^{2N,2M}$ on the mesh $Q^{N,M} \cup Q^{2N,2M}$. Then, the orders of global convergence are estimated in a standard way, using

$$P_\varepsilon^{N,M} := \log_2 \left(\frac{D_\varepsilon^{N,M}}{D_\varepsilon^{2N,2M}} \right).$$

The uniform two-mesh global differences and their corresponding uniform orders of global convergence are calculated by

$$D^{N,M} := \max_{\varepsilon \in S} D_\varepsilon^{N,M}, \quad P^{N,M} := \log_2 \left(\frac{D^{N,M}}{D^{2N,2M}} \right),$$

and $S = \{1, 2^{-2}, \ldots, 2^{-30}\}$ is the sample set of values taken for the singular perturbation parameter ε.

Example 1 Consider the following example

$$Lu := \varepsilon(u_t - u_{xx} - u_{yy}) + (1 + xy)u = 1, \quad (x, y) \in \Omega, t > 0,$$

$$u(x, y, t) = 0, \quad (x, y) \in \partial\Omega, \ t > 0,$$

with $\Omega = (0, 1)^2$ and $0 \le t \le 1$. The initial condition in this example is

$$g(x, y) = 4(1 - x^2)y(1 - y), \quad (x, y) \in \Omega, \tag{8}$$

which satisfies the condition $g_x(0, y) = 0$, but note that $\Delta g(0, 0) \ne 0$.

In Fig. 1 we display the numerical/analytical approximation of the solution u at the times $t = 0$ (initial condition), $t = t_1$ (first time step), $t = 0.25$ and $t = 1$ for $\varepsilon = 2^{-16}$ and $N = 128$ and $M = 32$. These subfigures display an abrupt change in the computed solution over the first time interval $t \in [0, t_1]$, $t_1 = 2^{-19} \ln 32$. This is further illustrated in Fig. 2 where the cross sections of the computed solution at $y = y_1$ and $x = x_1$ are displayed for $\varepsilon = 2^{-16}$ and $N = 128$ and $M = 32$. The surface solution in the subdomain $(x, y, t) \in [0, 1] \times \{y_1\} \times [0, \tau]$ and $(x, y, t) \in \{x_1\} \times [0, 1] \times [0, \tau]$ are shown and the evolution of the solution at $y = y_1$ (inside the boundary layer near $y = 0$) and $x = x_1$ (inside the boundary layer near $x = 0$)

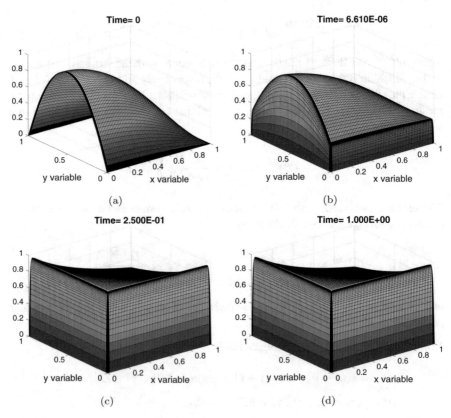

Fig. 1 Example 1: (**a**) Initial condition $u(x, y, 0)$, computed solutions U at (**b**) $t_1 = 2^{-19} \ln 32$, (**c**) $t = 0.25$ and (**d**) $t = 1$ for $\varepsilon = 2^{-16}$ and $N = 128$ and $M = 32$

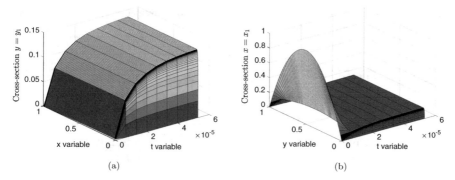

Fig. 2 Example 1: Cross sections of the computed solution U for (**a**) $y = y_1$ and (**b**) $x = x_1$

is observed. In the latter subfigure, the initial layer in the solution near $t = 0$ is quite noticeable.

In Table 1 we show the maximum two-mesh global differences for the values of $\varepsilon \in S$ and their orders of convergence. In the last row we give the uniform two-mesh global differences and their corresponding orders of convergence. The numerical results of this table suggest that the method is globally uniformly convergent with respect to the singular perturbation parameter ε. The computed orders of global convergence indicate that the numerical method converges with first-order when ε is large and with an order of the form $M^{-1} \ln M$ otherwise.

Example 2 The second example is similar to the Example 1 and the only difference between them is the initial condition. In this second example the initial condition is replaced by

$$g(x, y) = 4(1 - x)y(1 - y), \quad (x, y) \in \Omega, \tag{9}$$

and it satisfies $g_x(0, y) \neq 0$.

The maximum and uniform two-mesh global differences are given in Table 2 and they again suggest that the method converges uniformly and globally. Nevertheless, if we compare the orders of convergence from Tables 1 and 2, in the latter (i.e, when $g_x(0, y) \neq 0$) we observe a reduction in the orders of convergence for ε large. Nevertheless, similar orders of convergence are obtained in both examples when ε is small enough. This suggests that the condition $g_x(0, y) = 0$ may only marginally affect the uniform orders of convergence of the method.

Although $g_x(0, y) \neq 0$ in Example 2, we observe that the right-hand side of the scheme (7) is bounded when $\varepsilon \leq CM^{-1} \ln M$. This can be seen by noticing that

$$\frac{\varepsilon}{\sqrt{\pi t_k}} \leq \frac{\varepsilon}{\sqrt{\pi t_1}} = \frac{\sqrt{\varepsilon}}{2\sqrt{\pi M^{-1} \ln M}};$$

Table 1 Two-mesh global differences $D_\varepsilon^{N,M}$ and orders of convergence $P_\varepsilon^{N,M}$ associated with the approximation of the solution to Example 1

	N=32 M=8	N=64 M=16	N=128 M=32	N=256 M=64	N=512 M=128
$\varepsilon = 2^0$	**1.043E–01**	**6.180E–02**	**3.195E–02**	**1.677E–02**	**8.689E–03**
	0.756	0.952	0.930	0.949	
$\varepsilon = 2^{-2}$	6.390E–02	4.753E–02	2.567E–02	1.334E–02	6.959E–03
	0.427	0.889	0.944	0.939	
$\varepsilon = 2^{-4}$	3.056E–02	2.204E–02	1.490E–02	9.469E–03	5.730E–03
	0.471	0.565	0.654	0.725	
$\varepsilon = 2^{-6}$	6.169E–02	3.018E–02	1.950E–02	1.198E–02	7.148E–03
	1.031	0.630	0.703	0.745	
$\varepsilon = 2^{-8}$	7.887E–02	3.683E–02	2.179E–02	1.325E–02	7.853E–03
	1.099	0.757	0.718	0.754	
$\varepsilon = 2^{-10}$	8.776E–02	4.182E–02	2.225E–02	1.358E–02	8.032E–03
	1.069	0.910	0.712	0.758	
$\varepsilon = 2^{-12}$	9.650E–02	4.475E–02	2.228E–02	1.361E–02	8.042E–03
	1.109	1.006	0.711	0.759	
$\varepsilon = 2^{-14}$	9.886E–02	4.842E–02	2.228E–02	1.361E–02	8.041E–03
	1.030	1.120	0.711	0.759	
.	.	.	.	.	.
.	.	.	.	.	.
.	.	.	.	.	.
$\varepsilon = 2^{-30}$	9.370E–02	4.626E–02	2.227E–02	1.360E–02	8.038E–03
	1.018	1.055	0.711	0.759	
$D^{N,M}$	1.043E–01	6.180E–02	3.195E–02	1.677E–02	8.689E–03
$P^{N,M}$	0.756	0.952	0.930	0.949	

therefore the term (3) is bounded if $\varepsilon \leq CM^{-1} \ln M$. This argument cannot be applied if $M^{-1} \ll \varepsilon$ and this observation may explain why assuming the condition $g_x(0, y) = 0$ improves the orders of convergence for the case where ε is large compared to M^{-1}.

Example 3 The third example is given by

$$Lu := \varepsilon(u_t - u_{xx} - u_{yy}) + (1 + x + y + t)u$$

$$= 4(x(1 - x) + y(1 - y)), \quad (x, y, t) \in Q = \Omega \times (0, 1],$$

$$u(x, y, t) = 0, \quad (x, y) \in \partial\Omega, \quad t \in (0, 1],$$

$$u(x, y, 0) = g(x, y) = \sin\left(\frac{5\pi}{4}x + \frac{3\pi}{4}\right)\sin\left(\frac{5\pi}{4}y + \frac{3\pi}{4}\right), \quad (x, y) \in \Omega = (0, 1)^2.$$

Table 2 Two-mesh global differences $D_\varepsilon^{N,M}$ and orders of convergence $P_\varepsilon^{N,M}$ associated with the approximation to the solution of Example 2

	N=32 M=8	N=64 M=16	N=128 M=32	N=256 M=64	N=512 M=128
$\varepsilon = 2^0$	7.433E–02	**5.216E–02**	**3.260E–02**	**2.007E–02**	**1.276E–02**
	0.511	0.678	0.699	0.654	
$\varepsilon = 2^{-2}$	3.563E–02	3.573E–02	2.621E–02	1.732E–02	1.142E–02
	-0.004	0.447	0.598	0.601	
$\varepsilon = 2^{-4}$	3.989E–02	2.739E–02	1.746E–02	1.068E–02	6.386E–03
	0.542	0.649	0.710	0.742	
$\varepsilon = 2^{-6}$	5.621E–02	3.225E–02	2.056E–02	1.271E–02	7.566E–03
	0.801	0.650	0.694	0.748	
$\varepsilon = 2^{-8}$	7.650E–02	3.524E–02	2.203E–02	1.341E–02	7.949E–03
	1.118	0.678	0.716	0.755	
$\varepsilon = 2^{-10}$	8.731E–02	4.136E–02	2.228E–02	1.361E–02	8.045E–03
	1.078	0.892	0.712	0.758	
$\varepsilon = 2^{-12}$	9.647E–02	4.470E–02	2.229E–02	1.361E–02	8.044E–03
	1.110	1.004	0.711	0.759	
$\varepsilon = 2^{-14}$	**9.885E–02**	4.842E–02	2.228E–02	1.361E–02	8.041E–03
	1.030	1.120	0.711	0.759	
.	.	.	.	.	.
.	.	.	.	.	.
.	.	.	.	.	.
$\varepsilon = 2^{-30}$	9.370E–02	4.626E–02	2.227E–02	1.360E–02	8.038E–03
	1.018	1.055	0.711	0.759	
$D^{N,M}$	9.885E–02	5.216E–02	3.260E–02	2.007E–02	1.276E–02
$P^{N,M}$	0.922	0.678	0.699	0.654	

Observe that in this example

$$u(1, y, t) = u(x, 1, t) = 0, \quad u(x, 0, t) \neq 0, \ u(0, y, t) \neq 0,$$

and then this problem has an incompatibility between the initial condition $g(x, y)$ and the boundary condition along the faces $x = 0$ and $y = 0$.

The solution of Example 3 is first approximated without separating off any singular component related to the incompatible initial and boundary data. We use standard central differences and backward Euler method on the Shishkin mesh considered in the previous examples. The method fails to generate a global approximation to the solution; we have observed that the maximum two-mesh global differences are constant for all the values of N, M and ε. This is illustrated in Table 3.

Table 3 Example 3: Uniform two-mesh global differences using a standard numerical method on a Shishkin mesh, without separating off the singularity

	N=32	N=64	N=128	N=256	N=512
	M=8	M=16	M=32	M=64	M=128
$D^{N,M}$	3.534E–01	3.534E–01	3.535E–01	3.536E–01	3.536E–01

Therefore, the solution of Example 3 is decomposed as follows

$$u = s + w, \quad \text{where} \quad s(x, y, t) := g(x, y)e^{-b(0,0,0)t/\varepsilon} \text{erf}\left(\frac{x}{2\sqrt{t}}\right) \text{erf}\left(\frac{y}{2\sqrt{t}}\right)$$

and the function w is the solution of the problem

$$Lw = f(x, y, t) - (b(x, y, t) - b(0, 0, 0))s(x, y, t)$$

$$+ \left(G(x, y, t)\text{erf}\left(\frac{y}{2\sqrt{t}}\right) + \frac{2\varepsilon g_y(x, y)}{\sqrt{\pi t}}e^{-\frac{y^2}{4t}}\text{erf}\left(\frac{x}{2\sqrt{t}}\right)\right)e^{-b(0,0,0)t/\varepsilon}, \quad \text{in } Q,$$

$$\tag{10a}$$

$$w(x, y, t) = u(x, y, t) - s(x, y, t), \quad (x, y, t) \in \bar{Q}\backslash Q. \tag{10b}$$

Compared to the previous two examples, an additional singularity of the form $\text{erf}(y/(2\sqrt{t}))$ is now a multiple of the original problematic term G. In this example, the partial derivatives of the initial condition for $x = 0$ or $y = 0$ satisfy

$$g_x(x, 0) \neq 0, \ g_y(x, 0) \neq 0, \ g_x(0, y) \neq 0, \ \text{and} \ g_y(0, y) \neq 0.$$

The initial-boundary value problem (10) is approximated with the scheme (7) (with Lw given in (10a)) on the Shishkin mesh $Q^{N,M}$. The numerical results are given in Table 4. Observe, in this final example, that the uniform orders of convergence $P^{N,M}$ are also associated with the larger values of ε and the orders again indicate that the method converges globally for all the values of the parameter ε considered in the table.

In Fig. 3 we display the numerical approximation of u at the times $t = 0$ (initial condition) and $t = t_1$ (first time step) for the same values of the singular perturbation and discretization parameters; i.e., $\varepsilon = 2^{-16}$ and $N = 128$ and $M = 32$. We observe again an abrupt change in the solution as the difference in time between both figures is quite small. Our analytical/numerical method appears to accurately capture the singularity in the solution of this example.

Table 4 Two-mesh global differences $D_\varepsilon^{N,M}$ and orders of convergence $P_\varepsilon^{N,M}$ associated with the approximation to the solution of Example 3

	N=32 M=8	N=64 M=16	N=128 M=32	N=256 M=64	N=512 M=128
$\varepsilon = 2^0$	**2.343E–01**	**1.803E–01**	**1.046E–01**	6.211E–02	4.107E–02
	0.378	0.786	0.752	0.597	
$\varepsilon = 2^{-2}$	1.721E–01	1.584E–01	1.010E–01	**6.687E–02**	**4.355E–02**
	0.120	0.649	0.595	0.619	
$\varepsilon = 2^{-4}$	1.312E–01	1.064E–01	7.960E–02	5.583E–02	3.764E–02
	0.303	0.418	0.512	0.569	
$\varepsilon = 2^{-6}$	1.043E–01	7.490E–02	4.948E–02	3.172E–02	2.033E–02
	0.478	0.598	0.641	0.642	
$\varepsilon = 2^{-8}$	8.676E–02	5.808E–02	3.666E–02	2.223E–02	1.322E–02
	0.579	0.664	0.721	0.750	
$\varepsilon = 2^{-10}$	7.767E–02	5.325E–02	3.400E–02	2.097E–02	1.243E–02
	0.545	0.647	0.697	0.755	
$\varepsilon = 2^{-12}$	7.394E–02	5.176E–02	3.278E–02	2.022E–02	1.200E–02
	0.514	0.659	0.697	0.752	
$\varepsilon = 2^{-14}$	7.452E–02	5.181E–02	3.267E–02	1.982E–02	1.177E–02
	0.524	0.665	0.721	0.752	
.	.	.	.	.	.
.	.	.	.	.	.
.	.	.	.	.	.
$\varepsilon = 2^{-30}$	7.464E–02	5.192E–02	3.271E–02	1.980E–02	1.166E–02
	0.524	0.667	0.724	0.764	
$D^{N,M}$	2.343E–01	1.803E–01	1.046E–01	6.687E–02	4.355E–02
$P^{N,M}$	0.378	0.786	0.646	0.619	

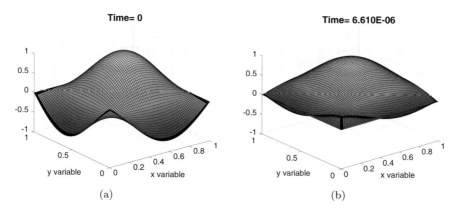

Fig. 3 Example 3: (**a**) Initial condition and (**b**) computed solution U at $t_1 = 2^{-19} \ln 32$ for $\varepsilon = 2^{-16}$ and $N = 128$ and $M = 32$

4 Conclusions

Based on the numerical results presented here, we conclude that the numerical/analytical approach in [4] may retain parameter-uniform convergence when it is extended to problems in two space dimensions. At a minimum, the extension to problems in two space dimensions merits a theoretical investigation.

Acknowledgements The research of J.L. Gracia has been partially supported by the Institute of Mathematics and Applications (IUMA), Spanish Ministry of Economy, project MTM2016-75139-R and the Diputación General de Aragón (E24-17R).

References

1. Farrell, P.A., Hegarty, A.F., Miller, J.J.H., O'Riordan, E., Shishkin, G.I.: Robust Computational Techniques for Boundary Layers. CRC Press, Boca Raton (2000)
2. Flyer, N., Fornberg, B.: Accurate numerical resolution of transients in initial-boundary value problems for the heat equation. J. Comput. Phys. **184**, 526–539 (2003)
3. Gracia, J.L., O'Riordan, E.: A singularly perturbed reaction-diffusion problem with incompatible boundary-initial data. In: Dimov, I., Farago, I., Vulkov, L. (eds.) Numerical Analysis and Its Applications: 5th International Conference, NAA 2012, Lozenetz, Bulgaria, June 15–20, 2012. Lecture Notes in Computer Science, Revised Selected Papers, vol. 8236, pp. 303–310. Springer, Heidelberg (2013)
4. Gracia, J.L., O'Riordan, E.: Parameter-uniform numerical methods for singularly perturbed parabolic problems with incompatible boundary-initial data. Appl. Numer. Math. **146**, 436–451 (2019)
5. Hemker, P.W., Shishkin, G.I.: Approximation of parabolic PDEs with a discontinuous initial condition. East-West J. Numer. Math. **1**, 287–302 (1993)
6. Hemker, P.W., Shishkin, G.I.: Discrete approximation of singularly perturbed parabolic PDEs with a discontinuous initial condition. Comput. Fluid Dyn. J. **2**, 375–392 (1994)
7. Kopteva, N.V., O'Riordan, E.: Shishkin meshes in the numerical solution of singularly perturbed differential equations. Int. J. Numer. Anal. Mod. **7**, 393–415 (2010)
8. Miller, J.J.H., O'Riordan, E.: The necessity of fitted operators and Shishkin meshes for resolving thin layer phenomena. CWI Q. **10**, 207–213 (1997)

Quasinorms in Semilinear Elliptic Problems

James Jackaman and Tristan Pryer

Abstract In this note we examine the a priori and a posteriori analysis of discontinuous Galerkin finite element discretisations of semilinear elliptic PDEs with polynomial nonlinearity. We show that optimal a priori error bounds in the energy norm are only possible for low order elements using classical a priori error analysis techniques. We make use of appropriate quasinorms that results in optimal energy norm error control.

We show that, contrary to the a priori case, a standard a posteriori analysis yields optimal upper bounds and does not require the introduction of quasinorms. We also summarise extensive numerical experiments verifying the analysis presented and examining the appearance of layers in the solution.

1 Introduction

Let $\Omega \subset \mathbb{R}^d$ with $d \geq 1$ be an open Lipschitz domain and consider the problem: find $u \in H_0^1(\Omega)$, such that

$$-\Delta u + |u|^{p-2} u = f \text{ in } \Omega$$
$$u = 0 \text{ on } \partial\Omega. \tag{1.1}$$

J. Jackaman
Department of Mathematics and Statistics, Memorial University of Newfoundland, St. John's, NL, Canada
e-mail: jjackaman@mun.ca

T. Pryer (✉)
Department of Mathematical Sciences, University of Bath, Bath, UK
e-mail: tmp38@bath.ac.uk

© Springer Nature Switzerland AG 2020
G. R. Barrenechea, J. Mackenzie (eds.), *Boundary and Interior Layers, Computational and Asymptotic Methods BAIL 2018*, Lecture Notes in Computational Science and Engineering 135,
https://doi.org/10.1007/978-3-030-41800-7_12

183

This class of equation is sometimes referred to as the Lane-Emden-Fowler equation and are related to problems with critical exponents [1]. Furthermore, they arise in the theory of boundary layers of viscous fluids [12].

We are particularly interested in the class of PDE (1.1) because of its application to the analysis of numerical schemes posed for the KdV-like equation

$$u_t - \left(|u|^{p-2} u\right)_x + u_{xxx} = 0. \tag{1.2}$$

Indeed, solutions of (1.2) posed over a 1-dimensional domain satisfy

$$0 = \frac{\mathrm{d}}{\mathrm{d}t} \mathscr{J}[u], \quad \text{with } \mathscr{J}[u] = \int_\Omega \frac{1}{2} |u_x|^2 + \frac{1}{p} |u|^p \, \mathrm{d}x \tag{1.3}$$

and energy minimisers of (1.3) satisfy (1.1) with $f = 0$ and appropriate boundary conditions. In [6, 7] a conservative Galerkin scheme was proposed for (1.2) and the a priori and a posteriori analysis of this scheme requires quasi-optimal approximation of the finite element solution of (1.1) and optimal a posteriori estimates. Hence our goal in this work is the derivation and a priori and a posteriori bounds of Galerkin discretisations of (1.1).

We proceed as follows: In Sect. 2 we introduce notation and the model problem. We give some insight as to its properties that we use in subsequent sections and propose a discontinuous Galerkin finite element approximation. In Sect. 3 we give a classical a priori analysis based on arguments in [2]. We show that in the energy norm, the analysis is suboptimal for high polynomial degrees and large values of p. In Sect. 4 we modify the notion of a quasinorm from the works of [10] to enable an optimal a priori error estimate to be shown. In Sect. 5 we derive an a posteriori estimate and finally, in Sect. 6, we showcase some numerical experiments.

2 Problem Setup

In this section we formulate the model problem, fix notation and give some basic assumptions. Weakly, we may consider the PDE (1.1) as: find $u \in H_0^1(\Omega)$, such that

$$\mathscr{A}(u, v) + \mathscr{B}(u; u, v) = \langle f, v \rangle \quad \forall v \in H_0^1(\Omega), \tag{2.1}$$

where $\langle \cdot, \cdot \rangle$ denotes the L^2 inner product and the bilinear form $\mathscr{A} : H_0^1(\Omega) \times H_0^1(\Omega) \to \mathbb{R}$ is given by

$$\mathscr{A}(u, v) := \int_\Omega \nabla u \cdot \nabla v \, \mathrm{d}x. \tag{2.2}$$

The semilinear form $\mathscr{B}$ is given by

$$\mathscr{B}(w; u, v) := \int_{\Omega} |w|^{p-2} uv \, dx. \tag{2.3}$$

It is straightforward to verify this problem admits a unique solution.

Proposition 2.1 (A Priori Bound 1) *Let $f \in H^{-1}(\Omega)$ and $u \in H_0^1(\Omega)$ solve (2.1). Then we have*

$$\frac{1}{2} \|\nabla u\|_{L^2(\Omega)}^2 + \|u\|_{L^p(\Omega)}^p \leq \|f\|_{H^{-1}(\Omega)}^2. \tag{2.4}$$

Proof Using a standard energy argument, take $v = u$ in (2.1), then

$$\|\nabla u\|_{L^2(\Omega)}^2 + \|u\|_{L^p(\Omega)}^p = \langle f, u \rangle \leq \|f\|_{H^{-1}(\Omega)} \|\nabla u\|_{L^2(\Omega)}$$

$$\leq \frac{1}{2} \left(\|f\|_{H^{-1}(\Omega)}^2 + \|\nabla u\|_{L^2(\Omega)}^2 \right), \tag{2.5}$$

as required. □

Proposition 2.2 (A Priori Bound 2) *Let $f \in L^q(\Omega)$, where $q = \frac{p}{p-1}$ and $u \in H_0^1(\Omega)$ solve (2.1) then we have*

$$\|\nabla u\|_{L^2(\Omega)}^2 + \frac{1}{q} \|u\|_{L^p(\Omega)}^p \leq \frac{1}{q} \|f\|_{L^q(\Omega)}^q. \tag{2.6}$$

Proof Again, take $v = u$ in (2.1), then

$$\|\nabla u\|_{L^2(\Omega)}^2 + \|u\|_{L^p(\Omega)}^p = \langle f, u \rangle \leq \|f\|_{L^q(\Omega)} \|u\|_{L^p(\Omega)} \leq \frac{1}{p} \|u\|_{L^p(\Omega)}^p + \frac{1}{q} \|f\|_{L^q(\Omega)}^q, \tag{2.7}$$

as required. □

Remark 2.3 (Behaviour of the a Priori Bounds in p) Note that the bounds given in Propositions 2.1 and 2.2 behave the same as p increases, since

$$\lim_{p \to \infty} \frac{1}{q} = 1, \tag{2.8}$$

however the bound given in Proposition 2.2 behave quite differently as p decreases, indeed

$$\lim_{p \to 1} \frac{1}{q} = 0. \tag{2.9}$$

We will only consider the case $p \geq 2$ in this work.

2.1 Discretisation

Let $\mathscr{T}$ be a regular subdivision of Ω into disjoint simplicial elements. We assume that the subdivision $\mathscr{T}$ is shape-regular [2, p.124], is $\overline{\Omega} = \cup_K \overline{K}$ and that the elemental faces are points (for $d = 1$), straight lines (for $d = 2$) or planar (for $d = 3$) segments; these will be, henceforth, referred to as *facets*. By Γ we shall denote the union of all $(d - 1)$-dimensional facets associated with the subdivision $\mathscr{T}$ including the boundary. Further, we set $\Gamma_i := \Gamma \backslash \partial\Omega$.

For a nonnegative integer k, we denote the set of all polynomials of total degree at most k by $\mathbb{P}^k(K)$. For $k \geq 1$, we consider the finite element space

$$V_h^k := \{ v \in L^2(\Omega) : v|_K \in \mathbb{P}^k(K), \forall K \}. \tag{2.10}$$

Further, let K^+, K^- be two (generic) elements sharing a facet $e := \partial K^+ \cap \partial K^- \subset \Gamma_i$ with respective outward normal unit vectors $\boldsymbol{n}^+$ and $\boldsymbol{n}^-$ on e. For a function $v : \Omega \to \mathbb{R}$ that may be discontinuous across Γ_i, we set $v^+ := v|_{e \subset \partial K^+}$, $v^- := v|_{e \subset \partial K^-}$, and we define the jump by

$$[v] := v^+ \boldsymbol{n}^+ + v^- \boldsymbol{n}^-;$$

if $e \in \partial K \cap \partial \Omega$, we set $[v] := v^+ \boldsymbol{n}$. Also, we define $h_K := \mathrm{diam}(K)$ and we collect them into the element-wise constant function $h : \Omega \to \mathbb{R}$, with $h|_K = h_K$, K, $h|_e = (h_{K^+} + h_{K^-})/2$ for $e \subset \Gamma_i$ and $h|_e = h_K$ for $e \subset \partial K \cap \partial \Omega$. We assume that the families of meshes considered in this work are locally quasi-uniform. Note that this restriction can be relaxed by following arguments as in [5].

For $s > 0$, we define the *broken* Sobolev space $\mathrm{H}^s(\mathscr{T})$, by

$$\mathrm{H}^s(\mathscr{T}) := \{ w \in \mathrm{L}^2(\Omega) : w|_K \in \mathrm{H}^s(K), K \in \mathscr{T} \},$$

along with the broken (element-wise) gradient and Laplacian $\nabla_h \equiv \nabla_h(\mathscr{T})$ and $\Delta_h \equiv \Delta_h(\mathscr{T})$.

We consider the interior penalty (IP) discontinuous Galerkin discretisation of (2.2), reading: find $u_h \in V_h^k$ such that

$$\mathscr{A}_h(u_h, v_h) + \mathscr{B}(u_h; u_h, v_h) = \langle f, v_h \rangle \quad \forall \, v_h \in V_h^k, \tag{2.11}$$

where

$$\mathscr{A}_h(u_h, v_h) = \int_\Omega \nabla_h u_h \cdot \nabla_h v_h \, \mathrm{d}x - \int_\Gamma \left([v_h] \cdot \{\!\!\{ \nabla u_h \}\!\!\} + [u_h] \cdot \{\!\!\{ \nabla v_h \}\!\!\} \right. \\ \left. - \sigma \, [u_h] \cdot [v_h] \right) \mathrm{d}s, \tag{2.12}$$

where $\sigma > 0$ is the, so-called, *discontinuity penalisation parameter* given by

$$\sigma := C_\sigma \frac{k^2}{h}. \tag{2.13}$$

Definition 2.5 (Mesh Dependent Norms) We introduce the mesh dependent H^1 norm to be

$$|w|_{dG}^2 := \|\nabla_h w\|_{\mathrm{L}^2(\Omega)}^2 + \|\sqrt{\sigma}\,[w]\|_{\mathrm{L}^2(\Gamma)}^2. \tag{2.14}$$

Note that the the bilinear form (2.2) satisfies boundedness and coercivity properties for C_σ chosen large enough [3, c.f.], that is

$$\begin{aligned}
\mathscr{A}_h(u_h, v_h) &\leq \tilde{C}_B \,|u_h|_{dG}\,|v_h|_{dG} \\
\mathscr{A}_h(u_h, u_h) &\geq \tilde{C}_C \,|u_h|_{dG}^2 \qquad \forall\, u_h, v_h \in V_h^k.
\end{aligned} \tag{2.15}$$

3   Classical A Priori Analysis

In this section we examine analysis based on classical arguments such as those used in [2] for the p-Laplacian.

Lemma 3.1 (Properties of $\mathscr{B}(\cdot;\,\cdot,\,\cdot)$, cf. [2, §5.3]) *There exist constants*

(1) $C_L > 0$ *such that*

$$\mathscr{B}(u - u_h; u - u_h, u - u_h) \leq C_L(\mathscr{B}(u; u, u - u_h) - \mathscr{B}(u_h; u_h, u - u_h)) \tag{3.1}$$

(2) $C_U > 0$ *such that*

$$\mathscr{B}(u; u, u - v_h) - \mathscr{B}(u_h; u_h, u - v_h) \leq C_U \,\|u - u_h\|_{\mathrm{L}^p(\Omega)}\,\|u - v_h\|_{\mathrm{L}^q(\Omega)}. \tag{3.2}$$

Theorem 3.2 *Let $u \in \mathrm{H}^2(\Omega) \cap \mathrm{H}_0^1(\Omega)$ solve (1.1) and $u_h \in V_h^k$ be the finite element approximation of (2.11) then for $k \geq 1$ we have*

$$|u - u_h|_{dG}^2 + \|u - u_h\|_{\mathrm{L}^p(\Omega)}^p \leq C \inf_{v_h \in V_h^k} \left(|u - v_h|_{dG}^2 + \|u - v_h\|_{\mathrm{L}^q(\Omega)}^q\right), \tag{3.3}$$

where $q = \frac{p}{p-1}$ is the Sobolev conjugate of p.

Proof Since u_h solves (2.11), we have, though Lemma 3.1 and Galerkin orthogonality

$$\tilde{C}_C \, |u - u_h|^2_{dG} + \frac{1}{C_L} \, \|u - u_h\|^p_{L^p(\Omega)}$$

$$\leq \mathscr{A}_h(u - u_h, u - u_h) + \frac{1}{C_L} \mathscr{B}(u - u_h; u - u_h, u - u_h)$$

$$\leq \mathscr{A}_h(u - u_h, u - u_h) + \mathscr{B}(u; u, u - u_h) - \mathscr{B}(u_h; u_h, u - u_h)$$

$$\leq \mathscr{A}_h(u - u_h, u - v_h) + \mathscr{B}(u; u, u - v_h) - \mathscr{B}(u_h; u_h, u - v_h) \quad \forall \, v_h \in V^k_h.$$
(3.4)

Note that

$$\mathscr{A}_h(u - u_h, u - v_h) \leq \frac{\tilde{C}^2_B}{2\tilde{C}_C} \, |u - v_h|^2_{dG} + \frac{\tilde{C}_C}{2} \, |u - u_h|^2_{dG}.$$
(3.5)

Further,

$$\mathscr{B}(u; u, u - v_h) - \mathscr{B}(u_h; u_h, u - v_h) \leq C_U \, \|u - u_h\|_{L^p(\Omega)} \, \|u - v_h\|_{L^q(\Omega)}$$

$$\leq \frac{1}{2C_L} \, \|u - u_h\|^p_{L^p(\Omega)}$$

$$+ \left(\frac{p}{2C_L}\right)^{-q/p} \frac{C_U}{q} \, \|u - v_h\|^q_{L^q(\Omega)}.$$
(3.6)

Substituting (3.5), (3.6) into (3.4) yields the desired result.   □

Corollary 3.3 *Choosing* $v_h = I_k u$, *the Clément interpolant of* u, *in Theorem 3.2 and under further smoothness requirements, that* $u \in W^{k+1,p}(\Omega)$, *we see that*

$$|u - u_h|^2_{dG} + \|u - u_h\|^p_{L^p(\Omega)} \leq C\left(h^{2k} \, |u|_{H^{k+1}(\Omega)} + h^{(k+1)q} \, |u|_{W^{k+1,p}(\Omega)}\right).$$
(3.7)

Remark 3.4 (Optimality of Corollary 3.3) Notice that the bound given in Corollary 3.3 depends upon $q = \frac{p}{p-1}$. Notice, as shown in Table 1, the energy error bounds are optimal only if $p = 2$ for all k or $k = 1$ for all p.

Table 1 In the following table we examine the optimality of the finite element approximation in the energy norm

k	p = 2	p = 3	p = 4	p = 5	p → ∞
1	2	3/2	4/3	5/4	1
2	3	9/4	2	15/8	3/2
3	4	3	8/3	5/2	2
4	5	15/4	10/3	25/8	5/2

The numerical values in the table correspond to $(k+1)q/2$ and are coloured green or red depending upon whether the bound is optimal, in the function approximation sense, or suboptimal respectively

Remark 3.5 (Dual Bounds) This lack of optimality propagates further when consider bounds based on duality approaches. Indeed, using the dual problem

$$-\Delta z + (p-1)\,u^{p-2}z = u - u_h \text{ in } \Omega$$
$$z = 0 \text{ on } \partial\Omega,$$
(3.8)

one can show

$$\|u - u_h\|_{L^2(\Omega)} \leq C\left(h\,|u - u_h|_{dG} + \|u - u_h\|^2_{L^p(\Omega)}\right).$$
(3.9)

We will not prove this here for brevity but, as illustrated in Table 2, the bound is optimal only when $p = 2$.

Table 2 In the following table we examine the optimality of the finite element approximation in the L^2 norm

k	p = 2	p = 3	p = 4	p = 5	p → ∞
1	2	4/3	1	4/5	0
2	3	8/3	2	3/2	0
3	4	4	8/3	2	0
4	5	19/4	10/3	5/2	0

The numerical values in the table correspond to $\min(k + 1, (k + 1)\,q/2 + 1, 4k/p, (2k + 2)/(p - 1))$ thus represent the convergence rate in L^2. They are coloured green or red depending upon whether the bound is optimal, in the function approximation sense, or suboptimal respectively. Notice that they L^2 norm estimate is only optimal for $p = 2$ for all k

4   A Priori Analysis Based on Quasi Norms

In this section we will examine the use of quasinorms to rectify the gap in the a priori analysis.

Definition 4.1 (Quasinorm) Let $v \in L^p(\Omega)$, $p \geq 2$, then for any $w \in L^p(\Omega)$ we define the quasinorm

$$\|v\|_{(w,p)}^2 := \int_\Omega |v|^2 (|w| + |v|)^{p-2} \, dx. \tag{4.1}$$

This satisfies the usual properties of a norm, in that

$$\|v\|_{(w,p)} \geq 0 \text{ and } \|v\|_{(w,p)} = 0 \iff v = 0. \tag{4.2}$$

However, the usual triangle inequality is replaced by

$$\|v_1 + v_2\|_{(w,p)} \leq C\big(\|v_1\|_{(w,p)} + \|v_2\|_{(w,p)}\big), \tag{4.3}$$

where $C = C(v_1, v_2, w, p)$

Remark 4.2 (Properties of the Quasinorm) As can be seen from the definition, the quasinorm is related to the L^p norm through

$$\|v\|_{L^p(\Omega)}^p \leq \|v\|_{(w,p)}^2 \leq C \|v\|_{L^p(\Omega)}^2, \tag{4.4}$$

for $v \in L^p(\Omega)$, $p \geq 2$ and any $w \in L^p(\Omega)$. The key property that the quasinorm satisfies that allows for optimal a priori treatment is that the semilinear form is coercive with respect to it, that is

$$\mathscr{B}(u; u, u - v) - \mathscr{B}(v; v, u - v) \geq \overline{C}_C \|u - v\|_{(u,p)}^2. \tag{4.5}$$

In addition, it is bounded [4] in that for any $\theta > 0$ there exists a $\gamma > 0$ such that

$$|\mathscr{B}(u; u, w) - \mathscr{B}(v; v, w)| \leq \overline{C}_B\Big(\theta^\gamma \|u - v\|_{(u,p)}^2 + \theta \|w\|_{(u,p)}^2\Big), \tag{4.6}$$

where

$$\gamma = \begin{cases} 1 & \text{if } \theta < 1 \\ \frac{1}{p-1} & \text{if } \theta \geq 1. \end{cases} \tag{4.7}$$

It was the lack of a sufficiently sharp boundedness property that led to suboptimality in the analysis presented in Sect. 3. The key observation to rectify the suboptimality is to measure the error in

$$|u - u_h|_{dG}^2 + \|u - u_h\|_{(u,p)}^2, \tag{4.8}$$

rather than the energy norm.

Henceforth, we will use the notation

$$C_C := \min\left(\overline{C}_C, \tilde{C}_C\right) \qquad C_B := \max\left(\overline{C}_B, \tilde{C}_B\right). \qquad (4.9)$$

Proposition 4.3 (A Priori Bound 3) *Let $f \in H^{-1}(\Omega)$ and $u \in H_0^1(\Omega)$ solve (2.1) then*

$$\|\nabla u\|_{L^2(\Omega)}^2 + 2^{3-p} \|u\|_{(u,p)}^2 \leq \|f\|_{H^{-1}(\Omega)}^2. \qquad (4.10)$$

Proof Notice that

$$\|\nabla u\|_{L^2(\Omega)}^2 + 2^{2-p} \|u\|_{(u,p)}^2 = \mathscr{A}(u,u) + \mathscr{B}(u;u,u) = \langle f, u \rangle$$
$$\leq \frac{1}{2}\left(\|f\|_{H^{-1}(\Omega)}^2 + \|\nabla u\|_{L^2(\Omega)}^2\right), \qquad (4.11)$$

as required. □

Theorem 4.4 *Let $u \in H^2(\Omega) \cap H_0^1(\Omega)$ solve (1.1) and u_h be the finite element approximation of (2.11) then for $k \geq 1$ we have*

$$|u - u_h|_{dG}^2 + \|u - u_h\|_{(u,p)}^2 \leq C \inf_{v_h \in V_h^k}\left(|u - v_h|_{dG}^2 + \|u - v_h\|_{(u,p)}^2\right). \qquad (4.12)$$

Proof Making use of the coercivity of $\mathscr{B}$ we have

$$C_C\left(|u - u_h|_{dG}^2 + \|u - u_h\|_{(u,p)}^2\right)$$
$$\leq \mathscr{A}_h(u - u_h, u - u_h) + \mathscr{B}(u;u,u-u_h) - \mathscr{B}(u_h;u_h,u-u_h) \qquad (4.13)$$
$$= \mathscr{A}_h(u - u_h, u - v_h) + \mathscr{B}(u;u,u-v_h) - \mathscr{B}(u_h;u_h,u-v_h) + \text{h.o.t},$$

for any $v_h \in V_h^k$, using Galerkin orthogonality. Now, through (4.6) and (2.15) we have

$$C_C\left(|u - u_h|_{dG}^2 + \|u - u_h\|_{(u,p)}^2\right) \leq \frac{C_C}{2}|u - u_h|_{dG} + \frac{C_B^2}{2C_C}|u - v_h|_{dG}$$
$$+ C_B\left(\theta^\gamma \|u - u_h\|_{(u,p)}^2 + \theta\|u - v_h\|_{(u,p)}^2\right). \qquad (4.14)$$

Choosing $\theta = \min\left(\frac{C_C}{2C_B}, \frac{1}{2}\right)$ then $\gamma = 1$. Rearranging the inequality yields the desired result. □

Lemma 4.5 *Let $p \geq 2$ and $v \in W^{k+1,p}(\Omega)$ then*

$$\inf_{v_h \in V_h^k} \|v - v_h\|_{(v,p)} \leq Ch^{k+1} |v|_{W^{k+1,p}(\Omega)}. \tag{4.15}$$

Proof Using the property of the quasinorm given in Remark 4.2 we have

$$\|v - v_h\|_{(v,p)}^2 \leq C \|v - v_h\|_{L^p(\Omega)}^2, \tag{4.16}$$

and the result follows from best approximation in $L^p(\Omega)$. $\square$

Corollary 4.6 *Under the conditions of Theorem 4.4 suppose that $u \in H^{k+1}(\Omega) \cap H_0^1(\Omega) \cap W^{k,p}(\Omega)$, then*

$$\left(|u - u_h|_{dG}^2 + \|u - u_h\|_{(u,p)}^2\right)^{1/2} \leq Ch^k \left(|u|_{H^{k+1}(\Omega)} + |u|_{W^{k,p}(\Omega)}\right). \tag{4.17}$$

Remark 4.7 (Optimality of Corollary 4.6) Notice that the bound given in Corollary 4.6 is optimal regardless of the choice of p for smooth enough u.

Remark 4.8 (Dual Bounds) By modifying the dual problem to

$$-\Delta z + (p-1) u^{p-2} z = (u - u_h)(|u| + |u - u_h|)^{p-2} \text{ in } \Omega$$
$$z = 0 \text{ on } \partial\Omega, \tag{4.18}$$

one can also show optimal a priori bounds for the quasinorm error.

5   A Posteriori Error Analysis

In this section we derive a reliable a posteriori estimator. To that end, we redefine $\mathscr{A}_h$ to extend it over $H^1(\Omega) \times H^1(\Omega)$ as:

$$\mathscr{A}_h(u_h v_h) = \int_\Omega \nabla_h u_h \cdot \nabla_h v_h \, dx - \int_\Gamma \left([v_h] \cdot \{P_{k-1}(\nabla u_h)\} + [u_h] \cdot \{P_{k-1}(\nabla v_h)\}\right.$$
$$\left. - \sigma [u_h] \cdot [v_h]\right) ds, \tag{5.1}$$

Proposition 5.1 (A Priori Bound 4) *Let $f \in H^{-1}(\Omega)$ and $u \in H_0^1(\Omega)$ solve (2.1) and $w \in H_0^1(\Omega)$ solve*

$$\mathscr{A}(w, v) + \mathscr{B}(w; w, v) = \langle f - \mathfrak{R}, v \rangle \quad \forall v \in H_0^1(\Omega), \tag{5.2}$$

for some $\mathfrak{R} \in H^{-1}(\Omega)$ *then*

$$\|\nabla u - \nabla w\|_{L^2(\Omega)}^2 + 2C_L \|u - w\|_{L^p(\Omega)}^p \leq \|\mathfrak{R}\|_{H^{-1}(\Omega)}^2 \tag{5.3}$$

Proof Through the definitions of u and w, we have the relation that

$$\mathscr{A}(u - w, v) + \mathscr{B}(u; u, v) - \mathscr{B}(w; w, v) = \langle \mathfrak{R}, v \rangle \quad \forall\, v \in H_0^1(\Omega). \tag{5.4}$$

Hence, choosing $v = u - w$

$$\|\nabla u - \nabla w\|_{L^2(\Omega)}^2 + C_L \|u - w\|_{L^p(\Omega)}^p \leq \langle \mathfrak{R}, u - w \rangle$$

$$\leq \frac{1}{2}\left(\|\mathfrak{R}\|_{H^{-1}(\Omega)}^2 + \|\nabla u - \nabla w\|_{L^2(\Omega)}^2\right), \tag{5.5}$$

as required. $\qquad\square$

To invoke the results of Proposition 5.1 we require an object $w \in H^1(\Omega)$. The dG solution $u_h \notin H^1(\Omega)$, so we make use of an appropriate postprocessor as an intermediate quantity.

Lemma 5.2 ([8]) *Let $\mathcal{N}$ denote the set of all Lagrange nodes of V_h^k, and $\mathscr{E}: V_h^k \to V_h^k \cap H_0^1(\Omega)$ be defined on the conforming Lagrange nodes $v \in \mathcal{N}$ by*

$$\mathscr{E}(v)(\nu) := \begin{cases} |\omega_\nu|^{-1} \displaystyle\sum_{K \in \omega_\nu} v|_K(\nu), & \nu \in \Omega; \\ 0, & \nu \in \partial\Omega, \end{cases}$$

with $\omega_\nu := \bigcup_{K \in \mathscr{T}: \nu \in \overline{K}} K$, the set of elements sharing the node $\nu \in \mathcal{N}$ and $|\omega_\nu|$ their cardinality. Then, the following bound holds

$$\sum_{K \in \mathscr{T}} |v - \mathscr{E}(v)|_{H^\alpha(K)}^2 \leq C_\alpha \sum_{e \in \Gamma} \left\| h^{1/2 - \alpha}[v] \right\|_{L^2(e)}^2, \tag{5.6}$$

with $\alpha = 0, 1$, $C_\alpha \equiv C_\alpha(k) > 0$ a constant independent of h, v and $\mathscr{T}$, but depending on the shape-regularity of $\mathscr{T}$ and on the polynomial degree k.

Proposition 5.3 *The reconstruction $\mathscr{E}(u_h)$ satisfies the perturbed PDE*

$$\mathscr{A}(\mathscr{E}(u_h), v) + \mathscr{B}(\mathscr{E}(u_h); \mathscr{E}(u_h), v) = \langle f - \mathfrak{R}, v \rangle \quad \forall\, v \in H_0^1(\Omega), \tag{5.7}$$

with

$$\langle \mathfrak{R}, v \rangle = \mathscr{A}_h(u_h - \mathscr{E}(u_h), v) + \mathscr{B}(u_h; u_h, v) - \mathscr{B}(\mathscr{E}(u_h); \mathscr{E}(u_h), v)$$

$$+ \langle f, v - v_h \rangle - \mathscr{A}_h(u_h, v - v_h) - \mathscr{B}(u_h; u_h, v - v_h) \quad \forall\, v_h \in V_h^k. \tag{5.8}$$

Theorem 5.4 *Let* $f \in H^{-1}(\Omega)$ *and* $u \in H_0^1(\Omega)$ *solve* (2.1) *and* $u_h \in V_h^k$ *solve* (2.11). *Further let* $\mathcal{E}(u_h) \in H_0^1(\Omega)$ *denote the reconstruction operator given in Lemma 5.2. Then,*

$$\|\nabla u - \nabla\mathcal{E}(u_h)\|_{L^2(\Omega)}^2 + 2C_L \|u - \mathcal{E}(u_h)\|_{L^p(\Omega)}^p \leq C \sum_{K \in \mathcal{T}} \left[\eta_R^2 + \sum_{e \in \partial K} \eta_J^2 \right],$$

(5.9)

where

$$\eta_R^2 := \left\| h\left(f + \Delta u_h - |u_h|^{p-2} u_h \right) \right\|_{L^2(K)}^2$$

$$\eta_J^2 := \left\| h^{1/2} [\nabla u_h] \right\|_{L^2(e)}^2 + \left\| h^{-1/2} [u_h] \right\|_{L^2(e)}^2 + \left\| h^{1/2} [u_h] \right\|_{L^2(e)}^2$$

(5.10)

Proof It suffices to determine an upper bound for $\|\mathfrak{R}\|_{H^{-1}(\Omega)}$. To that end, by Proposition 5.3

$$\langle \mathfrak{R}, v \rangle = \underbrace{\mathcal{A}_h(u_h - \mathcal{E}(u_h), v)}_{\mathscr{I}_1} + \underbrace{\mathcal{B}(u_h; u_h, v) - \mathcal{B}(\mathcal{E}(u_h); \mathcal{E}(u_h), v)}_{\mathscr{I}_2}$$

$$+ \underbrace{\langle f, v - v_h \rangle - \mathcal{A}_h(u_h, v - v_h) - \mathcal{B}(u_h; u_h, v - v_h)}_{\mathscr{I}_3},$$

(5.11)

and we proceed to bound the terms individually. Firstly,

$$\mathscr{I}_1 = \sum_{K \in \mathcal{T}} \int_K (\nabla u_h - \nabla\mathcal{E}(u_h)) \cdot \nabla v \, d\mathbf{x} - \sum_{e \in \Gamma} [u_h] \{ P_{k-1}(\nabla v) \} \, ds$$

$$\leq \sum_{K \in \mathcal{T}} \|\nabla u_h - \nabla\mathcal{E}(u_h)\|_{L^2(K)} \|\nabla v\|_{L^2(K)}$$

$$+ \sum_{e \in \Gamma} \left\| h^{-1/2} [u_h] \right\|_{L^2(e)} \left\| h^{1/2} \{ P_{k-1}(\nabla v) \} \right\|_{L^2(e)}$$

$$\leq \left(C_1^{1/2} + C_{dim,\mathcal{T}}^{1/2} C_{trace}^{1/2} \right) \left(\sum_{e \in \Gamma} \left\| h^{-1/2} [u_h] \right\|_{L^2(e)}^2 \right)^{1/2} \left(\sum_{K \in \mathcal{T}} \|\nabla v\|_{L^2(K)}^2 \right)^{1/2}$$

$$\leq C \left(\sum_{K \in \mathcal{T}} \sum_{e \in \partial K} \eta_J^2 \right)^{1/2} \|\nabla v\|_{L^2(\Omega)},$$

(5.12)

where $C_{dim,\mathcal{T}}^{1/2}$ is a constant depending on the dimension and the triangulation and C_{trace} is the constant from a trace estimate. The second term can be controlled by

$$
\begin{aligned}
\mathscr{I}_2 &= \int_\Omega \left(|u_h|^{p-2} u_h - |\mathscr{E}(u_h)|^{p-2} \mathscr{E}(u_h) \right) v \\
&\leq C(u_h, \mathscr{E}(u_h), p) \sum_{K \in \mathcal{T}} \|u_h - \mathscr{E}(u_h)\|_{L^2(K)} \|v\|_{L^2(K)} \, . \\
&\leq C(u_h, \mathscr{E}(u_h), p) \left(\sum_{K \in \mathcal{T}} \|u_h - \mathscr{E}(u_h)\|_{L^2(K)}^2 \right)^{1/2} \|v\|_{L^2(\Omega)} \\
&\leq C_P C(u_h, \mathscr{E}(u_h), p) C_0^{1/2} \left(\sum_{e \in \Gamma} \left\| h^{1/2} [u_h] \right\|_{L^2(e)}^2 \right)^{1/2} \|\nabla v\|_{L^2(\Omega)} \, ,
\end{aligned}
\tag{5.13}
$$

where C_P is the Poincaré constant. To finish, $\mathscr{I}_3$ is controlled by a standard a posteriori argument.

$$
\begin{aligned}
\mathscr{I}_3 &= \int_\Omega f(v - v_h) - \nabla_h u_h \cdot (\nabla v - \nabla_h v_h) - |u_h|^{p-2} u_h (v - v_h) \, dx \\
&\quad + \int_\Gamma [u_h] \cdot \{P_{k-1}(\nabla v - \nabla_h v_h)\} \, ds + [v - v_h] \{\nabla_h u_h\} - \sigma [u_h] \cdot [v - v_h] \, ds \\
&= \int_\Omega \left(f + \Delta_h u_h - |u_h|^{p-2} u_h \right) (v - v_h) \, dx - \int_\Gamma [\nabla u_h] \{v - v_h\} \, ds \\
&\quad + \int_\Gamma [u_h] \cdot \{P_{k-1}(\nabla v - \nabla_h v_h)\} - \sigma [u_h] \cdot [v - v_h] \, ds.
\end{aligned}
\tag{5.14}
$$

Splitting the integrals elementwise and making use of the Cauchy-Schwarz inequality we see

$$
\int_\Omega \left(f + \Delta_h u_h - |u_h|^{p-2} u_h \right)(v - v_h) \, dx \leq \sum_{K \in \mathcal{T}} \left\| h \left(f + \Delta_h u_h - |u_h|^{p-2} u_h \right) \right\|_{L^2(K)}
$$
$$
\times \left\| h^{-1}(v - v_h) \right\|_{L^2(K)}
\tag{5.15}
$$

Similarly, for the second,

$$
-\sum_{e \in \Gamma} \int_e [\nabla u_h] \{v - v_h\} \, ds \leq \sum_{K \in \mathcal{T}} \left[\sum_{e \in \partial K} \left\| h^{1/2}[\nabla u_h] \right\|_{L^2(e)} \left\| h^{-1/2} \{v - v_h\} \right\|_{L^2(e)} \right],
\tag{5.16}
$$

and third term

$$\sum_{e\in\Gamma}\int_e [u_h]\cdot\{\!\!\{P_{k-1}(\nabla v-\nabla v_h)\}\!\!\}\ ds \le \sum_{K\in\mathcal{T}}\Bigg[\sum_{e\in\partial K}\Big\|h^{-1/2}[u_h]\Big\|_{L^2(e)}$$

$$\times \Big\|h^{1/2}\{\!\!\{P_{k-1}(\nabla v-\nabla v_h)\}\!\!\}\Big\|_{L^2(e)}\Bigg].$$

(5.17)

For the final term

$$\sum_{e\in\Gamma}\int_e \sigma\,[u_h]\cdot[v-v_h]\ ds \le C_\sigma \sum_{K\in\mathcal{T}}\Bigg[\sum_{e\in\partial K}\Big\|h^{-1/2}[u_h]\Big\|_{L^2(e)}\Big\|h^{-1/2}[v-v_h]\Big\|_{L^2(e)}\Bigg]$$

(5.18)

Collecting (5.14)–(5.18) we have

$$\mathscr{I}_3 \le C\left(\sum_{K\in\mathcal{T}}\Bigg[\eta_R + \sum_{e\in\partial K}\eta_J\Bigg]\Phi(v-v_h)\right),$$

(5.19)

where

$$\Phi(w) = \max\left(\Big\|h^{-1}w\Big\|_{L^2(K)}, \max_{e\in\Gamma}\Big\|h^{-1/2}w\Big\|_{L^2(e)},\right.$$

$$\left.\max_{e\in\Gamma}\Big\|h^{1/2}\{\!\!\{P_{k-1}(\nabla w)\}\!\!\}\Big\|_{L^2(e)}, \max_{e\in\Gamma}\Big\|h^{-1/2}[w]\Big\|_{L^2(e)}\right).$$

(5.20)

Choosing $v_h = P_0 v$ and in view of the approximation properties and stability of the L^2 projector we have that

$$\Phi(v-v_h) \le C\,\|\nabla v\|_{L^2(\widehat{K})},$$

(5.21)

where $\widehat{K}$ denotes the patch of K. Using a discrete Cauchy-Schwarz inequality

$$\mathscr{I}_3 \le C\left(\sum_{K\in\mathcal{T}}\Bigg[\eta_R^2 + \sum_{e\in\partial K}\eta_J^2\Bigg]\right)^{1/2}\left(\sum_{K\in\mathcal{T}}\|\nabla v\|_{L^2(\widehat{K})}^2\right)^{1/2}$$

$$\le C\left(\sum_{K\in\mathcal{T}}\Bigg[\eta_R + \sum_{e\in\partial K}\eta_J^2\Bigg]\right)^{1/2}\|\nabla v\|_{L^2(\Omega)},$$

(5.22)

hence, making use of (5.12) and (5.13), we have

$$
\langle \mathfrak{R}, v \rangle = \mathcal{I}_1 + \mathcal{I}_2 + \mathcal{I}_3 \leq C \left(\sum_{K \in \mathcal{T}} \left[\eta_R + \sum_{e \in \partial K} \eta_J^2 \right] \right)^{1/2} \| \nabla v \|_{L^2(\Omega)}, \qquad (5.23)
$$

where the constant C depends only upon the shape regularity of the mesh, p and u_h. The result follows by dividing through by $\| \nabla v \|_{L^2(\Omega)}$ and taking the supremum over all possible $0 \neq v \in H_0^1(\Omega)$.   □

Corollary 5.5 *Making use of the triangle inequality, one may show under the conditions of Theorem 5.4 the following result holds:*

$$
|u - u_h|_{dG}^2 + 2 C_L \| u - u_h \|_{L^p(\Omega)}^p \leq C \sum_{K \in \mathcal{T}} \left[\eta_R^2 + \sum_{e \in \partial K} \eta_J^2 \right]. \qquad (5.24)
$$

6 Numerical Experiments

We now illustrate the performance of the scheme through a series of numerical experiments.

6.1 Test 1: Asymptotic Behaviour Approximating a Smooth Solution

As a first test, we consider the domain $\Omega = [0, 1]^2$. We fix f such that the exact solution is given by

$$
u(x, y) = \sin(\pi x) \sin(\pi y), \qquad (6.1)
$$

and approximate Ω through a uniformly generated, criss-cross triangular type mesh to test the asymptotic behaviour of the numerical approximation. The results are summarised in Fig. 1a–d, and confirm the theoretical findings in Sects. 4 and 5.

More specifically, we consider the case $k = 1, 2$, $p = 4, 8$ and show that convergence measured the L^p-norm, the (u, p) quasinorm and the dG norm are all optimal. Notice that the fact the L^p norm is optimal is contrary to the analysis. This is already well observed in other problems [9, 11]. In addition, the a posteriori estimator is of optimal rate with an effectivity index of just under 10.

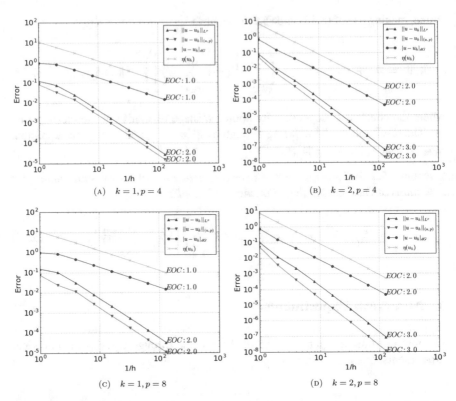

Fig. 1 Convergence plots for the dG scheme (2.11) for Test 1. We measure error norms involving the dG solution, u_h and the a posteriori estimator given in Corollary 5.5

6.2   Test 2: Behaviour of an Adaptive Scheme for Various p

We consider the domain $\Omega = [0, 1]^2$ and fix $f = 1000$ in this case there is no known solution. However, examining the energy functional (1.3) one can see that a minimiser has to 'balance' the L^2 norm of its derivative with the L^p norm of the function. For large p this almost translates into control of the ess sup which causes boundary layers to appear.

We approximate Ω through a uniformly generated, criss-cross triangular type initial mesh consisting of four elements. We run an adaptive algorithm of SOLVE, ESTIMATE, MARK, REFINE type, where SOLVE consists of solving the formulation (2.11), ESTIMATE is done through the evaluation of the estimator given in Corollary 5.5, MARK is a maximum strategy with 50% of the elements marked for refinement at each iteration and REFINE is a newest vertex bisection.

The results are summarised in Fig. 2a–d where we consider the case $k = 1$, $p = 4, 8, 12, 16, 20$ and examine the solution and underlying adaptive mesh.

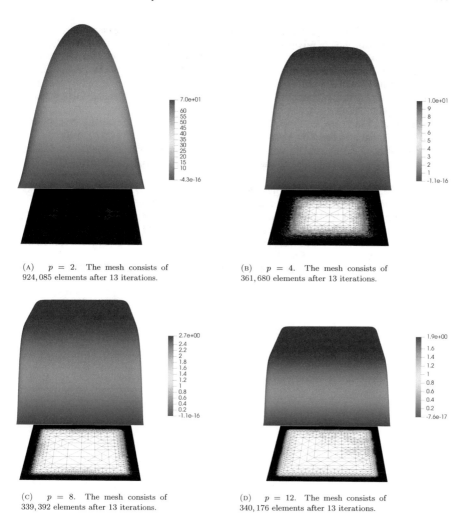

(A) $p = 2$. The mesh consists of
924, 085 elements after 13 iterations.

(B) $p = 4$. The mesh consists of
361, 680 elements after 13 iterations.

(C) $p = 8$. The mesh consists of
339, 392 elements after 13 iterations.

(D) $p = 12$. The mesh consists of
340, 176 elements after 13 iterations.

Fig. 2 Solutions of the dG scheme (2.11) for Test 2 and the underlying meshes generated through
the adaptive algorithm. (**a**) $p = 2$. The mesh consists of 924,085 elements after 13 iterations. (**b**)
$p = 4$. The mesh consists of 361,680 elements after 13 iterations. (**c**) $p = 8$. The mesh consists of
339,392 elements after 13 iterations. (**d**) $p = 12$. The mesh consists of 340,176 elements after 13
iterations

References

1. Clément, P., de Figueiredo, D.G., Mitidieri, E.: Quasilinear elliptic equations with critical
   exponents. Topol. Methods Nonlinear Anal. **7**(1), 133–170 (1996)
2. Ciarlet, P.G.: The Finite Element Method for Elliptic Problems. Studies in Mathematics and its
   Applications, vol. 4. North-Holland Publishing Co., Amsterdam (1978)
3. Ern, A., Guermond, J.-L.: Theory and Practice of Finite Elements. Applied Mathematical
   Sciences, vol. 159. Springer, New York (2004)

4. Ebmeyer, C., Liu, W.B.: Quasi-norm interpolation error estimates for the piecewise linear finite element approximation of p-Laplacian problems. Numer. Math. **100**(2), 233–258 (2005)
5. Georgoulis, E., Makridakis, C., Pryer, T.: Babuška-Osborn techniques in discontinuous Galerkin methods: L^2-norm error estimates for unstructured meshes. arXiV preprint: https://arxiv.org/abs/1704.05238 (2018)
6. Jackaman, J., Pryer, T.: Conservative Galerkin methods for dispersive Hamiltonian problems. arXiv preprint (2018)
7. Jackaman, J., Papamikos, G., Pryer, T.: The design of conservative finite element discretisations for the vectorial modified KdV equation. Appl. Numer. Math. **137**, 230–251 (2018)
8. Karakashian, O.A., Pascal, F.: A posteriori error estimates for a discontinuous Galerkin approximation of second-order elliptic problems. SIAM J. Numer. Anal. **41**(6), 2374–2399 (electronic) (2003)
9. Katzourakis, N., Pryer, T.: On the numerical approximation of p-Biharmonic and ∞-Biharmonic functions. Numer. Methods Partial Differ. Equ. **35**(1), 155–180 (2019)
10. Liu, W.B., Barrett, J.W.: Finite element approximation of some degenerate monotone quasilinear elliptic systems. SIAM J. Numer. Anal. **33**(1), 88–106 (1996)
11. Pryer, T.: On the finite element approximation of Infinity-Harmonic functions. Proc. R. Soc. Edinb. Sect. A: Math. Phys. Sci. (2018)
12. Wong, J.S.W.: On the generalized Emden–Fowler equation. SIAM Rev. **17**(2), 339–360 (1975)

Centred Splash of a Vertical Jet on a Horizontal Rotating Disc: Recent Findings and Resolving Controversies Over the Hydraulic Jump

Bernhard Scheichl

Abstract Highlights of the asymptotic and numerical analysis of the steady axisymmetric swirl flow of a Newtonian liquid over a spinning disc and generated by a jet, impacting perpendicularly onto the latter in the direction of gravity, are presented. Ubiquitous in engineering applications and involving a myriad of disparate velocity and length scales, thus extreme aspect ratios, this flow configuration is an archetypical one for the application of dimensional reasoning and matched asymptotic expansions in fluid dynamics. Particular interest lies on the recent advances in the rigorous description of the thin developed layer relatively far from the jet, which is essentially parametrised by a suitably defined Rossby number, the influence of gravity on the thin film and its interplay with a finitely remote disc edge. The latter upstream influence explains the phenomenon of the hydraulic jump in developed flow. The clarification of long-standing and more recent controversies around this concludes the analysis.

1   Origin and Statement of the Problem

Thin liquid films, involving the hydraulic jump as an important phenomenon, are ubiquitous in fluids engineering. The axisymmetric non-rotating jump can be reproduced readily in a kitchen sink. From the viewpoint of applications, however, the flow over a rotating disc attracts even more interest. This we study asymptotically.

In the following, we tacitly refer to the sketch of the assumed flow configuration in Fig. 1 and the subsequent assumptions. The horizontal and vertical directions are given by respectively the plane aligned with a rigid, impervious, perfectly

B. Scheichl (✉)
Institute of Fluid Mechanics and Heat Transfer, Technische Universität Wien, Vienna, Austria

AC2T research GmbH (Austrian Excellence Center for Tribology), Wiener Neustadt, Austria
e-mail: bernhard.scheichl@tuwien.ac.at; bernhard.scheichl@ac2t.at

© Springer Nature Switzerland AG 2020
G. R. Barrenechea, J. Mackenzie (eds.), *Boundary and Interior Layers, Computational and Asymptotic Methods BAIL 2018*, Lecture Notes in Computational Science and Engineering 135,
https://doi.org/10.1007/978-3-030-41800-7_13

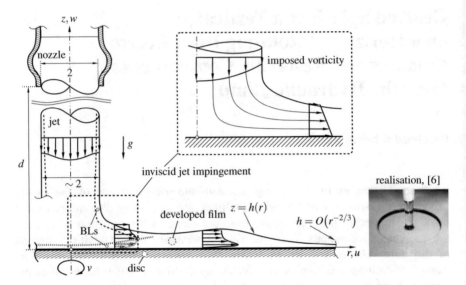

Fig. 1 Viable flow configuration for generating a radially spreading liquid film (not to scale, quantities non-dimensional), realisation exhibiting a strong hydraulic jump (taken from [6], reprinted with permission)

smooth disc and its axis of rotation. This coincides with that of a liquid jet with circular cross-section and exiting a nozzle, positioned sufficiently far above the disc, as driven by gravity acting vertically. The liquid is a Newtonian fluid of uniform density, (kinematic) viscosity, and surface tension with respect to the gaseous environment at rest and under constant pressure. Its so generated fluid flow past the disc is stationary (laminar) and axisymmetric throughout. Two radii come in the shortlist as reference lengths for the flow: that of the nozzle, typical of the jet unperturbed by its impingement, and the radius of the disc defining its edge. Since the second is only of concern if elliptic effects are, i.e. the upstream influence by gravity and/or surface tension, the first is preferred. The cross-section-averaged flow velocity at the nozzle exit then serves as the reference flow speed, and all variables are non-dimensional accordingly; subscripts indicate (unambiguously) partial derivatives.

We introduce polar coordinates r, z in respectively the radial and vertical directions, having their origin in the disc centre where the flow stagnates, and the dependent quantities $[u, v, w, p](r, z)$ and $h(r)$ denoting the radial, azimuthal, and vertical flow components, the difference p of the pressure in the fluid and the gaseous environment, and the (in case of jet contraction by gravity, double-valued) vertical height of the liquid layer respectively. Then the key groups

$$d \gg 1, \quad r_e \gg 1, \quad v \ll 1, \quad g \ll 1, \quad \sigma \ll 1, \quad 0 \le \alpha_0 < \infty \qquad \text{(1a–f)}$$

essentially parametrise the flow. Herein, $z = d$ and $r = r_e$ indicate the positions of the nozzle orifice and the disc edge, and the other quantities denote reciprocal Reynolds, Froude, and Weber numbers and a reciprocal squared Rossby number respectively (in the order of their appearance). The insinuated asymptotic limits are corroborated by the definitions of these quantities and the range of values they usually assume in engineering practice. In addition, we require v to satisfy the sharper constraint

$$vd \ll 1. \tag{2}$$

We also note that the well-known Rayleigh–Plateau instability is safely avoided, thus the jet sufficiently smooth and stable, as long as d^2 is sufficiently large as compared to the Weber number $1/\sigma$.

The Navier–Stokes equations governing this free-surface flow comprise the continuity equation satisfied identically by a streamfunction ψ,

$$r[u, w] = [\psi_z, -\psi_r], \tag{3}$$

and the momentum equations for the r-, azimuthal, and z-directions respectively,

$$uu_r + wu_z - v^2/r = v\{[(ru)_r/r]_r + u_{zz}\} - p_r, \tag{4}$$

$$uv_r + wv_z + uv/r = v\{[(rv)_r/r]_r + v_{zz}\}, \tag{5}$$

$$uw_r + ww_z = v\{(rw_r)_r/r + w_{zz}\} - p_z - g. \tag{6}$$

The third contributions to in (4) and (5) are the centripetal and Coriolis acceleration in the Eulerian frame. Usual boundary conditions (BCs) account for axial symmetry, no-penetration of and no-slip on the disc, and conservation of the volume flow rate;

$$r = 0: \ \psi = w_r = 0, \quad z = 0: \ \psi = u = 0, \quad v = \sqrt{\alpha_0}\, r, \quad z = h(r): \ \psi = \tfrac{1}{2}. \tag{7a–d}$$

Having in mind Bonnet's expression for the mean curvature κ of the free surface of the liquid, $\kappa = -(2r)^{-1}\mathrm{d}\big[rh'(1+h'^2)^{-1/2}\big]/\mathrm{d}r$, we arrive at the additional dynamic BCs holding at the latter,

$$[2h'(w_z - u_r) + (1 - h'^2)(u_z + w_r)]_{z=h} = 0, \tag{8}$$

$$[v_z - h'(v_r - v/r)]_{z=h} = 0, \tag{9}$$

$$2v[h'^2 u_r - h'(u_z + w_r) + w_z]_{z=h} = (1 + h'^2)(p|_{z=h} - 2\sigma\kappa). \tag{10}$$

These express vanishing total stress components in respectively the median, azimuthal, and surface-normal directions. The first two represent free-slip conditions, the latter takes into account the capillary (Laplace) pressure jump. Finally, the appropriate upstream or initial conditions (ICs) and downstream conditions then read

$$[\psi, u, v, w, p](r, d) = [\psi_o(r), 0, 0, -\psi_o'/r, \sigma], \quad h(1) = d, \quad h(r_e) = h_e.$$
$$(11a\text{–}c)$$

Here $\psi_o(r)$ subject to (7d) *models* the flow profile at the nozzle outlet, where the pressure is given by the capillary hoop stress in agreement with (10); prescribing a film height h_e at the disc edge, however, must be reviewed critically: see Sect. 3.

For a more elaborate discussion of the set of the full governing equations (1)–(11) we refer to the forerunner study [7]. The present one revisits its essence, focusses on some interesting intricacies of the analysis therein, and is intended to represent its supplement and a noteworthy extension. Specifically, the reader is provided with a survey of the current progress in the asymptotic description of the inherent hydraulic jump. Here also the recent and partially analogous results in [11] deserve a mention, of course.

2   Radially Spreading Thin Film

By (2), the bulk of the slender falling jet remains inviscid so that it is described by classical inviscid-flow theory. Due to the negligibly small capillary pressure according to (1e), Bernoulli's theorem yields $(u^2 + w^2)/2 - \bar{g}(d - z) \sim B(\psi)$ with the Bernoulli function $B(\psi)$ specified via (11a) and $\bar{g} := gd$. The appealing idea of controlling the radially spreading thin film (in industrial applications) by controlling the imposed vorticity $B'(\psi)$ via a targeted design of the nozzle shape suggests to prevent the jet from being contracted markedly, according to Toricelli's law, and essentially unaffected by the specific form of w_o at its impingement. Hence, $\bar{g}$ should not be to large. We therefore stipulate a least-degenerate limit: $\bar{g} = O(1)$.

Prescribing a downstream condition consistent with the global flow picture and thus allowing for strict forward flow as $r \gg 1$ closes the problem of an inviscid free jet and its deflection where $z = O(r)$. From (3), $w/u \ll 1$ in the far field, which implies $u \sim \sqrt{2(B + \bar{g})}$ where ψ takes on similarity form: say, $\psi \sim f(\eta)/2$ with $\eta := z/h(r) = O(1)$, thus $u \sim f'(\eta)/(2rh)$. Finally, by solving

$$f' = C\sqrt{2[B(f/2) + \bar{g}]}, \quad f(0) = 0, \quad f(1) = 1, \quad 2rh \sim C = const, \quad (12)$$

we determine f and the eigenvalue C governing the radial flattening of the layer. Two model cases, i.e. of zero and constant vorticity, are of specific importance: (i) the irrotational (uniform) flow just beneath the nozzle exit, $w_o = -1$, yielding $B \equiv 1/2$, $f = \eta$ as $C = 1/\sqrt{1 + 2\bar{g}}$; (ii) the classical (and more realistic) Hagen–Poiseuille profile $w_o = 2(r^2 - 1)$, giving $B = 2 - 4\psi$, $f = \sqrt{4 + 2\bar{g}}\,C(\eta - \eta^2) + \eta^2$, $C = \sqrt{1 + \bar{g}/2} - \sqrt{\bar{g}/2}$. In [7], jet bending is accounted for more complex representations of w_o under the variation of $\bar{g}$ and jet contraction, even affected by capillarity.

Viscous boundary layers (BLs) form along the free surface and on the disc encompassing its centre. The second type merges with the predominantly inviscid flow considered above to eventually form the developed thin film spreading outwards. As long as submerged in the region of jet deflection, this rotary BL flow is parametrised by α_0 and exhibits a radial component matching that of the inviscid flow on its top. This accelerates from being of $O(r)$ near stagnation to $O(1)$. The following order-of-magnitude estimates give an account of the aforementioned merge, where ν serves as the basic perturbation parameter. At first, (3) and (4) entail

$$urh \sim 1, \quad u^2/r \sim \nu u/h^2 \qquad (13a\text{–}c)$$

for the whole developed layer; (4) and (5) with (7c)

$$u^2/r \sim v^2/r \sim \nu u/\delta_K^2, \quad v \sim r\sqrt{\alpha_0}, \quad \delta_K := \nu^{1/2}\alpha_0^{-1/4} \qquad (14a\text{–}c)$$

for a BL with thickness δ_K where rotation is dominantly at play (generalisation of the classical self-preserving von-Kármán-type BL underneath a stagnant flow, see [9]). We now advantageously consider three flow regimes.

(I) $r \sim r_\nu := \nu^{-1/3}$: there the dominant effect of viscous shear extends to the whole layer as long as rotation is so weak that $u \sim 1$ (from (13) with $u \sim 1$).

(II) $r \sim r_c := 1/\sqrt{\alpha_0}$: there the centrifugal force induces a u-component of $O(1)$ in the BL (from (14b) and $u \sim v \sim r\sqrt{\alpha_0}$).

(III) $r \sim r_r := \alpha_0^{-1/8}\nu^{-1/4}$: there the effect of rotation becomes important across the entire layer (from (13a) and (14) for $\delta_K \sim h$).

From this the useful categorisation (A)–(D) ensues, differing by the magnitude of r where rotation becomes effective:

(A) $\alpha_0 \ll \nu^{2/3}$, $(r_c \gg) r_r \gg r_\nu$ (slowly rotating film): relatively far downstream of the formation of a developed film.

(B) $\alpha_0 \sim \nu^{2/3}$, $r_c \sim r_r \sim r_\nu$ (moderately rotating film): where the developed film emerges (least-degenerate case).

(C) $\alpha_0 \gg \nu^{2/3}$, $r_c \ll r_r (\ll r_\nu)$ (rapidly rotating film): where the BL is still much thinner than the film. That is, the BL evolves towards a von-Kármán BL ($u \sim v \gg 1$) that has entrained the whole film for $r \sim r_r$.

(D) $1 \ll \alpha_0 \ll \nu^{-2}$ (very rapidly rotating film): according to (14b), the BL has already assumed von-Kármán form beneath jet impingement.

Finally, if r_r becomes of $O(1)$, i.e. α as large as ν^{-2}, the BL rotates so fast that it already sucks in the impinging jet and prevents it from being deflected in an essentially inviscid fashion. This scenario renders the current flow description invalid.

As suggested by the generic case (B) and item (II), we scrutinise the regime of competing inertial, centrifugal, and viscous forces, characterised by the coordinate stretching $(R, Z) := (\nu^{1/3}r, 2\nu^{-1/3}z) = O(1)$. Accordingly, we introduce the similarity parameter $\alpha := v^{-2/3}\alpha_0 = O(1)$, so that the flow regimes (A)–(D) and the associated terminology are readily identified below. Substituting the appropriate expansions $[\psi, u, v, w] \sim [\Psi/2, U, \alpha R V, v^{-2/3}W](R, Z)$, $h \sim v^{1/3}H(R)/2$ into (3)–(6) shows that the rescaled flow quantities satisfy a parabolic shallow-water problem as pressure forces remain negligibly small. This governs perfectly super-critical flow controlled by α and the vorticity introduced by the jet:

$$R[U, W] = [\Psi_Z, -\Psi_R], \tag{15}$$

$$UU_R + WU_Z - \alpha R V^2 = U_{ZZ}, \tag{16}$$

$$UV_R + WV_Z + 2UV/R = V_{ZZ} \tag{17}$$

are supplemented with BCs given by (7b–d) and the leading-order forms of (8), (9) and ICs capturing the overlap of the thin-film region with that of inviscid jet deflection;

$$Z = 0: \ \Psi = U = 0, \ V = 1, \quad Z = H(R): \ \Psi = 1, \ U_Z = V_Z = 0, \tag{18}$$

$$R \to 0: \ [U, V, H] \sim [f'(\zeta)\,\mathrm{sgn}(\zeta), 1 - \mathrm{sgn}(\zeta), C/R], \quad \zeta := Z/H(R). \tag{19}$$

By the absence of an upstream influence, a downstream condition as (11c) is ignored, i.e. the disc considered as infinitely large. Adopting an advanced Keller–Box scheme and advantageously utilising the coordinates R, ζ for marching downstream achieves highly accurate numerical solutions of (15)–(19). However, the ICs complying with the no-slip conditions in (18) render the problem singular, and its regularisation involves the BL structure discussed above. This is recovered for $R \to 0$ but itself becomes singular and splits into several flow regions in the limit $\alpha \to \infty$. In agreement with the scenarios (C) and (D), its full resolution involves the subtleties of matching exponentially varying terms (under investigation). In our numerical solutions for situation (i) with $\bar{g} \to 0$ above, an adaptive refining of the discretisation introduces the BL in an automated way. (This strategy and hence the numerical resolution for $R \ll 1$ are obviously compromised for rather large values of α.)

Equation (16) decouples from (17) for vanishingly small disc spin ($\alpha \to 0$). In the case $\alpha = v \equiv 0$ the subsequent analysis bears substantial analogies to but differs in subtle details from its counterpart applied to a planar layer; cf. [2, 4, 7]. These originate in the radial divergence of the flow, solely expressed by the factors r in (3).

As confirmed by Fig. 2a, for rather high disc spin, the radially increasing centrifugal force accelerates the film flow throughout. If α is lowered, here if $\sqrt{\alpha}$ falls below $\simeq 0.723$, the viscous shear force first decelerates flow before the centrifugal one takes over. Hence, as a remarkable feature of the flow, H then undergoes a maximum. This becomes more pronounced and shifted outwards the lower the disc spin is. Eventually, for zero disc spin, Watson's classical self-similar flow, see [12], is attained quite rapidly. This predicts an unbounded increase of the layer: $H/R^2 \to \bar{C} := \pi/(3\sqrt{3}) \simeq 0.6046\,(R \to \infty)$. In the limit $\alpha \to 0$, we restore

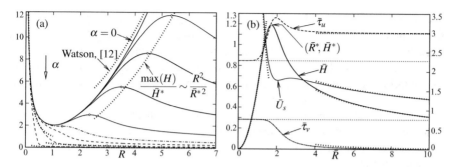

Fig. 2 Numerical solutions for strictly supercritical flow; *dotted* asymptotes refer to Watson's self-preserving flow, inviscid jet bending ($H \sim C/R$), and fully developed flow, cf. (22): (**a**) H vs. R, *solid*: $\sqrt{\alpha} = 0, 0.01, 0.02, 0.05, 0.2$, *dash-dotted*: $\sqrt{\alpha} = 0.723$, *dashed*: $\sqrt{\alpha} = 2, 10, 100$; (**b**) $\bar{H}$, $\bar{U}_s$ (left ordinate) and $\bar{\tau}_u, \bar{\tau}_v$ (right ordinate) vs. $\bar{R}$: canonical representations for slowly rotating flow

(15)–(18) in full to leading order when we put a hat on all quantities therein and α set to unity as we consider the following expansion referring to a slowly rotating film,

$$[\Psi, U, V, W, H] \sim [\bar{\Psi}, \alpha^{3/8}\bar{U}, \bar{V}, \alpha^{1/4}\bar{W}, \alpha^{-1/4}\bar{H}], \quad (\bar{R}, \bar{Z}) := (\alpha^{1/8}R, \alpha^{1/4}Z). \tag{20}$$

Then generic conditions of matching with Watson's flow far upstream replace (19);

$$\bar{R} \to 0: \quad [\bar{\Psi}, \bar{V}, \bar{H}] \sim [f_W(\bar{\zeta}), g_W(\bar{\zeta}), \bar{C}\bar{R}^2], \quad \zeta := \bar{Z}/\bar{H}(\bar{R}), \tag{21}$$

with f_W, g_W denoting Watson's self-similar solution and the raised azimuthal component, governed by (16) for $\alpha = 0$ and (17). Consequently, this modification of the above problem describes the spread of the rotating film in universal form.

For its numerical solution, where $\bar{R}$, $\bar{\zeta}$ serve as independent variables, we refer to Fig. 2b: $[\bar{\tau}_u, \bar{\tau}_v] := [\bar{\Psi}_{\bar{\zeta}\bar{\zeta}}, -\bar{V}_{\bar{\zeta}}](\bar{R}, 0)$ are the suitably scaled wall shear stresses; $\bar{U}_s$ is the surface speed $\bar{U}(R, H)$; the height maximum occurs for $(\bar{R}, \bar{H}) = (\bar{R}^*, \bar{H}^*) \simeq (1.700, 1.191)$, confirmed for even rather moderate values of α; see Fig. 2a.

By viscous diffusion, the predominant centrifugal–shear balance for large values of $\bar{R}$ drives the flow towards its fully developed terminal state where the fluid rotates like a solid body ($\bar{V} \sim 1$), as spotted in Fig. 2. One finds that

$$[\bar{U}\bar{R}^{1/3}, \bar{R}^{8/3}(\bar{V} - 1), \bar{H}\bar{R}^{2/3}] \sim 3^{1/3}[3^{1/3}(\bar{\zeta} - \bar{\zeta}^2/2), -2\bar{\zeta} + \bar{\zeta}^3 - \bar{\zeta}^4/4, 1], \tag{22}$$

where the reminder term is a power series in $\bar{R}^{-8/3}$. This entails the similarity laws $UR^{1/3} = O(\alpha^{1/3})$ and $HR^{2/3} \sim (3/\alpha)^{1/3}$ at play for $R \gg \alpha^{-1/8}$, hence $U \ll \alpha^{3/8}$

and $H \ll \alpha^{-1/4}$. These relationships state $r \gg r_r$, $u \ll r_r\sqrt{\alpha_0}$, $h \ll \delta_K$, according to (III) and (14b,c) above. Then scenario (C) ensures their validity far downstream from film formation for very large values of α. In that case, the von-Kármán BL comprises the whole layer and avoids the formation of its height maximum. During its evolution, the accordingly overshooting U-profiles are gradually transformed into the parabolic one predicted by (22).

The otherwise negligibly weak streamline curvature due to plate-normal momentum transfer and possibly its interplay with the capillary and hydrostatic pressure variations, see (6), perturb the current flow structure by virtue of a strong viscous–inviscid interaction emerging near the real disc edge ($r = r_e$). This accomplishes the gradual transition from the no- to a free-slip condition, cf. [8]. Inspection of (6) and (4) discloses that the associated upstream effect leads to a global failure of the above structure once h has increased such that gh has become of $O(u^2)$ and the developed liquid layer undergoes a viscous hydraulic jump.

3   Upstream Influence: The Toroidal Viscous Hydraulic Jump

In classical hydraulic engineering, the occurrence of the hydraulic jump is referred to a discontinuity separating two states of an inviscid flow in the shallow-water limit, i.e. with the pressure being hydrostatic; see [10]. Thus, the irreversible energetic losses are first construed as being concentrated in the infinitesimally thin interface. This, at a first glance obvious idealisation has led to some fundamental insight and to the celebrated analogy to inviscid gas dynamics, with the Froude number taking the place of the Mach number and the jump that of a shock wave. However, as a closer look shows, it suffers from the severe conceptional shortcoming that shocks manifest exact weak solutions of the Euler equations whereas here their contraction process provokes an unbounded growth of the wall-normal flow component, violating the initially quasi-one-dimensional approach. Therefore, a rational treatment of the jump phenomenon recognises the typical sudden increase of the elevation of the layer as inevitably smeared ("bore") over a finite distance, accounting for its dissipative structure and vanishing if the Reynolds number grows beyond all bounds. The complete description of a marginally weak (transcritical) hydraulic jump within this asymptotic framework of triple-deck theory, namely as an eigensolution settling in a nonlinear fashion essentially by an interplay of gravity and viscous forces only dominant in a thin sublayer adjacent to the wall, was first accomplished in [5].

3.1   The Weakly Elliptic Problem

As a consequence of the aforesaid, in the current setting governed by (1) and (A)–(D) in Sect. 2, the viscous hydraulic "jump" can arise for slowly rotating developed flow. It thus typifies a smooth solution of the shallow-water problem including the hydrostatic pressure gradient on the global characteristic length, the disc radius r_e.

In order to extend the above canonical flow representation for $\alpha_0 \ll \nu^{2/3}$ towards both the case of zero disc spin and the inclusion of the pressure gradient, we formally replace α in (20) by some small perturbation parameter β. With the aid of (6) and (10) and the previous transformations between the original variables and the current ones, this entails the following generalisation of (16):

$$\bar{U}\bar{U}_{\bar{R}} + \bar{W}\bar{U}_{\bar{Z}} - \bar{\alpha}\bar{R}\,\bar{V}^2 = \bar{U}_{\bar{Z}\bar{Z}} - \bar{P}'(\bar{R}), \quad \bar{P}(\bar{R}) := G\bar{H} - S\bar{H}'', \quad (23)$$

$$\bar{\alpha} := \alpha/\beta, \quad G := g\nu^{1/3}/(2\beta), \quad S := \sigma\nu/(2\beta^{3/4}), \quad \bar{R} \le \bar{R}_e := r_e\nu^{1/3}\beta^{1/8}. \,(24)$$

For physically realistic flows, it is expedient to assume $S \ll G$. (However, capillarity can be at play to leading order as streamline curvature and correspondingly drastic changes of $\bar{H}$ render the shallow-water approximation invalid near the disc edge; cf. [8].) We consider a least-degenerate distinguished limit by taking the remaining parameters $\bar{\alpha}$, G, $\bar{R}_e$ as of $O(1)$, where the rescaled disc radius $\bar{R}_e$ controls the inevitable upstream influence if $G > 0$ in an awkward manner elucidated below. Then (23) supplemented with (15), (23), (17), (18), and (21) (overbars added) governs the viscous hydraulic jump. Notably, the film height steepens more abruptly as the typical Froude number $1/G$ becomes smaller. The associated onset of the jump was first appreciated rigorously in [2]; the global structure for its two-dimensional counterpart in [4]. Here some novel insight, tied in with disc rotation, is put forward.

There are three sensible possibilities to define β in (24):

(a) $\beta := \alpha$ means a pronounced hydrostatic effect by the increased film height if $G = O(1)$ or $g = O(\alpha_0/\nu)$;
(b) $\beta := g\nu^{1/3}/(2G)$, $G = const$ measures the impact of rotation on the jump;
(c) $\beta := r_e^{-8}\nu^{-8/3}$ or $\bar{R}_e = 1$ as the most obvious, physically motivated, choice, morphing the original scaling given by the nozzle flow into $\bar{\alpha} = \bar{\alpha}_c := \alpha_0 r_e^8 \nu^2$, $G = G_c := g\,r_e^8\nu^3/2$, $S = S_c := \sigma\,r_e^6\nu^3/2$ (r_e is the characteristic length scale).

The choices (a) and (b) resort to the dominant parabolic structure of the problem and make evident that a viscous jump can only be expected if r_e is not much smaller than r_r, according to the items (III) and (A) in Sect. 2, or $g^{-1/8}\nu^{-3/8}$.

Inspection of (23) for $G > 0$ shows that the marching problem ignoring a downstream condition is ill-posed as a scale shortening arising for $\bar{R} \to 0$ causes the emergence of a sublayer where $\bar{\zeta} = O(G\,\bar{R}^{8/3})$ and convection, pressure and viscous shear predominate. In turn, irregular expansions of strong superexponential variations superimpose the regular ones controlled by Watson's solution. For instance,

$$\frac{\bar{H}}{\bar{C}\bar{R}^2} \sim 1 + \bar{R}^8\left[A\exp\left(-\frac{\Omega}{G^3\bar{R}^{24}}\right) + O(1)\right], \quad \Omega := \frac{3^{17/2}}{2^8\,\pi^{15/2}}\frac{\Gamma(\frac{1}{3})^{12}}{\Gamma(\frac{5}{6})^{15}} \simeq 184.094\,.$$

$$(25)$$

Herein, positive/negative values of the constant A, presently determined by the resolution of our marching scheme, refer to a compressive/expansive branch-off from Watson's self-preserving flow. This inevitably has our marching solutions terminate at some $\bar{R} = \bar{R}_t$, say, in the form of the universal singularity of expansive type discussed extensively in [4]. However, as the associated scale shortening must initiate the aforementioned invalidity of the shallow-water approximation, $\bar{R}_t$ is identified with $\bar{R}_e$ and thus fixes the value of A. Most important, h_e in (11c) can *not* be prescribed but is part of the solution as that singularity is to be met at $\bar{R} = \bar{R}_e$. Hence, a compressive bifurcation having a strength A parametrised by α, G, and $\bar{R}_e$ such that the flow re-expands as $\bar{R} \to \bar{R}_e$ effects the jump. This subtle intrinsic ellipticity or upstream influence generated at the plate edge unequivocally settles the known existing controversies on its position and strength and accompanying flow separation.

The marching procedure for tracking the bifurcation is computationally much less expensive than iterative elliptic schemes including the singularity (as a transient one, see [4]). Efforts are under way to implement (25), allowing for a proper control of A such that a jump indeed forms. Also, the transformation invariance of (23) enables a (preliminary) systematic study; the subscripts c indicate item (c), see Fig. 3:

$$[\bar{R}, \bar{Z}, \bar{H}, \bar{U}, \bar{\alpha}, G] = [\bar{R}_e \bar{R}_c, \ \bar{R}_e^2 \bar{Z}_c, \ \bar{R}_e^2 \bar{H}_c, \ \bar{U}_c/\bar{R}_e^3, \ \bar{\alpha}_c/\bar{R}_e^8, \ G_c/\bar{R}_e^7]. \qquad (26)$$

Viscous diffusion can facilitate a lubrication limit of (23) governed by the order-of-magnitude estimates $\bar{U}\bar{R}\bar{H} \sim 1$, see (13a), $\bar{V} \sim 1$, $\bar{U}^2/\bar{R} \ll G\bar{H}/\bar{R} \sim \bar{\alpha}\bar{R} \sim \bar{U}/\bar{H}^2$. Then $\bar{R} \gg \bar{H}^{1/2}$, i.e. also $\bar{R} \gg \bar{\alpha}^{-1/8}$ and $G \gg \bar{\alpha}$ from integration of (23)

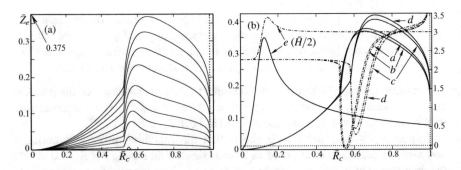

Fig. 3 Jumps by (25): (**a**) $(G, \bar{\alpha}, \bar{R}_e) = (1, 0.027, \simeq 2.008)$, $(\bar{\alpha}_c, G_c) \simeq (7.145, 131.77)$, $\bar{\Psi} = 0$ (encircling separation bubble), 0.002, 0.02, 0.05, 0.1, 0.2, 0.4, 0.6, 0.8, 1; (**b**) $\bar{H}_c$, $\bar{\tau}_u$ as in Fig. 2b, a: $(G, \bar{\alpha}, \bar{R}_e) = (1, 0, \simeq 1.985)$, b: see (**a**), c: $(2, 0.5, \simeq 1.694)$, d: $(8, 2, \simeq 1.413)$, e: $(1, 9, 10)$ (regular)

subject to (18):

$$\bar{U} \sim \frac{3\bar{\alpha}^{1/2}}{G^{1/8}\hat{R}\hat{H}}\left(\bar{\zeta} - \frac{\bar{\zeta}^2}{2}\right), \quad \hat{R} - \frac{d\hat{H}}{d\hat{R}} = \frac{3}{\hat{R}\hat{H}^3}, \quad (\hat{R}, \hat{H}) := \left(\frac{\bar{\alpha}^{1/2}}{G^{3/8}}\bar{R}, G^{1/4}\bar{H}\right).$$

(27)

Hence, $\hat{H}'(\hat{R}) < 0$, $\hat{H}^4 \sim -12 \ln \hat{R}$ in the gravity-dominated upstream limit $\hat{R} \to 0$, which is also discerned for $G \ll 1$ and $\bar{R} = O(1)$, cf. [4], and $\hat{H} \sim (3/\hat{R}^2)^{1/3}$ in the rotation-controlled downstream limit $\hat{R} \to \infty$ of fully developed flow, see (22). As the first cannot be maintained for arbitrarily large values of $\hat{R}$, either the expansive singularity terminates downstream marching or the centrifugal force enters the lubrication approximation further downstream: increased disc rotation diminishes the jump strength and finally suppresses the edge singularity, as the result in Fig. 3b for $G = 1$, $\bar{\alpha} = 9$ confirms and is currently studied in full depth and breadth.

3.2 On the Role of Surface Tension

We multiply (23) with $\bar{R}\bar{U}$, conveniently add $\bar{R}\bar{W}\bar{P}_{\bar{Z}} \equiv 0$, and integrate the so obtained equation over some fraction $\mathcal{V}$ of the whole layer bounded by vertical up- and downstream sections at $\bar{R} = \bar{R}_u$ and $\bar{R} = \bar{R}_d$ $(> \bar{R}_u)$, say. With (15) and the aid of the divergence theorem and a "jump" operator applied to some quantity $Q(\bar{R}, \bar{Z})$,

$$[\![Q]\!] := \bar{R}_d \int_0^{\bar{H}(\bar{R}_d)} Q(\bar{R}_d, \bar{Z})\,d\bar{Z} - \bar{R}_u \int_0^{\bar{H}(\bar{R}_u)} Q(\bar{R}_u, \bar{Z})\,d\bar{Z}, \quad (28)$$

we recover the shallow-water approximation of the mechanical-power budget for $\mathcal{V}$,

$$\tfrac{1}{2}[\![\bar{U}^3]\!] + G[\![\bar{H}]\!] - S[\![\bar{H}'']\!] = \int_{\bar{R}_u}^{\bar{R}_d}\int_0^{\bar{H}(\bar{R})} (\bar{U}\bar{V}^2 - \bar{R}\bar{U}_{\bar{Z}}^2)\,d\bar{Z}\,d\bar{R} \quad (29)$$

as $[\![\bar{U}\bar{R}\bar{U}]\!] = 0$; see (13a): the net flux of the inertial (kinetic and potential) energy as well as the capillary pressure through the confining cross-sections equals the power exerted by the centrifugal (body) force and the negative viscous dissipation in $\mathcal{V}$.

Notably, as seen from the derivation of (29), surface tension *has no impact* on the conversion of mechanical energy under the absence of the Marangoni effect and the associated tangential surface stress. Here the following rationale is illuminating. To this end, we for the moment relax a basic assumption by allowing $S = S(\bar{R})$, i.e. for the Marangoni effect. Then $S'(\bar{R})$ equals the thereby generated Marangoni stress, τ_M, exerted in radial direction on the liquid just below the surface. Multiplying this tangential stress equilibrium for an infinitesimally short surface filament in the shallow-water limit with the surface speed $\bar{U}_s$ yields upon integration by parts

$$(S\,\bar{U}_s)(\bar{R}_u) - (S\,\bar{U}_s)(\bar{R}_d) - \int_{\bar{R}_u}^{\bar{R}_d} \bar{U}_s\,S'\,\mathrm{d}\bar{R} = M := \int_{\bar{R}_u}^{\bar{R}_d} \tau_M\,\mathrm{d}\bar{R}. \tag{30}$$

This relationship states that the powers of externally (first two contributions and $-M$) and internally (third term) acting stresses add up to zero for an infinitesimally thin control cell $\mathcal{C}$ encompassing the considered fraction of the inertia-less free surface, in full agreement with the conventional statement about the conservation of mechanical energy. Conversely, if the control volume $\mathcal{V}$ just excludes $\mathcal{C}$, one would have to add M on the right-hand side of (29) (the capillary pressure acting onto the surface is powerless). Adding this and (30) recombines these two control cells to the original one, $\mathcal{V}$: since the surface powers M and $-M$ cancel, (29) is recovered with the first three terms in (30) added on its right-hand side. Accordingly, these denote the externally and internally effected powers of surface tension. By (30), these sum up to M, thus furnishing the correct extension of (29) if $S' = \tau_M \neq 0$.

Unfortunately, these quite basic findings have apparently been confounded in the very recent (intensely debated) study [1]. The mechanical-power balance therein, see Eq. (5.2) in [1], contains erroneously (amongst other doubtful terms) the external power of surface tension, of $O(\sigma u r)$. Its difference, giving the first two terms in (30), is referred to as "the increase of the surface area". In fact, the cancelling third term in (30) describes the stretching of the surface due to the infinitesimally small relative speed $\bar{U}_s(\bar{R} + \mathrm{d}\bar{R}) - \bar{U}_s(\bar{R})$ of neighbouring fluid particles. However, the power balance in [1] lacks this and the capillary contributions to (29). The latter are of $O(\sigma u r h h'')$ (where (13a) holds). This discloses the queried appearance of the external capillary power as inconsistent with the underlying slender-layer approximation or $|h'| \ll 1$, in [1] postulated tacitly (and unquestioned) even when the control volume is radially contracted. In our opinion, the hydrostatic contributions of $O(g u r h^2)$, see (29), then misled the authors of [1] to their central statement, namely that the local Weber number $We := u^2 h/\sigma$ balances the local Froude number $Fr := u^2/(gh)$ in leading order, obviously in conflict with the correct estimate $\sigma h'' \sim gh$ or $We/Fr \sim |h''| \ll 1$ expressing a hydrostatic–capillary force balance; cf. [3]. Therefore, the authors' novel assertion that "hydraulic jumps result form energy loss due to surface tension" where "gravity plays no significant role" and their associated interpretation of the (admittedly, intriguing) experimental findings must be questioned seriously in the light of their incorrect adoption of the energy balance.

4 Further Outlook

It is the ultimate goal of an advanced asymptotic theory to clarify the transition of the flow passing the edge towards a developed downfall parabola for $G = O(1)$, which completes the overall picture of the viscous hydraulic jump and is topic of our ongoing research. So far, this task has been accomplished partially under the assumption $G \ll 1$, namely when the effect of streamline curvature by disc-normal momentum transfer dominates locally over the gravitational one; see [8]. The rational description of the free-surface eddy or "roller" emerging during the hydraulic jump for sufficiently small Weber numbers immediately upstream of separation—and observed experimentally, but in [4] also numerically by an ad-hoc inclusion of streamline curvature into the classical shallow-water problem—attracts particular interest. Once the laminar hydraulic jump is fully understood under a variation of the Froude and Weber numbers, the laminar–turbulent transition accompanying it provides a most formidable challenge to be mastered with asymptotic techniques.

Acknowledgements The author is grateful to the Austrian Research Promotion Agency (FFG) for granting this research within the *COMET K2* programme (grant no.: 849109, project acronym: *XTribology*).

References

1. Bhagat, R.K., Jha, N.K., Linden, P.F., Wilson, D.I.: On the origin of the circular hydraulic jump in a thin liquid film. J. Fluid Mech. **851**, R5-1–R5-11 (2018). https://doi.org/10.1017/jfm.2018.558
2. Bowles, R.I., Smith, F.T.: The standing hydraulic jump: theory, computations and comparisons with experiments. J. Fluid Mech. **242**, 145–168 (1992). https://doi.org/10.1017/s0022112092002313
3. Bush, J.W.M., Aristoff, J.M.: The influence of surface tension on the circular hydraulic jump. J. Fluid Mech. **489**, 229–238 (2013). https://doi.org/10.1017/S0022112003005159
4. Higuera, F.J.: The hydraulic jump in a viscous laminar flow. J. Fluid Mech. **274**, 69–92 (1994). https://doi.org/10.1017/s0022112094002041
5. Kluwick, A., Cox, E.A., Exner, A., Grinschgl, Ch.: On the internal structure of weakly nonlinear bores in laminar high Reynolds number flow. Acta Mech. **210**(1–2), 135–157 (2010). https://doi.org/10.1007/s00707-009-0188-x
6. Martens, E.A., Watanabe, S., Bohr, T.: Model for polygonal hydraulic jumps. Phys. Rev. E **85**(3), 036316-1–036316-14 (2012). https://doi.org/10.1103/PhysRevE.85.036316
7. Scheichl, B., Kluwick, A.: Laminar spread of a circular liquid jet impinging axially on a rotating disc. J. Fluid Mech. **864**, 449–489 (2019). https://doi.org/10.1017/jfm.2018.1009
8. Scheichl, B., Bowles, R.I., Pasias, G.: Developed liquid film passing a trailing edge under the action of gravity and capillarity. J. Fluid Mech. **850**, 924–953 (2018). https://doi.org/10.1017/jfm.2018.464
9. Tifford, A.N. & Chu, S.T.: On the flow around a rotating disk in a uniform stream. J. Aeron. Sci. **19**(4), 284–285 (1952). https://doi.org/10.2514/8.2255
10. Valiani, A., Caleffi, V.: Free surface axially symmetric flows and radial hydraulic jumps (Technical Note). J. Hydraul. Eng. **142**(4), 06015025-1–06015025-6 (2016). https://doi.org/10.1061/(ASCE)HY.1943-7900.0001104
11. Wang, Y., Khayat, R.E.: Impinging jet flow and hydraulic jump on a rotating disk. J. Fluid Mech. **839**, 525–560 (2018). https://doi.org/10.1017/jfm.2018.43
12. Watson, E.J.: The radial spread of a liquid jet over a horizontal plane. J. Fluid Mech. **20**(3), 481–499 (1964). https://doi.org/10.1017/S0022112064001367

Homogenisation of Parabolic/Hyperbolic Media

Sebastian Franz and Marcus moppi Waurick

Abstract We consider an evolutionary problem with rapidly oscillating coefficients. This causes the problem to change frequently between a parabolic and a hyperbolic state. We prove convergence of the homogenisation process in the unit square and present a numerical method to deal with approximations of the resulting equations. A numerical study finalises the contribution.

MSC (2010) 35M10, 35B35, 35B27, 65M12, 65M60

1 Introduction

In the present article, we discuss an academic example of a partial differential equation with highly oscillatory change of type. Highly oscillatory coefficients of partial differential equations are the central object of study in homogenisation theory, see e.g. [1, 4, 18] for some standard references. In real-world applications a change of type can be observed, when discussing a solid-fluid interaction model. In these kind of models, the solid is modelled by a (hyperbolic) elasticity equation and the fluid is of parabolic type.

S. Franz
Institute of Scientific Computing, TU Dresden, Dresden, Germany
e-mail: sebastian.franz@tu-dresden.de

M. moppi Waurick (✉)
University of Strathclyde, Glasgow, Scotland, UK
e-mail: marcus.waurick@strath.ac.uk

© Springer Nature Switzerland AG 2020
G. R. Barrenechea, J. Mackenzie (eds.), *Boundary and Interior Layers,
Computational and Asymptotic Methods BAIL 2018*, Lecture Notes in
Computational Science and Engineering 135,
https://doi.org/10.1007/978-3-030-41800-7_14

An example of the equations to be studied is the following system of equations in the unit square $\Omega = (0, 1)^2$

$$\partial_t u - \Delta u = f_{\mathrm{w}} \text{ on } \Omega_{\mathrm{w}},$$

$$\partial_t^2 u - \Delta u = f_{\mathrm{b}} \text{ on } \Omega_{\mathrm{b}}.$$

Ω should be thought of being a chessboard like structure with Ω_{w} being the white areas and Ω_{b} being the black areas. u satisfies natural transmission conditions on the interfaces. Our aim is to study the limit of the white and black squares' diameters tending to zero.

We shall present a convergence estimate for this homogenisation problem as well as a numerical study. Equations with change of type (ranging from elliptic to parabolic to hyperbolic) can be treated with the notion of so-called evolutionary equations, which are due to Picard [12], see also [13]. The notion of evolutionary equations is an abstract class of equations formulated in a Hilbert space setting and comprises partial differential-algebraic problems and may further be described as implicit evolution equation, hence the name 'evolutionary equations'.

More precisely, given a Hilbert space H and bounded linear operators $M_0, M_1 \in L(H)$ as well as a skew-self-adjoint operator A in H, we consider the problem of finding $U : \mathbb{R} \to H$ for some given right-hand side $F : \mathbb{R} \to H$ such that

$$(\partial_t M_0 + M_1 + A)U = F, \tag{1}$$

where ∂_t denotes the time derivative. The solution theory for this equation is set up in an exponentially weighted Hilbert space describing space-time. We shall specify the ingredients in the next section.

For evolutionary equations, a numerical framework has been developed in [9]. In particular, this numerical treatment allows for equations with change of type. Other related numerical approaches for these kinds of problems use the language of Friedrich systems, see e.g. [7, 10]. We refer to the introductory part in [9] for a detailed discussion and relation to numerical methods for Friedrich systems.

Qualitatively, problems with highly oscillatory change of type (varying from elliptic/parabolic/hyperbolic) have been considered in [15] in a one-dimensional setting. For a higher-dimensional setting of highly-oscillatory type in the context of Maxwell's equations, we refer to [16]. For a solid-fluid interaction homogenisation with oscillations between hyperbolic and parabolic parts we refer to [6].

For a general systematic account of similar quantitative homogenisation results (without change of type), we refer to [2, 17]. Quantitative results for equations with highly oscillatory change of type have been obtained in [3, 8]. In the latter reference, we have employed results and techniques stemming from [5] to transfer operator-norm estimates on (static) problems posed on $\mathbb{R}^n$ to corresponding estimates for periodic time-dependent problems on the one-dimensional unit cell.

The present contribution is very much in line with the approach presented in [8]. The major difference, however, is the transference to a higher-dimensional setting.

For the sake of the argument, we restrict ourselves to two spatial dimensions. The higher-dimensional case is then adopted without further difficulties.

We shortly comment on the organisation of this paper. We start by presenting the analytical background in the next section. In this section, we shall also derive the necessary convergence estimates for the homogenisation problem.

Our numerical approach will be provided in Sect. 3. We conclude the article with a small case study.

2 Analytical Background

In this section, we rephrase and summarise some results from [5]. The key ingredients are [5, Theorem 3.9] as well as [5, Proposition 3.16].

First of all, we properly define the operators involved. Let $\Omega = (0, 1)^2$. Then we define

$$\widetilde{\mathrm{grad}} \colon C^1_\#(\Omega) \subseteq L^2(\Omega) \to L^2(\Omega)^2, \phi \mapsto (\partial_j \phi)_{j \in \{1,2\}},$$

where $C^1_\#(\Omega) := \{\phi|_\Omega; \phi \in C^1(\mathbb{R}^2), \phi(\cdot) = \phi(\cdot + k) \quad (k \in \mathbb{Z}^2)\}$. Note that $\widetilde{\mathrm{grad}}$ is densely defined as $C^1_c(\Omega) \subseteq C^1_\#(\Omega)$. We define $\mathrm{div}_\# := -\widetilde{\mathrm{grad}}^*$. It is easy to see, that $C^1_\#(\Omega)^2 \subseteq \mathrm{dom}(\mathrm{div}_\#)$ and so $\mathrm{grad}_\# := -\mathrm{div}_\#^*$ is a well-defined operator extending $\widetilde{\mathrm{grad}}$. Note that it can be shown that

$$\mathrm{dom}(\mathrm{grad}_\#) = H^1_\#(\Omega) := \{\phi|_\Omega; \phi \in H^1_{\mathrm{loc}}(\mathbb{R}^2), \phi(\cdot) = \phi(\cdot + k) \quad (k \in \mathbb{Z}^2)\}.$$

Next, let $s_0, s_1 \colon \mathbb{R}^2 \to \mathbb{C}$ be measurable, bounded, $(0, 1)^2$-periodic functions satisfying $s_0(x) = s_0(x)^* \geq 0$ for all $x \in \mathbb{R}^2$ and

$$\rho_0 s_0(x) + \Re s_1(x) \geq c$$

for some $\rho_0 \geq 0$ and $c > 0$ and all $x \in \mathbb{R}^2$.

We define $M_0 \in L(L^2(\Omega)^3)$ by

$$M_0(\phi)_{j \in \{1,2,3\}} := \begin{pmatrix} s_0 & 0 \\ 0 & 1 \end{pmatrix} \begin{pmatrix} \phi_1 \\ (\phi_j)_{j \in \{2,3\}} \end{pmatrix} := \begin{pmatrix} (\Omega \ni x \mapsto s_0(x)\phi_1(x)) \\ (\phi_j)_{j \in \{2,3\}} \end{pmatrix},$$

and M_1 similarly replacing s_0 by s_1. Note that we have

$$\rho_0 M_0 + \Re M_1 \geq c$$

in the sense of positive definiteness; furthermore M_0 is selfadjoint. A straight forward application of [12, Solution Theory] leads to the following result. We recall that ∂_t is the distributional derivative with respect to the first variable in the space

$$L^2_\rho(H) := \{f \in L^2_{\text{loc}}(\mathbb{R}; H); \int_{\mathbb{R}} \|f(t)\|^2_H \exp(-2vt)dt < \infty\}$$

with maximal domain, that is,

$$\partial_t : H^1_\rho(H) \subseteq L^2_\rho(H) \to L^2_\rho(H), \phi \mapsto \phi'.$$

It will be obvious from the context, which ρ and which Hilbert space H is chosen. In the next theorem, we have $H = L^2(\Omega)^3$.

Theorem 1 ([12]) *Let $\rho \geq \rho_0$. Then*

$$S := \overline{\partial_t M_0 + M_1 + \begin{pmatrix} 0 & \text{div}_\# \\ \text{grad}_\# & 0 \end{pmatrix}}^{-1} \in L(L^2_\rho(L^2(\Omega)^3))$$

and $\|S\| \leq 1/c$.

Remark 2 Note that it can be shown [15, Remark 2.3] that if $F \in \text{dom}(\partial_t)$ then $SF \in \text{dom}(\partial_t) \cap \text{dom}(\begin{pmatrix} 0 & \text{div}_\# \\ \text{grad}_\# & 0 \end{pmatrix})$. Moreover, let $F = (f, 0) \in L^2_\rho(L^2(\Omega) \oplus L^2(\Omega)^2)$. Then $U = (u, v) = SF$ satisfies the following two equations

$$\partial_t s_0 u + s_1 u + \text{div}_\# v = f$$

$$\partial_t v = -\text{grad}_\# u.$$

Substituting the second equation into the first one, we obtain

$$\partial^2_t s_0 u + \partial_t s_1 u - \text{div}_\# \text{grad}_\# u = \partial_t f, \tag{2}$$

which is a damped wave equation, if $s_0 > 0$ everywhere. The conditions imposed on s_0 and s_1, however, also allow for regions, where $s_0 = 0$ (or $s_0 = 0$ entirely). On these regions, the equation is a heat-type equation. If there are regions where either s_0 or s_1 vanish (but not both on the same region), the resulting equation is of mixed type.

Coming back to the example from the introduction, we set $s_0 = \chi_b$ to be the characteristic function of Ω_b and $s_1 = \chi_w$ the characteristic function of Ω_w, where Ω_b and Ω_w are open, disjoint sets such that $\chi_w + \chi_b = 1$ almost everywhere. On Ω_b Eq. (2) reads

$$\partial^2_t u - \text{div}_\# \text{grad}_\# u = \chi_b \partial_t f,$$

similarly on Ω_w we obtain

$$\partial_t u - \operatorname{div}_\# \operatorname{grad}_\# u = \chi_w \partial_t f.$$

In order to understand the condition on the interface, we just read off $u \in$ $\operatorname{dom}(\operatorname{grad}_\#)$; which is in particular a continuity-type condition on the interface (if the latter was smooth enough). Note that in order to solve the equation in the first place, *no* regularity assumption on either Ω_b or Ω_w (or their boundary) was needed. We emphasise, that transmission conditions are not necessary for the formulation of the equations but are rather a consequence of U being a solution to the equation; see also [15, Remark 3.2].

Next, we aim to study the limit behaviour of $\mathcal{S}_N$, which is given as $\mathcal{S}$ but with $s_0(N\cdot)$ and $s_1(N\cdot)$ respectively replacing s_0 and s_1. In particular, our aim is to establish the following theorem. For this, we define

$$H_\rho^k(H) := \operatorname{dom}(\partial_t^k)$$

endowed with the graph norm of ∂_t^k acting as an operator from $L_\rho^2(H)$ into itself. It can be shown that given $\rho > \rho_0$ that ∂_t is continuously invertible in $L_\rho^2(H)$; so that $u \mapsto \|\partial_t^k u\|$ is equivalent to the graph norm on $H_\rho^k(H)$.

Theorem 3 *Let $\rho > \rho_0$. There exists $\kappa \geq 0$ such that for all $N \in \mathbb{N}$ and $f \in H_\rho^2(L^2(\Omega))$ we have*

$$\left\| \left(\partial_t \begin{pmatrix} s_0(N\cdot) & 0 \\ 0 & 1 \end{pmatrix} + \begin{pmatrix} s_1(N\cdot) & 0 \\ 0 & 0 \end{pmatrix} + \begin{pmatrix} 0 & \operatorname{div}_\# \\ \operatorname{grad}_\# & 0 \end{pmatrix} \right)^{-1} \begin{pmatrix} f \\ 0 \end{pmatrix} \right.$$

$$\left. - \left(\partial_t \begin{pmatrix} \langle s_0 \rangle & 0 \\ 0 & 1 \end{pmatrix} + \begin{pmatrix} \langle s_1 \rangle & 0 \\ 0 & 0 \end{pmatrix} + \begin{pmatrix} 0 & \operatorname{div}_\# \\ \operatorname{grad}_\# & 0 \end{pmatrix} \right)^{-1} \begin{pmatrix} f \\ 0 \end{pmatrix} \right\|_{L_\rho^2(L^2(\Omega)^3)}$$

$$\leq \frac{\kappa}{N} \|\partial_t^2 f\|_{L_\rho^2(L^2(\Omega))},$$

where

$$\langle s_j \rangle := \int_\Omega s_j(x)dx \quad (j \in \{0, 1\}).$$

In order to prove this theorem, we need to introduce the Fourier–Laplace transformation: Let H be a Hilbert space. For $\phi \in C_c(\mathbb{R}; H)$ we define

$$\mathcal{L}_\rho \phi(\xi) := \frac{1}{\sqrt{2\pi}} \int_\mathbb{R} \phi(t) \exp(-it\xi - \rho t)dt.$$

A variant of Plancherel's theorem yields that $\mathcal{L}_\rho$ extends to a unitary operator from $L^2_\rho(H)$ into $L^2(H)$. A remarkable property of $\mathcal{L}_\rho$ is that

$$\partial_t = \mathcal{L}^*_\rho(im + \rho)\mathcal{L}_\rho,$$

where m is the multiplication by argument operator in $L^2(\mathbb{R}; H)$ with maximal domain; see [11, Corollary 2.5]. Thus, applying the Fourier–Laplace transformation to the norms on either side of the inequality in Theorem 3, we deduce that it suffices to show that there exists $\kappa \geq 0$ such that for all $N \in \mathbb{N}$, $z \in \mathbb{C}_{\Re \geq \rho}$ and $f \in L^2(\Omega)$ we have

$$\left\| \left(z \begin{pmatrix} s_0(N\cdot) & 0 \\ 0 & 1 \end{pmatrix} + \begin{pmatrix} s_1(N\cdot) & 0 \\ 0 & 0 \end{pmatrix} + \begin{pmatrix} 0 & \mathrm{div}_\# \\ \mathrm{grad}_\# & 0 \end{pmatrix} \right)^{-1} \begin{pmatrix} f \\ 0 \end{pmatrix} \right.$$

$$\left. - \left(z \begin{pmatrix} \langle s_0 \rangle & 0 \\ 0 & 1 \end{pmatrix} + \begin{pmatrix} \langle s_1 \rangle & 0 \\ 0 & 0 \end{pmatrix} + \begin{pmatrix} 0 & \mathrm{div}_\# \\ \mathrm{grad}_\# & 0 \end{pmatrix} \right)^{-1} \begin{pmatrix} f \\ 0 \end{pmatrix} \right\|_{L^2(\Omega)^3} \tag{3}$$

$$\leq \frac{\kappa}{N} \| z^2 f \|_{L^2(\Omega)}.$$

This inequality will be shown using the results of [5]. For this we need some auxiliary statements.

Lemma 4 *The space* $\mathrm{ran}(\mathrm{grad}_\#)$ *is closed on* $L^2(\Omega)^2$.

Proof Since Ω has continuous boundary, we get that $H^1(\Omega)$ embeds compactly into $L^2(\Omega)$. Since $\mathrm{grad}_\# \subseteq \mathrm{grad}$, where $\mathrm{grad} \colon H^1(\Omega) \subseteq L^2(\Omega) \to L^2(\Omega)^2$ is the distributional gradient and $\mathrm{grad}_\#$ is closed, we obtain that $H^1_\#(\Omega)$ is compactly embedded into $L^2(\Omega)$, as well. It is now standard to show that $\mathrm{ran}(\mathrm{grad}_\#) \subseteq L^2(\Omega)^2$ is closed, see e.g. [14, Lemma 4.1(b)].

Using Lemma 4, we define

$$\iota \colon \mathrm{ran}(\mathrm{grad}_\#) \hookrightarrow L^2(\Omega)^2, \phi \mapsto \phi$$

and obtain that

$$\iota^* \colon L^2(\Omega)^2 \to \mathrm{ran}(\mathrm{grad}_\#)$$

is the (surjective) orthogonal projection according to the decomposition $L^2(\Omega)^2 = \ker(\mathrm{div}_\#) \oplus \mathrm{ran}(\mathrm{grad}_\#)$.

Proposition 5 ([5, Proposition 3.8]) *Let* $f \in L^2(\Omega)$. *Then the following conditions are equivalent:*

(i) $u \in \mathrm{dom}(\mathrm{div}_\# \, \mathrm{grad}_\#)$ *satisfies*

$$- \mathrm{div}_\# \, \mathrm{grad}_\# \, u + z^2 s_0 + z s_1 u = f$$

(ii) $u \in \text{dom}(\text{grad}_\#)$ and $q \in \text{dom}(\text{div}_\#)$ satisfy

$$
\left(\begin{pmatrix} z s_0 + s_1 & 0 \\ 0 & z \end{pmatrix} + \begin{pmatrix} 0 & \text{div}_\# \\ \text{grad}_\# & 0 \end{pmatrix} \right) \begin{pmatrix} u \\ q \end{pmatrix} = \begin{pmatrix} z^{-1} f \\ 0 \end{pmatrix}.
$$

(iii) $u \in \text{dom}(\text{grad}_\#)$ and $q \in \text{dom}(\text{div}_\#) \cap \text{ran}(\text{grad}_\#)$ satisfy

$$
\left(\begin{pmatrix} z s_0 + s_1 & 0 \\ 0 & z \end{pmatrix} + \begin{pmatrix} 0 & \text{div}_\# \, \iota \\ \iota^* \, \text{grad}_\# & 0 \end{pmatrix} \right) \begin{pmatrix} u \\ q \end{pmatrix} = \begin{pmatrix} z^{-1} f \\ 0 \end{pmatrix}.
$$

Proof The equivalence of 1 and 2 follows from [5, Proposition 3.8] by multiplying 1 by z^{-1} and by putting $\varepsilon = 1$, $\theta = 0$, $n = 1$, $s = z s_0 + s_1$ and $a = z^{-1}$ in [5, Proposition 3.8]. The implication from 3 to 1 follows upon realising that $\text{div}_\# \, \text{grad}_\# = \text{div}_\# \, \iota \iota^* \, \text{grad}_\#$. Thus, it remains to establish that 2 is sufficient for 3. For this implication, however, that the second equation in 2, implies that $zq \in \text{ran}(\text{grad}_\#)$ and, hence, $q \in \text{ran}(\text{grad}_\#)$. Therefore, $zq = \iota^* zq = z\iota^* q$.

Next, we introduce the Floquet–Bloch or Gelfand transformation:

Definition 6 Let $N \in \mathbb{N}$, $f : \mathbb{R}^2 \to \mathbb{C}$. Then define

$$
\mathcal{V}_N f(\theta, y) := \frac{1}{N} \sum_{k \in \{0, \dots, N-1\}^2} f(y + k) e^{-i\theta \cdot k}
$$

$$
(y \in [0, 1)^2, \theta \in \{2\pi k / N; k \in \{0, \dots, N-1\}^2\})
$$

and for $f \in L^2(0, 1)$

$$
\mathcal{T}_N f := \frac{1}{N} f \left(\frac{\cdot}{N} \right).
$$

As in [8] one can show the following result:

Theorem 7

(a) The mapping $\mathcal{V}_N : L^2_\#((0, N)^2) \to L^2(0, 1)^{N^2}$ given by

$$
f \mapsto (\mathcal{V}_N f(2\pi k / N, \cdot))_{k \in \{0, \dots, N-1\}^2}
$$

is unitary, where $L^2_\#((0, N)^2)$ denotes the set of $(0, N)^2$-periodic $L^2_{\text{loc}}(\mathbb{R}^2)$ functions endowed with the scalar product from $L^2((0, N)^2)$.

(b) The mapping $G_N := \mathcal{V}_N \mathcal{T}_N$ is unitary.

The mapping G_N in the previous theorem is also called the Floquet–Bloch or Gelfand transformation. With this transformation at hand, we are in the position to transform the inequality in (3) into an equivalent form such that [5, Section 2] is applicable. The reason is the following representation:

Proposition 8 ([5] and [8]) *Let* $N \in \mathbb{N}$, $k \in \{0, \ldots, N-1\}^2$ *and* $\theta := 2\pi k/N$, $f \in L^2(\Omega)$. *Then we have*

$$
\left(G_N \left(\left(\begin{matrix} zs_0(N\cdot) + s_1(N\cdot) \ 0 \\ 0 \ z \end{matrix} \right) + \left(\begin{matrix} 0 \ \mathrm{div}_{\#}\, \iota \\ \iota^* \,\mathrm{grad}_{\#} \ 0 \end{matrix} \right) \right)^{-1} \left(\begin{matrix} f \\ 0 \end{matrix} \right) G_N^* \right)_k
$$
$$
= \left(\left(\begin{matrix} zs_0(\cdot) + s_1(\cdot) \ 0 \\ 0 \ z \end{matrix} \right) + \frac{1}{N} \left(\begin{matrix} 0 \ \mathrm{div}_\theta\, \iota_\theta \\ \iota_\theta^* \,\mathrm{grad}_\theta \ 0 \end{matrix} \right) \right)^{-1} \left(\begin{matrix} (G_N f G_N^*)_k \\ 0 \end{matrix} \right),
$$

where div_θ *and* grad_θ *as well as* ι_θ *are given as in [5, Section 3].*

Proof Let $(u, q) := \left(\left(\begin{matrix} zs_0(N\cdot) + s_1(N\cdot) \ 0 \\ 0 \ z \end{matrix} \right) + \left(\begin{matrix} 0 \ \mathrm{div}_{\#}\, \iota \\ \iota^* \,\mathrm{grad}_{\#} \ 0 \end{matrix} \right) \right)^{-1} \left(\begin{matrix} f \\ 0 \end{matrix} \right)$. By Proposition 5, we have that

$$
-\mathrm{div}_{\#} \,\mathrm{grad}_{\#}\, u + z^2 s_0(\cdot)u + z s_1(\cdot)u = zf, \quad zq = -\mathrm{grad}_{\#}\, u.
$$

Then, by the argument just after [5, Proposition 3.5] (use an adapted version of [5, Proposition 3.5], where the Gelfand transform used there is replaced by the discrete version introduced here), it follows that $u_k := (G_N u G_N^*)_k$ satisfies

$$
-\frac{1}{N^2} \,\mathrm{div}_\theta \,\mathrm{grad}_\theta\, u_k + z^2 s_0(\cdot)u_k + z s_1(\cdot)u_k = z(G_N f G_N^*)_k =: z f_k.
$$

Applying G_N to $zq = -\mathrm{grad}_{\#}\, u$, we obtain

$$
z(G_N q)_k = -\mathrm{grad}_\theta (G_N u)_k = -\iota_\theta^* \,\mathrm{grad}_\theta (G_N u)_k,
$$

which yields the assertion.

Now, along the lines of [5, Section 3] it is possible to show the following result, which eventually implies Theorem 3.

Theorem 9 ([5, Proof of Theorem 3.1; Eq (14)] and [8, Proof of Theorem 3.10]) *There exists* $\kappa \geq 0$ *such that for all* $N \in \mathbb{N}$, $k \in \{0, \ldots, N-1\}^2$ *with* $\theta = 2\pi k/N$ *and* $f \in L^2(\Omega)$ *we have*

$$
\left\| \left(\left(\begin{matrix} zs_0(\cdot) + s_1(\cdot) \ 0 \\ 0 \ z \end{matrix} \right) + \frac{1}{N} \left(\begin{matrix} 0 \ \mathrm{div}_\theta\, \iota_\theta \\ \iota_\theta^* \,\mathrm{grad}_\theta \ 0 \end{matrix} \right) \right)^{-1} \right.
$$
$$
\left. - \left(\left(\begin{matrix} z\langle s_0 \rangle + \langle s_1 \rangle \ 0 \\ 0 \ z \end{matrix} \right) + \frac{1}{N} \left(\begin{matrix} 0 \ \mathrm{div}_\theta\, \iota_\theta \\ \iota_\theta^* \,\mathrm{grad}_\theta \ 0 \end{matrix} \right) \right)^{-1} \right\| \leq \frac{\kappa}{N} \|z^2 f\|_{L^2(\Omega)}.
$$

With this theorem, the assertion of Theorem 3 follows upon applying the inverse Gelfand transformation first and afterwards the inverse Fourier–Laplace transformation; see also [8, Proof of Theorem 3.10] for the precise argument.

Remark 10 We shall comment on the differences of the one-dimensional case in [8] to the two-dimensional case (or higher-dimensional cases) discussed here. In order to capture the higher-dimensional setting, it is crucial to somehow project away the kernel of the divergence, which in dimensions greater than one is infinite-dimensional. This is the central objective for Proposition 5. Moreover, a suitable variant of the transformation G_N accounting for the dimension of the problem has to be used.

3 Numerical Method

In this whole section, we address solving the equation

$$\left(\partial_t M_0 + M_1 + \begin{pmatrix} 0 & \mathrm{div}_\# \\ \mathrm{grad}_\# & 0 \end{pmatrix}\right) U = F \tag{4}$$

by a numerical method to illustrate the homogenisation process. The analytical results (Theorem 1 and Remark 2) state that given $F \in H^1_\rho(H)$, the solution U of (4) yields $U \in L^2_\rho(H^1_\#(\Omega) \times H(\mathrm{div}_\#, \Omega))$ such that $M_0 U \in H^1_\rho(H)$ with $H = L^2(\Omega)^3$. We will use a discontinuous Galerkin method in time and a conforming Galerkin method in space. For that let $0 = t_0 < t_1 < \cdots < t_M = T$ be a mesh for the time interval $[0, T]$ using M equidistant intervals $I_m = (t_{m-1}, t_m)$ of length $\tau = t_m - t_{m-1} = \frac{T}{M}, m \in \{1, \ldots, M\}$. The method could also be defined on a non-uniform mesh in time with the obvious changes. For the discretisation of $\bar\Omega = [0, 1]^2$ we use an equidistant tensor-product mesh with K^2 mesh-cells $\sigma_{ij} = (x_{i-1}, x_i) \times (y_{j-1}, y_j)$, where $x_i = \frac{i}{K}, i \in \{0, \ldots, K\}$ and $y_j = \frac{j}{K}$, $j \in \{0, \ldots, K\}$. Note that the mesh-size in both space dimensions is $h = \frac{1}{K}$. Again a non-equidistant tensor product mesh with different mesh-sizes in the different dimensions is also possible.

We will approximate $U = (u, v)$ using piecewise polynomials, globally discontinuous in time and piecewise polynomials, globally continuous (H^1-conforming) in space for u and globally $H(\mathrm{div})$-conforming for v. Thus our discrete space is given by

$$\mathcal{U}^{h,\tau} := \left\{ U \in H_\rho([0, T]; H) : U|_{I_m} \in \mathcal{P}_q(I_m, \mathcal{V}_u(\Omega) \times \mathcal{V}_v(\Omega)), m \in \{1, \ldots, M\} \right\},$$

where the spatial spaces are

$$\mathcal{V}_u(\Omega) := \left\{ u \in H_{\#}^1(\Omega) : u|_{\sigma_{ij}} \in \mathcal{Q}_p(\sigma_{ij}), \; 0 \le i, j \le K \right\},$$

$$\mathcal{V}_v(\Omega) := \left\{ v \in H_{\#}(\mathrm{div}, \Omega) : v|_{\sigma_{ij}} \in \mathcal{RT}_{p-1}(\sigma_{ij}), \; 0 \le i, j \le K \right\}.$$

Here, $\mathcal{P}_q(I_m, H)$ is the space of polynomials of degree up to q on the interval I_m with values in H and $\mathcal{Q}_p(\sigma_{ij})$ is the space of polynomials with total degree up to p on the cell $\sigma_{ij} \subseteq \Omega$. Furthermore, $\mathcal{RT}_{p-1}(\sigma_{ij})$ is the Raviart–Thomas space on σ_{ij}, defined by

$$\mathcal{RT}_{p-1}(\sigma_{ij}) = (\mathcal{Q}_{p-1}(\sigma_{ij}))^n + x \mathcal{Q}_{p-1}(\sigma_{ij}).$$

Finally, the "#" denotes periodic boundary conditions. This means, that $w \in \mathcal{V}_u(\Omega)$ fulfils

$$w(0, \zeta) = w(1, \zeta), \; w(\zeta, 0) = w(\zeta, 1), \quad \text{for any } \zeta \in [0, 1]$$

and $w \in \mathcal{V}_v(\Omega)$ fulfils using the outer normal $\boldsymbol{n}$ on $\partial\Omega$

$$(\boldsymbol{n} \cdot w)(0, \zeta) = -(\boldsymbol{n} \cdot w)(1, \zeta), \; (\boldsymbol{n} \cdot w)(\zeta, 0) = -(\boldsymbol{n} \cdot w)(\zeta, 1), \quad \text{for any } \zeta \in [0, 1].$$

With these notions at hand, we can now properly specify the numerical method. For any given right-hand side $F \in \mathcal{U}^{h,\tau}$ and initial condition $x_0 \in H$, find $\mathcal{U} \in \mathcal{U}^{h,\tau}$, such that for all $\Phi \in \mathcal{U}^{h,\tau}$ and $m \in \{1, 2, \ldots, M\}$ it holds

$$Q_m \left[(\partial_t M_0 + M_1 + A)\mathcal{U}, \Phi \right]_\rho + \langle M_0 [\![\mathcal{U}]\!]_{m-1}^{x_0}, \Phi_{m-1}^+ \rangle = Q_m \left[F, \Phi \right]_\rho. \tag{5}$$

Here, we denote by

$$[\![\mathcal{U}]\!]_{m-1}^{x_0} := \begin{cases} \mathcal{U}(t_{m-1}+) - \mathcal{U}(t_{m-1}-), & m \in \{2, \ldots, M\} \\ \mathcal{U}(t_0+) - x_0, & m = 1, \end{cases}$$

the jump at t_{m-1}, by $\Phi_{m-1}^+ := \Phi(t_{m-1}+)$ the right-sided trace and by

$$Q_m [a, b]_\rho := \frac{\tau_m}{2} \sum_{i=0}^q \omega_i^m \langle a(t_{m,i}), b(t_{m,i}) \rangle$$

a right-sided weighted Gauß–Radau quadrature formula on I_m approximating

$$\langle a, b \rangle_{\rho, m} := \int_{t_{m-1}}^{t_m} \langle a(t), b(t) \rangle \exp(-2\rho(t - t_{m-1})) \mathrm{d}t,$$

see [9] for further details.

We can cite the convergence results from [9] which were for Dirichlet boundary conditions. The proof needs only marginal modifications to hold for the periodic case too. We introduce two measures for the error. The first one measures the error in an L^∞-L^2 sense, while the second one is a discrete version of the $L^2_\rho(H)$-norm:

$$E^2_{\sup}(a) := \sup_{t\in[0,T]} \langle M_0 a(t), a(t) \rangle,$$

$$E^2_Q(a) := e^{2\rho T} \sum_{m=1}^{M} Q_m [a, a]_\rho \, e^{-2\rho t_{m-1}}.$$

Note that $E_Q(a) = \|a\|$ for $a \in \mathcal{U}^{h,\tau}$.

Theorem 11 *We assume for the solution U of Example (4) the regularity*

$$U \in H^1_\rho(H^p_\#(\Omega) \times H^p_\#(\Omega)^2) \cap H^{q+3}_\rho(L^2(\Omega) \times L^2(\Omega)^2)$$

as well as

$$AU \in L^2_\rho(H^p_\#(\Omega) \times H^p_\#(\Omega)^2).$$

Then we have for the error of the numerical solution $U^{h,\tau}$ of (5) with a generic constant C

$$E^2_{\sup}(U - U^{h,\tau}) + E^2_Q(U - U^{h,\tau}) \leq C e^{2\rho T} (\tau^{2(q+1)} + T h^{2p}).$$

Note that the spatial regularity is only needed in each cell σ_{ij} of the spatial mesh as local interpolation error estimates are used.

4 Numerical Study

All computations were done in SOFE (https://github.com/SOFE-Developers/SOFE), a finite element suite for Matlab/Octave.

For our numerical study let us assume an equidistant rectangular background mesh covering Ω with nodes $(\tilde{x}_i = \frac{i}{N}, \tilde{y}_j = \frac{j}{N}), i, j \in \{0, \dots, N\}$ for an even number $N \in \mathbb{N}$. This background mesh will be used in defining the oscillating coefficients.

Our rough coefficient problem is given by

$$\left(\partial_t \begin{pmatrix} \epsilon_N & 0 \\ 0 & 1 \end{pmatrix} + \begin{pmatrix} 1 - \epsilon_N & 0 \\ 0 & 0 \end{pmatrix} + \begin{pmatrix} 0 & \mathrm{div}_\# \\ \mathrm{grad}_\# & 0 \end{pmatrix} \right) U_N = \begin{pmatrix} f \\ 0 \end{pmatrix}, \tag{6}$$

where the coefficient function ϵ_N is defined as

$$\epsilon_N(x, y) := \begin{cases} 1, & \exists i, j \in \mathbb{N}_0 : (x, y) \in (\tilde{x}_i, \tilde{x}_{i+1}) \times (\tilde{y}_j, \tilde{y}_{j+1}) \text{ and } i + j \text{ is even,} \\ 0, & \text{otherwise.} \end{cases}$$

The corresponding homogenised problem is then

$$\left(\partial_t \begin{pmatrix} \frac{1}{2} & 0 \\ 0 & 1 \end{pmatrix} + \begin{pmatrix} \frac{1}{2} & 0 \\ 0 & 0 \end{pmatrix} + \begin{pmatrix} 0 & \text{div}_\# \\ \text{grad}_\# & 0 \end{pmatrix} \right) U_{\text{hom}} = \begin{pmatrix} f \\ 0 \end{pmatrix}. \tag{7}$$

The theoretical results of Sects. 2 and 3 provide the following expected convergence behaviour

$$\|U_N - U_{\text{hom}}\|_{H^1_\rho(\mathbb{R}, H)} \le CN^{-1},$$

$$E_Q(U^{h,\tau}_{\text{hom}} - U_{\text{hom}}) \le C(h^p + \tau^{q+1}), \quad E_Q(U^{h,\tau}_N - U_N) \le C(h^p + \tau^{q+1})$$

for smooth solutions U_{hom} and U_N. In general we cannot expect the solutions to be very smooth. Thus, for our experiments we only chose a polynomial order $p = 2$ in space and $q = 1$ in time. Setting furthermore $h = \tau = 1/(2N)$ (and thus $K = 2N$, $M = 3N$ for $T = 3/2$ which couples the numerical mesh and the background mesh) we combine the above expected estimates and obtain

$$E_Q(U^{h,\tau}_N - U_{\text{hom}}) \le E_Q(U^{h,\tau}_N - U_N) + E_Q(U_N - U_{\text{hom}})$$

$$\le E_Q(U^{h,\tau}_N - U_N) + C\|U_N - U\|_{H^1_\rho(\mathbb{R}, H)} \le CN^{-1},$$

where the second inequality comes from Sobolev's embedding theorem (see e.g. [11, Lemma 5.2]). Note that one should always use an integer-multiple of N for the number K of mesh cells in order to catch the jumps in the coefficients of U_N.

Let us finalise the definition of our problem by setting $T = 3/2$ and the right-hand sides

$$f(t, x, y) = \begin{cases} 1, & t \in (0, 1) \text{ and } \max\{|2x - 1|, |2y - 1|\} \le \frac{1}{4}, \\ 0, & \text{otherwise.} \end{cases}$$

Thus f is one in the time-space cube $(0, 1) \times [1/4, 3/4]^2$ and otherwise zero. Figure 1 shows (numerical approximations by our method of) the solutions U_4, U_8, U_{16} and U_{hom} at different times. In the first row the rough coefficients can be seen quite nicely, while the solution becomes smooth very quickly (lower rows). Furthermore, already for a very coarse background mesh of $N = 16$ the solutions U_N and U_{hom} are very similar. This visualises the homogenisation process.

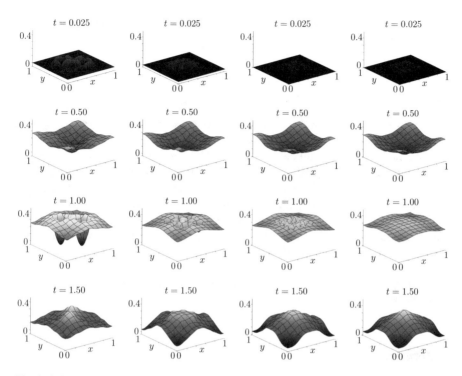

Fig. 1 Solutions U_4, U_8, U_{16} and U_{hom} (left to right) at $t = 0.025$, 0.5, 1 and 1.5 (top to bottom)

Table 1 Convergence results for $\tilde{U}_{\text{hom}} - U_N^{h,\tau}$ and $\tilde{U}_N - U_N^{h,\tau}$ of problem (6) using $h = \tau = \frac{1}{2N}$

N	$E_{\text{sup}}(\tilde{U}_{\text{hom}} - U_N^{h,\tau})$		$E_Q(\tilde{U}_{\text{hom}} - U_N^{h,\tau})$		$E_{\text{sup}}(\tilde{U}_N - U_N^{h,\tau})$		$E_Q(\tilde{U}_N - U_N^{h,\tau})$	
2	7.175e−02		2.778e−02		5.046e−02		1.336e−02	
4	4.391e−02	0.71	1.969e−02	0.50	2.346e−02	1.11	6.692e−03	1.00
8	2.256e−02	0.96	8.802e−03	1.16	1.171e−02	1.00	3.165e−03	1.08
16	1.038e−02	1.12	4.186e−03	1.07	6.063e−03	0.95	1.507e−03	1.07
32	5.081e−03	1.03	2.005e−03	1.06	3.172e−03	0.93	6.633e−04	1.18
64	2.383e−03	1.09	9.445e−04	1.09	1.590e−03	1.00	3.012e−04	1.14

In Table 1 we see the results for a simulation using polynomial degrees $p = q+1 = 2$. As no exact solutions to (6) and (7) are known, we use reference solutions $\tilde{U}_N$ and $\tilde{U}_{\text{hom}}$ computed with the same approach and polynomial degree $p = 3$ on a mesh with 256 cells in each space dimension and 384 cells in time dimension. The reference solution mesh is therefore twice as fine as the finest one used in the simulation.

Note that we also provided the experimental orders of convergence (eoc), calculated for errors E_n and E_{2n} by

$$\mathrm{eoc}_n = \frac{\ln \frac{E_n}{E_{2n}}}{\ln 2}.$$

We observe first order convergence of the numerical solution $U_N^{h,\tau}$ towards U_{hom} and towards U_N. The first result confirms the reasoning at the beginning of this section. The second numerical result can be explained by two different facts. First, for our simulation we have coupled the numerical mesh and the background mesh such that both converge simultaneously. Second, numerical simulations investigating the convergence for a fixed N, that means fixed coefficients, do also not show second order convergent errors as expected by Theorem 11 and thus directing to a non-smoothness of the solution U_N. Considering the oscillating coefficients and discontinuous f this reduction is to be expected.

Acknowledgements We thank the anonymous referee for their thorough report helping to considerably improve the manuscript.

References

1. Bensoussan, A., Lions, J.-L., Papanicolaou, G.: Asymptotic analysis for periodic structures. In: Studies in Mathematics and its Applications, vol. 5. North-Holland Publishing Co., Amsterdam (1978)
2. Birman, M.Sh., Suslina, T.A.: Periodic second-order differential operators. Threshold properties and averaging. Algebra i Analiz **15**(5), 1–108 (2003)
3. Cherednichenko, K., Waurick, M.: Resolvent estimates in homogenisation of periodic problems of fractional elasticity. J. Differ. Equ. **264**(6), 3811–3835 (2018)
4. Cioranescu, D., Donato, P.: An introduction to homogenization. In: Oxford Lecture Series in Mathematics and its Applications, vol. 17. The Clarendon Press/Oxford University Press, New York (1999)
5. Cooper, S., Waurick, M.: Fibre homogenisation (2017). arXiv: 1706.00645
6. Dasser, S.: Méthode de pénalisation pour l'homogénéisation d'un problème de couplage fluide-structure. C. R. Acad. Sci. Paris Sér. I Math. **320**(6), 759–764 (1995)
7. Ern, A., Guermond, J.-L.: Discontinuous Galerkin methods for Friedrichs' systems. I. General theory. SIAM J. Numer. Anal. **44**(2), 753–778 (2006)
8. Franz, S., Waurick, M.: Resolvent estimates and numerical implementation for the homogenisation of one-dimensional periodic mixed type problems. ZAMM **98**(7), 1284–1294 (2018)
9. Franz, S., Trostorff, S., Waurick, M.: Numerical methods for changing type systems. IMA J. Numer. Anal. (2018). https://doi.org/10.1093/imanum/dry007
10. Jensen, M.: Discontinuous Galerkin methods for Friedrichs systems with irregular solutions. PhD thesis, University of Oxford, Oxford (2004)
11. Kalauch, A., Picard, R., Siegmund, S., Trostorff, S., Waurick, M.: A Hilbert space perspective on ordinary differential equations with memory term. J. Dyn. Differ. Equ. **26**(2), 369–399 (2014)
12. Picard, R.: A structural observation for linear material laws in classical mathematical physics. Math. Methods Appl. Sci. **32**, 1768–1803 (2009)

13. Picard, R., McGhee, D.: Partial differential equations: a unified Hilbert space approach. In: Expositions in Mathematics, vol. 55. DeGruyter, Berlin (2011)
14. ter Elst, A.F.M., Gorden, G., Waurick, M.: The Dirichlet-to-Neumann operator for divergence form problems. Ann. Mat. Pura Appl. (2018). https://doi.org/10.1007/s10231-018-0768-2
15. Waurick, M.: Stabilization via homogenization. Appl. Math. Lett. **60**, 101–107 (2016)
16. Waurick, M.: Nonlocal H-convergence. Cal. Var. Partial Differ. Equ. (2018). https://doi.org/10.1007/s00526-018-1436-5
17. Zhikov, V.V., Pastukhova, S.E.: On operator estimates in homogenization theory. Uspekhi Mat. Nauk **71**(3(429)), 27–122 (2016)
18. Zhikov, V.V., Kozlov, S.M., Oleinik, O.A., T'en Ngoan, K.: Averaging and G-convergence of differential operators. Russian Math. Surv. **34**(5), 69–147 (1979)

Isogeometric Analysis for Singularly Perturbed Problems in 1-D: A Numerical Study

Klio Liotati and Christos Xenophontos

We perform numerical experiments on one-dimensional singularly perturbed problems of reaction-convection-diffusion type, using isogeometric analysis. In particular, we use a Galerkin formulation with B-splines as basis functions. The question we address is: how should the knots be chosen in order to get uniform, exponential convergence in the maximum norm? We provide specific guidelines on how to achieve precisely this, for three different singularly perturbed problems.

1 Introduction

We consider second order singularly perturbed problems (SPPs) in one-dimension, of reaction-convection-diffusion type, whose solution contains boundary layers (see, e.g. [7]). The approximation of SPPs has received a lot of attention in the last few decades, mainly using finite differences (FDs) and finite elements (FEs) on *layer adapted meshes* (see, e.g. [5]). Various formulations and results are available in the literature, both theoretical and computational [5]. One method that has not, to our knowledge, been applied to general SPPs is *Isogeometric Analysis* (IGA). Since the introduction of IGA by T. R. Hughes et. al. [4], the method has been successfully applied to a large number of problem classes. Even though much attention has been given to convection-dominated problems [1], the method has not been applied, as far as we know, to a typical singularly perturbed problem, such as (3)–(4) ahead.

K. Liotati · C. Xenophontos (✉)
Department of Mathematics and Statistics, University of Cyprus, Nicosia, Cyprus
e-mail: xenophontos@ucy.ac.cy

© Springer Nature Switzerland AG 2020
G. R. Barrenechea, J. Mackenzie (eds.), *Boundary and Interior Layers,
Computational and Asymptotic Methods BAIL 2018*, Lecture Notes in
Computational Science and Engineering 135,
https://doi.org/10.1007/978-3-030-41800-7_15

Our goal in this article is to study the application of IGA to SPPs and in particular the approximation of the solution to (3)–(4) ahead. We use a Galerkin formulation with B-splines as basis functions and select appropriate knot vectors, such that as the polynomial degree increases, the error in the approximation, measured in the maximum norm, decays exponentially, independently of any singular perturbation parameter(s). This is the analog of performing p refinement in the FEM.

The rest of the paper is organized as follows: in Sect. 2 we give a brief introduction to IGA, as described in [1]. In Sect. 3 we present the model problem and its regularity. In Sect. 4 we give the Galerkin formulation and construct the discrete problem. Finally, Sect. 5 shows the results of our numerical computations and Sect. 6 gives our conclusions.

With $I \subset \mathbf{R}$ an interval with boundary ∂I and measure $|I|$, we will denote by $C^k(I)$ the space of continuous functions on I with continuous derivatives up to order k. We will use the usual Sobolev spaces $W^{k,m}(I)$ of functions on Ω with $0, 1, 2, \ldots, k$ generalized derivatives in $L^m(I)$, equipped with the norm and seminorm $\|\cdot\|_{k,m,I}$ and $|\cdot|_{k,m,I}$, respectively. When $m = 2$, we will write $H^k(I)$ instead of $W^{k,2}(I)$, and for the norm and seminorm, we will write $\|\cdot\|_{k,I}$ and $|\cdot|_{k,I}$, respectively. The usual $L^2(I)$ inner product will be denoted by $\langle \cdot, \cdot \rangle_I$, with the subscript omitted when there is no confusion. We will also use the space

$$H_0^1(I) = \left\{ u \in H^1(I) : u|_{\partial \Omega} = 0 \right\}.$$

The norm of the space $L^\infty(I)$ of essentially bounded functions is denoted by $\|\cdot\|_{\infty,I}$. Finally, the letters C, c will be used for generic positive constants, independent of any discretization or singular perturbation parameters.

2   Isogeometric Analysis

In this study we use B-splines as basis functions and follow [1] closely. To this end let $\mathcal{E} = \{\xi_1, \xi_2, \ldots, \xi_{n+p+1}\}$ be a **knot vector**, where $\xi_i \in \mathbf{R}$ is the ith knot, $i = 1, 2, \ldots, N + p + 1$, p is the polynomial order and N is the number of basis functions used to construct the B-spline. The numbers in $\mathcal{E}$ are non-decreasing and may be repeated, in which case we are talking about a *non-uniform* knot vector. If the first and last knot values appear $p + 1$ times, the knot vector is called *open* (see [1] for more details). With a knot vector $\mathcal{E}$ in hand, the B-spline basis functions are defined recursively, starting with piecewise constants ($p = 0$):

$$B_{i,0}(\xi) = \begin{cases} 1, & \xi_i \leq \xi < \xi_{i+1} \\ 0, & otherwise \end{cases}.$$

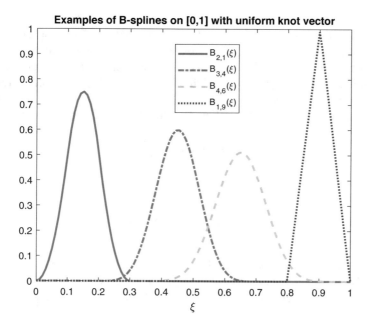

Fig. 1 Examples of B-splines using the knot vector $\Xi = [0, 0.1, 0.2 \ldots 1]$

For $p = 1, 2, \ldots$, they are defined by the *Cox-de Boor recursion formula* [2, 3]

$$B_{i,p}(\xi) = \frac{\xi - \xi_i}{\xi_{i+p} - \xi_i} B_{i,p-1}(\xi) + \frac{\xi - \xi_i}{\xi_{i+p} - \xi_i} B_{i,p-1}(\xi).$$

In Fig. 1 we show some of the above B-splines, obtained with the uniform knot vector $\Xi = [0, 0.1, 0.2 \ldots 1]$ for various polynomial degrees p.

We also mention the recursive formula for obtaining the derivative of a B-spline [1]:

$$\frac{d}{d\xi} B_{i,p}(\xi) = \frac{p}{\xi_{i+p} - \xi_i} B_{i,p-1}(\xi) - \frac{p}{\xi_{i+p+1} - \xi_{i+1}} B_{i+1,p-1}(\xi).$$

We will be considering open knot vectors, having possibly repeated entries (other than the endpoints). If we assume we have $\xi_1, \ldots, \xi_m$ distinct knots, each having multiplicity r_i, then

$$\Xi = [\underbrace{\xi_1, \ldots, \xi_1}_{r_1 \text{ times}}, \underbrace{\xi_2, \ldots, \xi_2}_{r_2 \text{ times}}, \ldots, \underbrace{\xi_m, \ldots, \xi_m}_{r_m \text{ times}}]$$

and there holds $\sum_{i=1}^{m} r_i = N + p + 1$. Since we are using open knots, we have $r_1 = r_m = p + 1$. The regularity of the B-spline at each knot ξ_i is determined by r_i, in that the B-spline has $p - r_i$ continuous derivatives at ξ_i. For this reason, we define $k_i = p - r_i + 1$ as a measure of the regularity at the knot ξ_i and set $\boldsymbol{k} = [k_1, \ldots, k_m]$. Note that $k_1 = k_m = 0$ due to the fact we are using an open knot vector.

B-splines form a partition of unity and they span the space of piecewise polynomials of degree p on the subdivision $\{\xi_1, \ldots, \xi_m\}$. Each basis function is positive and has support in $[\xi_i, \xi_{i+p+1}]$. In the sections that follow, we will approximate the solution to the BVP under consideration, using the space

$$S_{\boldsymbol{k}}^p = span \left\{ B_{k,p} \right\}_{k=1}^N, \tag{1}$$

with dimension

$$N = \dim \left(S_{\boldsymbol{k}}^p \right) = mp - \sum_{i=1}^m k_i. \tag{2}$$

We point out that we are using a uniform polynomial degree p, while we allow for the regularity at each knot to (possibly) vary. A more general approach would be to allow p to vary as well. We will refer to N as the number of degrees of freedom, DOF.

3   The Model Problem

We will apply isogeometric analysis to the following model SPP: Find u such that

$$- \varepsilon_1 u''(x) + \varepsilon_2 b(x) u'(x) + c(x) u(x) = f(x) \text{ in } I = (0, 1), \tag{3}$$

$$u(0) = u(1) = 0, \tag{4}$$

where $0 < \varepsilon_1, \varepsilon_2 \le 1$ are given parameters that can approach zero and the functions b, c, f are given and sufficiently smooth. We assume that there exist constants β, γ, ρ, independent of $\varepsilon_1, \varepsilon_2$, such that $\forall x \in \bar{I}$

$$b(x) \ge \beta \ge 0, \quad c(x) \ge \gamma > 0, \quad c(x) - \frac{\varepsilon_2}{2} b'(x) \ge \rho > 0. \tag{5}$$

The structure of the solution to (3) depends on the roots of the characteristic equation associated with the differential operator. For this reason, we let $\lambda_0(x), \lambda_1(x)$ be the solutions of the characteristic equation and set

$$\mu_0 = - \max_{x \in [0,1]} \lambda_0(x), \quad \mu_1 = \min_{x \in [0,1]} \lambda_1(x),$$

Table 1 Different regimes based on the relationship between ε_1 and ε_2

		μ_0	μ_1
Convection-diffusion	$\varepsilon_1 << \varepsilon_2 = 1$	1	ε_1^{-1}
Convection-reaction-diffusion	$\varepsilon_1 << \varepsilon_2^2 << 1$	ε_2^{-1}	$\varepsilon_2/\varepsilon_1$
Reaction-diffusion	$1 >> \varepsilon_1 >> \varepsilon_2^2$	$\varepsilon_1^{-1/2}$	$\varepsilon_1^{-1/2}$

or equivalently,

$$\mu_{0,1} = \min_{x \in [0,1]} \frac{\mp \varepsilon_2 b(x) + \sqrt{\varepsilon_2^2 b^2(x) + 4\varepsilon_1 c(x)}}{2\varepsilon_1}. \tag{6}$$

The values of μ_0, μ_1 determine the strength of the boundary layers and since $|\lambda_0(x)| < |\lambda_1(x)|$ the layer at $x = 1$ is stronger than the layer at $x = 0$. Essentially, there are three regimes [5], as shown in Table 1.

We assume that b, c, f are analytic functions satisfying, for some positive constants $\gamma_f, \gamma_c, \gamma_b$ independent of $\varepsilon_1, \varepsilon_2$, and $\forall\, n = 0, 1, 2, \ldots$

$$\left\| f^{(n)} \right\|_{\infty, I} \leq Cn!\gamma_f^n, \quad \left\| c^{(n)} \right\|_{\infty, I} \leq Cn!\gamma_c^n, \quad \left\| b^{(n)} \right\|_{\infty, I} \leq Cn!\gamma_b^n.$$

Then, it was shown in [9] that, in the case of constant coefficients,[1] the solution u to (3), (4) can be decomposed into a smooth part u_S, a boundary layer part at the left endpoint u_{BL}^-, a boundary layer part at the right endpoint u_{BL}^+, and a (negligible) remainder, viz.

$$u = u_S + u_{BL}^- + u_{BL}^+ + u_R,$$

with

$$\left| u_S^{(n)}(x) \right| \leq CK^n n!,$$

$$\left| \left(u_{BL}^- \right)^{(n)}(x) \right| \leq CK_1^n \mu_0^n e^{-\ell\mu_0 x}, \quad \left| \left(u_{BL}^+ \right)^{(n)}(x) \right| \leq CK_2^n \mu_1^n e^{-\ell\mu_1(1-x)},$$

$$\| u_R \|_{\infty, \partial I} + \| u_R \|_{0, I} + \varepsilon_1 \left\| u_R' \right\|_{0, I} \leq C \max\{e^{-\delta\varepsilon_2/\varepsilon_1}, e^{-\delta/\varepsilon_2}\},$$

for all $x \in \bar{I}$, where the constants $C, K, K_1, K_2, \delta > 0$ depend *only* on the data. Figure 2 shows the behavior of the solution to (3)–(4), in all three regimes.

If $\varepsilon_1, \varepsilon_2$ are not small (see Sect. 5 for precise conditions), then no boundary layers are present and approximating u may be done using a fixed mesh (of say

---

[1] We expect the same result to hold true for variable coefficients.

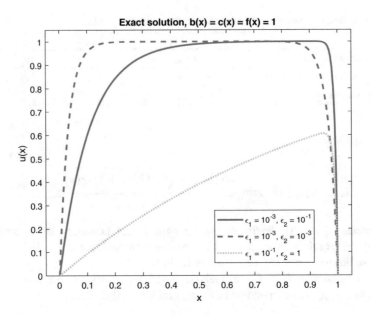

Fig. 2 Exact solution for different values of $\varepsilon_1, \varepsilon_2$

one element) and increasing p. (For IGA, the knot vector could simply be $\varXi = [0, \ldots, 0, 1, \ldots, 1]$.) If, on the other hand, $\varepsilon_1, \varepsilon_2$ are small then classical techniques fail and the mesh must be chosen carefully. The challenge lies in approximating the typical boundary layer function $\exp(-x/\varepsilon)$. In the context of FDs and FEs, the mesh points **must** depend on ε, as is well documented in the literature under the name *layer-adapted* meshes [5]. We expect something similar to hold for IGA, in the sense that the knot vector *must* depend on ε. We will illustrate this in Sect. 5.

4   The Galerkin Formulation and the Discrete Problem

Isogeometric analysis may be combined with a number of formulations; we choose to use Galerkin's approach, i.e. we multiply (3) by a suitable test function, integrate by parts and use the boundary conditions (4). The resulting variational formulation reads: Find $u \in H_0^1(I)$ such that

$$\mathcal{B}(u, v) = \mathcal{F}(v) \quad \forall\, v \in H_0^1(I), \tag{7}$$

where

$$\mathcal{B}(u, v) = \varepsilon_1 \langle u', v' \rangle_I + \varepsilon_2 \langle bu', v \rangle_I + \langle cu, v \rangle_I, \quad \mathcal{F}(v) = \langle f, v \rangle_I. \tag{8}$$

The bilinear form $\mathcal{B}(\cdot, \cdot)$ given by (8) is *coercive* (due to (5)) with respect to the *energy norm*

$$\|u\|_{E,I}^2 := \varepsilon_1 \, |u|_{1,I}^2 + \|u\|_{0,I}^2 ,$$

i.e.,

$$\mathcal{B}(u, u) \geq \|u\|_{E,I}^2 \quad \forall \, u \in H_0^1(I). \tag{9}$$

Next, we restrict our attention to a finite dimensional subspace $S \subset H_0^1(I)$, that will be selected shortly, and obtain the discrete version of (7) as: find $u_N \in S$ such that

$$\mathcal{B}(u_N, v) = \mathcal{F}(v) \quad \forall \, v \in S. \tag{10}$$

The space S is chosen as $S = S_k^p$, given by (1). Thus, we may write the approximate solution as

$$u_N = \sum_{k=0}^N \alpha_k B_{k,p},$$

with $\vec{\alpha} = [\alpha_1, \ldots, \alpha_N]^T$ unknown coefficients, and substitute in (10) to obtain the linear system of equations

$$\underbrace{(\varepsilon_1 A_1 + \varepsilon_2 A_2 + A_0)}_{M \in \mathbf{R}^{N \times N}} \vec{\alpha} = \vec{F}, \tag{11}$$

where

$$[A_1]_{i,j} = \int_I B_{i,p}'(\xi) B_{j,p}'(\xi) d\xi \ , \ [A_2]_{i,j} = \int_I B_{i,p}'(\xi) B_{j,p}(\xi) d\xi,$$

$$[A_0]_{i,j} = \int_I B_{i,p}(\xi) B_{j,p}(\xi) d\xi \ , \ [\vec{F}]_{i=} \int_I B_{i,p}(\xi) f(\xi) d\xi,$$

for $i, j = 1, \ldots, N$. The linear system (11) has a unique solution, due to the fact that the coefficient matrix M in (11) is non-singular. To see this, let $0 \neq v \in S_k^p$ be arbitrary and write it as $v = \sum_{k=0}^N \beta_k B_{k,p}$, with the coefficients β_k not all zero.

From (9), we have

$$0 < \left\| \sum_{k=0}^{N} \beta_k B_{k,p} \right\|_{E,I}^{2} \leq \mathcal{B} \left(\sum_{k=0}^{N} \beta_k B_{k,p}, \sum_{\ell=0}^{N} \beta_\ell B_{\ell,p} \right)$$

$$\leq \sum_{k=0}^{N} \sum_{\ell=0}^{N} \beta_k \mathcal{B} \left(B_{k,p}, B_{\ell,p} \right) \beta_\ell = \vec{\beta}^T M \vec{\beta},$$

which shows that M is positive definite, hence invertible.

We close this section by mentioning that in our implementation of the method, the entries in the matrices in (11), i.e. integrals of B-splines, are computed numerically to any desired accuracy (using MATLAB's `integrate` command).

5   Numerical Results

In this section we present the results of numerical computations for three examples with known exact solution—this makes our results reliable. We will 'mimic' the FEM recommendations for such problems (see, e.g., [9] and the references therein), and select our open knot vector (for the interval $I = (0, 1)$) as follows:

With μ_0, μ_1 given by (6), if $p\mu_1^{-1} \geq 1/2$ then

$$\Xi = [\underbrace{0, \ldots, 0}_{p+1 \text{ times}}, \underbrace{1, \ldots, 1}_{p+1 \text{ times}}\}]. \tag{12}$$

If $p\mu_0^{-1} < 1/2$ then, for **reaction-diffusion**

$$\Xi = [\underbrace{0, \ldots, 0}_{p+1 \text{ times}}, p_{\max}\varepsilon_1^{1/2}, 1 - p_{\max}\varepsilon_1^{1/2}, \underbrace{1, \ldots, 1}_{p+1 \text{ times}}\}], \tag{13}$$

for **convection-diffusion**

$$\Xi = [\underbrace{0, \ldots, 0}_{p+1 \text{ times}}, 1 - p_{\max}\varepsilon_1, \underbrace{1, \ldots, 1}_{p+1 \text{ times}}\}], \tag{14}$$

and for **reaction-convection-diffusion**

$$\Xi = [\underbrace{0, \ldots, 0}_{p+1 \text{ times}}, p_{\max}\mu_0^{-1}, 1 - p_{\max}\mu_1^{-1}, \underbrace{1, \ldots, 1}_{p+1 \text{ times}}\}], \tag{15}$$

where $p = 2, \ldots, p_{\max}$ is the polynomial degree, which we change to improve accuracy. The number of degrees of freedom in each case is given by $DOF = 3p$ and we take $p_{max} = 9$ for the computations.

We will be measuring the percentage relative error in the maximum norm,

$$Error = 100 \times \frac{\|u - u_N\|_{\infty, I}}{\|u\|_{\infty, I}},$$

which we will estimate as follows:

$$Error \approx 100 \times \max_{k=1,\ldots,K} |u(x_k) - u_N(x_k)| / \max_{k=1,\ldots,K} |u(x_k)|,$$

where $\{x_k\}_{k=1}^{K} \in I$ are points in $(0, 1)$, chosen uniformly in the layer region and outside—we use $K = 400$ in each region for our experiments below. We choose to use the maximum norm as an error measure, because the energy norm is 'not balanced' for reaction-diffusion problems (see [6] and the references therein).

The examples that follow cover all three regimes, and try to answer the question of how the method performs as $\varepsilon_1, \varepsilon_2 \to 0$. To reduce the error, we increase the dimension of the space by increasing p, hence strictly speaking, we are performing p-refinement. (In the FEM literature this has been referred to as hp-refinement [8].)

Example 1 We consider (3), (4) with $b(x) = 0$, $c(x) = f(x) = 1$, which makes the problem reaction-diffusion with $\mu_0 = \mu_1 = \varepsilon_1^{-1/2}$. Figure 3 shows the percentage

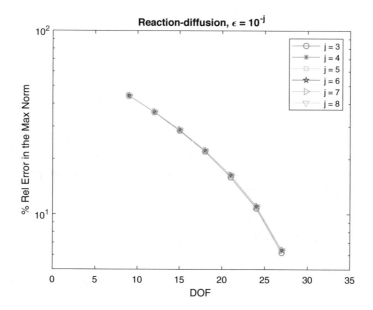

Fig. 3 Maximum norm convergence for Example 1, using the knot vector (13)

Table 2 Percentage relative error in the maximum norm for Example 1, using the knot vector (13)

ε_1 :	10^{-6}	10^{-8}	10^{-10}	10^{-12}	10^{-14}	10^{-16}
DOF						
6	44.05	44.19	44.21	44.22	44.21	44.21
9	35.96	36.09	36.11	36.12	36.11	36.11
12	28.48	28.78	28.81	28.82	28.82	28.82
15	21.96	22.23	22.25	22.26	22.26	22.26
18	15.92	16.31	16.34	16.35	16.35	16.35
21	10.75	11.01	11.04	11.04	11.04	11.04
24	6.18	6.37	6.39	6.39	6.39	6.39
27	2.42	2.41	2.42	2.41	2.41	2.41

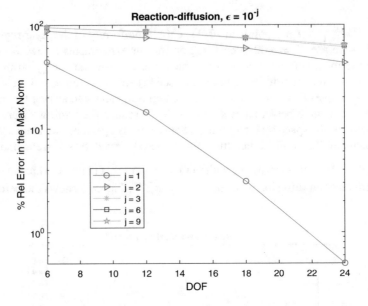

Fig. 4 Maximum norm convergence for Example 1, using the knot vector (12)

relative error measured in the maximum norm, versus the number of degrees of freedom DOF (cf. (2)) in a semi-log scale, and Table 2 lists the errors. The fact that we see straight lines indicates the exponential convergence of the method, while the robustness is verified since the straight lines coincide. We also show, in Fig. 4, the results of using a mesh that does *not* depend on ε_1; in particular we use (12), $\varepsilon_1 = 10^{-j}$, $j = 1, 2, 3, 6, 9, 12$ and we increase p. As can be seen from the figure, for large ε_1 the method yields good results, but as $\varepsilon_1 \to 0$, the results deteriorate and we (basically) have no convergence.

Example 2 We next consider (3), (4) with $b(x) = c(x) = f(x) = \varepsilon_2 = 1$, which makes the problem convection-diffusion with $\mu_0 = 1$, $\mu_1 = \varepsilon_1^{-1}$. In Fig. 5 we show the error in the maximum norm versus the number of degrees of freedom, in a semi-

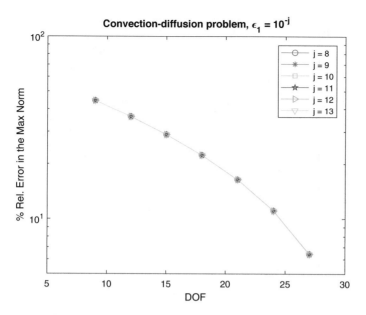

Fig. 5 Maximum norm convergence for Example 2, using the knot vector (14)

Table 3 Percentage relative error in the maximum norm for Example 2, using the knot vector (14)

ε_1 :	10^{-8}	10^{-9}	10^{-10}	10^{-11}	10^{-12}	10^{-13}
DOF						
6	44.44	44.45	44.45	44.44	44.44	44.44
9	36.23	36.22	36.22	36.22	36.22	36.21
12	28.91	28.90	28.91	28.91	28.90	28.89
15	22.32	22.32	22.33	22.32	22.32	22.30
18	16.40	16.41	16.40	16.41	16.40	16.38
21	11.08	11.09	11.08	11.08	11.08	11.05
24	6.41	6.41	6.41	6.42	6.41	6.38
27	2.42	2.42	2.42	2.42	2.42	2.40
30	1.92	1.92	1.92	1.93	1.93	1.95

log scale, for different values of ε_1. The errors are listed in Table 3. Once again we observe robust exponential convergence. The same result as Example 1 is obtained when we use the knot vector (12), hence this is not shown here.

Example 3 We finally consider (3), (4) with $b(x) = c(x) = f(x) = 1$, and choose the values of $\varepsilon_1, \varepsilon_2$ to satisfy $\varepsilon_1 \ll \varepsilon_2^2$, so that the problem becomes convection-reaction-diffusion with $\mu_0 = \varepsilon_2^{-1}$, $\mu_1 = \varepsilon_2 \varepsilon_1^{-1}$. In Fig. 6 we show the convergence of the method, measured in the maximum norm, for different values of $\varepsilon_1, \varepsilon_2$. Table 4 shows the actual errors. We observe robust exponential convergence in this final case as well.

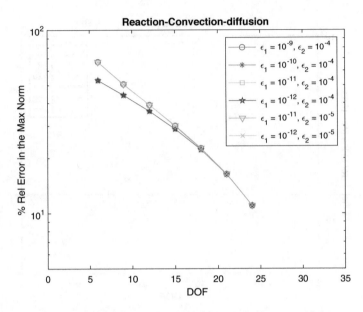

Fig. 6 Maximum norm convergence for Example 3, using the knot vector (15)

Table 4 Percentage relative error in the maximum norm for Example 3, using the knot vector (15)

DOF	$\varepsilon_1 = 10^{-9}$ $\varepsilon_2 = 10^{-4}$	$\varepsilon_1 = 10^{-10}$ $\varepsilon_2 = 10^{-4}$	$\varepsilon_1 = 10^{-11}$ $\varepsilon_2 = 10^{-4}$	$\varepsilon_1 = 10^{-12}$ $\varepsilon_2 = 10^{-4}$	$\varepsilon_1 = 10^{-11}$ $\varepsilon_2 = 10^{-5}$	$\varepsilon_1 = 10^{-12}$ $\varepsilon_2 = 10^{-5}$
6	66.98	53.17	53.18	53.20	66.97	66.74
9	50.68	44.18	44.17	44.19	50.71	50.41
12	39.09	36.07	36.08	36.09	39.11	39.10
15	30.01	28.77	28.78	28.79	30.05	29.84
18	22.61	22.22	22.23	2.22	22.65	22.29
21	16.35	16.31	16.30	16.32	16.41	16.36
24	11.00	11.02	11.03	11.01	11.05	11.04
27	6.37	6.38	6.38	6.38	6.39	6.39
30	2.43	2.43	2.43	2.42	2.41	2.41

6 Conclusions

In this article we studied the performance of IGA for one-dimensional reaction-convection-diffusion problems with two small parameters. We observed that if the knot vector is chosen appropriately and depending on the singular perturbation parameter(s), then p-refinement yields robust, exponential rates of convergence. The theoretical justification of what we have observed will appear in [10].

As a next step, we intend to study the two-dimensional analogs, as well as higher order operators. In particular, we are investigating the use of IGA to fourth order

SPPs in two-dimensions, where their use is one of the few available choices for obtaining an approximation, in curvilinear two-dimensional domains.

References

1. Cottrell, J.A., Hughes, T.R., Basilevs, Y.: Isogeometric Analysis: Toward Integration of CAD and FEA. Wiley and Sons, Chichester (2009)
2. Cox, M.G.: The numerical evaluation of B-splines. Technical report, National Physics Laboratory DNAC 4 (1971)
3. De Boor, C.: On calculation with B-splines. J. Approx. Theory **6**, 50–62 (1972)
4. Hughes, T.R., Cottrell, J.A., Basilevs, Y.: Isogeometric analysis: CAD, finite elements, NURBS, exact geometry and mesh refinement. Comput. Methods Appl. Mech. Eng. **194**, 4125–4195 (2005)
5. Linß, T.: Layer-Adapted Meshes for Reaction-Convection-Diffusion Problems. Lecture Notes in Mathematics, vol. 1985. Springer, Berlin (2010)
6. Melenk, J.M., Xenophontos, C.: Robust exponential convergence of hp-FEM in balanced norms for singularly perturbed reaction-diffusion equations. Calcolo **53**, 105–132 (2016)
7. Roos, H.-G., Stynes, M., Tobiska, L.: Robust Numerical Methods for Singularly Perturbed Differential Equations. Springer Series in Computational Mathematics, 2nd edn., vol. 24. Springer, Berlin (2008). Convection-diffusion-reaction and flow problems
8. Schwab, C., Suri, M.: The p and hp version of the finite element method for problems with boundary layers. Math. Comput. **65**, 1404–1429 (1996)
9. Sykopetritou, I.: An hp finite element method for a second order singularly perturbed boundary value problem with two small parameters, M.Sc Thesis, Department of Mathematics & Statistics, University of Cyprus (2018)
10. Xenophontos, C., Sykopetritou, I.: Isogeometric analysis for singularly perturbed problems in 1-D: error estimates. Electron. Trans. Numer. Anal. **20**, 1–25 (2020)

Editorial Policy

1. Volumes in the following three categories will be published in LNCSE:

i)   Research monographs
ii) Tutorials
iii) Conference proceedings

Those considering a book which might be suitable for the series are strongly advised to contact the publisher or the series editors at an early stage.

2. Categories i) and ii). Tutorials are lecture notes typically arising via summer schools or similar events, which are used to teach graduate students. These categories will be emphasized by Lecture Notes in Computational Science and Engineering. **Submissions by interdisciplinary teams of authors are encouraged.** The goal is to report new developments – quickly, informally, and in a way that will make them accessible to non-specialists. In the evaluation of submissions timeliness of the work is an important criterion. Texts should be well-rounded, well-written and reasonably self-contained. In most cases the work will contain results of others as well as those of the author(s). In each case the author(s) should provide sufficient motivation, examples, and applications. In this respect, Ph.D. theses will usually be deemed unsuitable for the Lecture Notes series. Proposals for volumes in these categories should be submitted either to one of the series editors or to Springer-Verlag, Heidelberg, and will be refereed. A provisional judgement on the acceptability of a project can be based on partial information about the work: a detailed outline describing the contents of each chapter, the estimated length, a bibliography, and one or two sample chapters – or a first draft. A final decision whether to accept will rest on an evaluation of the completed work which should include

– at least 100 pages of text;
– a table of contents;
– an informative introduction perhaps with some historical remarks which should be accessible to readers unfamiliar with the topic treated;
– a subject index.

3. Category iii). Conference proceedings will be considered for publication provided that they are both of exceptional interest and devoted to a single topic. One (or more) expert participants will act as the scientific editor(s) of the volume. They select the papers which are suitable for inclusion and have them individually refereed as for a journal. Papers not closely related to the central topic are to be excluded. Organizers should contact the Editor for CSE at Springer at the planning stage, see *Addresses* below.

In exceptional cases some other multi-author-volumes may be considered in this category.

4. Only works in English will be considered. For evaluation purposes, manuscripts may be submitted in print or electronic form, in the latter case, preferably as pdf- or zipped ps-files. Authors are requested to use the LaTeX style files available from Springer at http://www.springer.com/gp/authors-editors/book-authors-editors/manuscript-preparation/5636 (Click on LaTeX Template → monographs or contributed books).

For categories ii) and iii) we strongly recommend that all contributions in a volume be written in the same LaTeX version, preferably LaTeX2e. Electronic material can be included if appropriate. Please contact the publisher.

Careful preparation of the manuscripts will help keep production time short besides ensuring satisfactory appearance of the finished book in print and online.

5. The following terms and conditions hold. Categories i), ii) and iii):

Authors receive 50 free copies of their book. No royalty is paid.
Volume editors receive a total of 50 free copies of their volume to be shared with authors, but no royalties.

Authors and volume editors are entitled to a discount of 40 % on the price of Springer books purchased for their personal use, if ordering directly from Springer.

6. Springer secures the copyright for each volume.

Addresses:

Timothy J. Barth
NASA Ames Research Center
NAS Division
Moffett Field, CA 94035, USA
barth@nas.nasa.gov

Michael Griebel
Institut für Numerische Simulation
der Universität Bonn
Wegelerstr. 6
53115 Bonn, Germany
griebel@ins.uni-bonn.de

David E. Keyes
Mathematical and Computer Sciences
and Engineering
King Abdullah University of Science
and Technology
P.O. Box 55455
Jeddah 21534, Saudi Arabia
david.keyes@kaust.edu.sa

and

Department of Applied Physics
and Applied Mathematics
Columbia University
500 W. 120 th Street
New York, NY 10027, USA
kd2112@columbia.edu

Risto M. Nieminen
Department of Applied Physics
Aalto University School of Science
and Technology
00076 Aalto, Finland
risto.nieminen@aalto.fi

Dirk Roose
Department of Computer Science
Katholieke Universiteit Leuven
Celestijnenlaan 200A
3001 Leuven-Heverlee, Belgium
dirk.roose@cs.kuleuven.be

Tamar Schlick
Department of Chemistry
and Courant Institute
of Mathematical Sciences
New York University
251 Mercer Street
New York, NY 10012, USA
schlick@nyu.edu

Editor for Computational Science
and Engineering at Springer:

Martin Peters
Springer-Verlag
Mathematics Editorial IV
Tiergartenstrasse 17
69121 Heidelberg, Germany
martin.peters@springer.com

Lecture Notes
in Computational Science
and Engineering

24. T. Schlick, H.H. Gan (eds.), *Computational Methods for Macromolecules: Challenges and Applications.*

25. T.J. Barth, H. Deconinck (eds.), *Error Estimation and Adaptive Discretization Methods in Computational Fluid Dynamics.*

26. M. Griebel, M.A. Schweitzer (eds.), *Meshfree Methods for Partial Differential Equations.*

27. S. Müller, *Adaptive Multiscale Schemes for Conservation Laws.*

28. C. Carstensen, S. Funken, W. Hackbusch, R.H.W. Hoppe, P. Monk (eds.), *Computational Electromagnetics.*

29. M.A. Schweitzer, *A Parallel Multilevel Partition of Unity Method for Elliptic Partial Differential Equations.*

30. T. Biegler, O. Ghattas, M. Heinkenschloss, B. van Bloemen Waanders (eds.), *Large-Scale PDE-Constrained Optimization.*

31. M. Ainsworth, P. Davies, D. Duncan, P. Martin, B. Rynne (eds.), *Topics in Computational Wave Propagation.* Direct and Inverse Problems.

32. H. Emmerich, B. Nestler, M. Schreckenberg (eds.), *Interface and Transport Dynamics.* Computational Modelling.

33. H.P. Langtangen, A. Tveito (eds.), *Advanced Topics in Computational Partial Differential Equations.* Numerical Methods and Diffpack Programming.

34. V. John, *Large Eddy Simulation of Turbulent Incompressible Flows.* Analytical and Numerical Results for a Class of LES Models.

35. E. Bänsch (ed.), *Challenges in Scientific Computing - CISC 2002.*

36. B.N. Khoromskij, G. Wittum, *Numerical Solution of Elliptic Differential Equations by Reduction to the Interface.*

37. A. Iske, *Multiresolution Methods in Scattered Data Modelling.*

38. S.-I. Niculescu, K. Gu (eds.), *Advances in Time-Delay Systems.*

39. S. Attinger, P. Koumoutsakos (eds.), *Multiscale Modelling and Simulation.*

40. R. Kornhuber, R. Hoppe, J. Périaux, O. Pironneau, O. Wildlund, J. Xu (eds.), *Domain Decomposition Methods in Science and Engineering.*

41. T. Plewa, T. Linde, V.G. Weirs (eds.), *Adaptive Mesh Refinement – Theory and Applications.*

42. A. Schmidt, K.G. Siebert, *Design of Adaptive Finite Element Software.* The Finite Element Toolbox ALBERTA.

43. M. Griebel, M.A. Schweitzer (eds.), *Meshfree Methods for Partial Differential Equations II.*

44. B. Engquist, P. Lötstedt, O. Runborg (eds.), *Multiscale Methods in Science and Engineering.*

45. P. Benner, V. Mehrmann, D.C. Sorensen (eds.), *Dimension Reduction of Large-Scale Systems.*

46. D. Kressner, *Numerical Methods for General and Structured Eigenvalue Problems.*

47. A. Boriçi, A. Frommer, B. Joó, A. Kennedy, B. Pendleton (eds.), *QCD and Numerical Analysis III.*

48. F. Graziani (ed.), *Computational Methods in Transport.*

49. B. Leimkuhler, C. Chipot, R. Elber, A. Laaksonen, A. Mark, T. Schlick, C. Schütte, R. Skeel (eds.), *New Algorithms for Macromolecular Simulation.*

50. M. Bücker, G. Corliss, P. Hovland, U. Naumann, B. Norris (eds.), *Automatic Differentiation: Applications, Theory, and Implementations.*

51. A.M. Bruaset, A. Tveito (eds.), *Numerical Solution of Partial Differential Equations on Parallel Computers.*

52. K.H. Hoffmann, A. Meyer (eds.), *Parallel Algorithms and Cluster Computing.*

53. H.-J. Bungartz, M. Schäfer (eds.), *Fluid-Structure Interaction.*

54. J. Behrens, *Adaptive Atmospheric Modeling.*

55. O. Widlund, D. Keyes (eds.), *Domain Decomposition Methods in Science and Engineering XVI.*

56. S. Kassinos, C. Langer, G. Iaccarino, P. Moin (eds.), *Complex Effects in Large Eddy Simulations.*

57. M. Griebel, M.A Schweitzer (eds.), *Meshfree Methods for Partial Differential Equations III.*

58. A.N. Gorban, B. Kégl, D.C. Wunsch, A. Zinovyev (eds.), *Principal Manifolds for Data Visualization and Dimension Reduction.*

59. H. Ammari (ed.), *Modeling and Computations in Electromagnetics: A Volume Dedicated to Jean-Claude Nédélec.*

60. U. Langer, M. Discacciati, D. Keyes, O. Widlund, W. Zulehner (eds.), *Domain Decomposition Methods in Science and Engineering XVII.*

61. T. Mathew, *Domain Decomposition Methods for the Numerical Solution of Partial Differential Equations.*

62. F. Graziani (ed.), *Computational Methods in Transport: Verification and Validation.*

63. M. Bebendorf, *Hierarchical Matrices.* A Means to Efficiently Solve Elliptic Boundary Value Problems.

64. C.H. Bischof, H.M. Bücker, P. Hovland, U. Naumann, J. Utke (eds.), *Advances in Automatic Differentiation.*

65. M. Griebel, M.A. Schweitzer (eds.), *Meshfree Methods for Partial Differential Equations IV.*

66. B. Engquist, P. Lötstedt, O. Runborg (eds.), *Multiscale Modeling and Simulation in Science.*

67. I.H. Tuncer, Ü. Gülcat, D.R. Emerson, K. Matsuno (eds.), *Parallel Computational Fluid Dynamics 2007.*

68. S. Yip, T. Diaz de la Rubia (eds.), *Scientific Modeling and Simulations.*

69. A. Hegarty, N. Kopteva, E. O'Riordan, M. Stynes (eds.), *BAIL 2008 – Boundary and Interior Layers.*

70. M. Bercovier, M.J. Gander, R. Kornhuber, O. Widlund (eds.), *Domain Decomposition Methods in Science and Engineering XVIII.*

71. B. Koren, C. Vuik (eds.), *Advanced Computational Methods in Science and Engineering.*

72. M. Peters (ed.), *Computational Fluid Dynamics for Sport Simulation.*

73. H.-J. Bungartz, M. Mehl, M. Schäfer (eds.), *Fluid Structure Interaction II - Modelling, Simulation, Optimization.*

74. D. Tromeur-Dervout, G. Brenner, D.R. Emerson, J. Erhel (eds.), *Parallel Computational Fluid Dynamics 2008.*

75. A.N. Gorban, D. Roose (eds.), *Coping with Complexity: Model Reduction and Data Analysis.*

76. J.S. Hesthaven, E.M. Rønquist (eds.), *Spectral and High Order Methods for Partial Differential Equations.*

77. M. Holtz, *Sparse Grid Quadrature in High Dimensions with Applications in Finance and Insurance.*

78. Y. Huang, R. Kornhuber, O.Widlund, J. Xu (eds.), *Domain Decomposition Methods in Science and Engineering XIX.*

79. M. Griebel, M.A. Schweitzer (eds.), *Meshfree Methods for Partial Differential Equations V.*

80. P.H. Lauritzen, C. Jablonowski, M.A. Taylor, R.D. Nair (eds.), *Numerical Techniques for Global Atmospheric Models.*

81. C. Clavero, J.L. Gracia, F.J. Lisbona (eds.), *BAIL 2010 – Boundary and Interior Layers, Computational and Asymptotic Methods.*

82. B. Engquist, O. Runborg, Y.R. Tsai (eds.), *Numerical Analysis and Multiscale Computations.*

83. I.G. Graham, T.Y. Hou, O. Lakkis, R. Scheichl (eds.), *Numerical Analysis of Multiscale Problems.*

84. A. Logg, K.-A. Mardal, G. Wells (eds.), *Automated Solution of Differential Equations by the Finite Element Method.*

85. J. Blowey, M. Jensen (eds.), *Frontiers in Numerical Analysis - Durham 2010.*

86. O. Kolditz, U.-J. Gorke, H. Shao, W. Wang (eds.), *Thermo-Hydro-Mechanical-Chemical Processes in Fractured Porous Media - Benchmarks and Examples.*

87. S. Forth, P. Hovland, E. Phipps, J. Utke, A. Walther (eds.), *Recent Advances in Algorithmic Differentiation.*

88. J. Garcke, M. Griebel (eds.), *Sparse Grids and Applications.*

89. M. Griebel, M.A. Schweitzer (eds.), *Meshfree Methods for Partial Differential Equations VI.*

90. C. Pechstein, *Finite and Boundary Element Tearing and Interconnecting Solvers for Multiscale Problems.*

91. R. Bank, M. Holst, O. Widlund, J. Xu (eds.), *Domain Decomposition Methods in Science and Engineering XX.*

92. H. Bijl, D. Lucor, S. Mishra, C. Schwab (eds.), *Uncertainty Quantification in Computational Fluid Dynamics.*

93. M. Bader, H.-J. Bungartz, T. Weinzierl (eds.), *Advanced Computing.*

94. M. Ehrhardt, T. Koprucki (eds.), *Advanced Mathematical Models and Numerical Techniques for Multi-Band Effective Mass Approximations.*

95. M. Azaïez, H. El Fekih, J.S. Hesthaven (eds.), *Spectral and High Order Methods for Partial Differential Equations ICOSAHOM 2012.*

96. F. Graziani, M.P. Desjarlais, R. Redmer, S.B. Trickey (eds.), *Frontiers and Challenges in Warm Dense Matter.*

97. J. Garcke, D. Pflüger (eds.), *Sparse Grids and Applications – Munich 2012.*

98. J. Erhel, M. Gander, L. Halpern, G. Pichot, T. Sassi, O. Widlund (eds.), *Domain Decomposition Methods in Science and Engineering XXI.*

99. R. Abgrall, H. Beaugendre, P.M. Congedo, C. Dobrzynski, V. Perrier, M. Ricchiuto (eds.), *High Order Nonlinear Numerical Methods for Evolutionary PDEs - HONOM 2013.*

100. M. Griebel, M.A. Schweitzer (eds.), *Meshfree Methods for Partial Differential Equations VII.*

122. A. Gerisch, R. Penta, J. Lang (eds.), *Multiscale Models in Mechano and Tumor Biology*. Modeling, Homogenization, and Applications.

123. J. Garcke, D. Pflüger, C.G. Webster, G. Zhang (eds.), *Sparse Grids and Applications - Miami 2016*.

124. M. Schäfer, M. Behr, M. Mehl, B. Wohlmuth (eds.), *Recent Advances in Computational Engineering*. Proceedings of the 4th International Conference on Computational Engineering (ICCE 2017) in Darmstadt.

125. P.E. Bjørstad, S.C. Brenner, L. Halpern, R. Kornhuber, H.H. Kim, T. Rahman, O.B. Widlund (eds.), *Domain Decomposition Methods in Science and Engineering XXIV*. 24th International Conference on Domain Decomposition Methods, Svalbard, Norway, February 6–10, 2017.

126. F.A. Radu, K. Kumar, I. Berre, J.M. Nordbotten, I.S. Pop (eds.), *Numerical Mathematics and Advanced Applications – ENUMATH 2017*.

127. X. Roca, A. Loseille (eds.), *27th International Meshing Roundtable*.

128. Th. Apel, U. Langer, A. Meyer, O. Steinbach (eds.), *Advanced Finite Element Methods with Applications*. Selected Papers from the 30th Chemnitz Finite Element Symposium 2017.

129. M. Griebel, M. A. Schweitzer (eds.), *Meshfree Methods for Partial Differencial Equations IX*.

130. S. Weißer, BEM-based Finite Element *Approaches on Polytopal Meshes*.

131. V. A. Garanzha, L. Kamenski, H. Si (eds.), *Numerical Geometry, Grid Generation and Scientific Computing*. Proceedings of the 9th International Conference, NUMGRID2018/Voronoi 150, Celebrating the 150th Anniversary of G. F. Voronoi, Moscow, Russia, December 2018.

132. H. van Brummelen, A. Corsini, S. Perotto, G. Rozza (eds.), *Numerical Methods for Flows*.

133. ——

134. S. J. Sherwin, D. Moxey, J. Peiro, P. E. Vincent, C. Schwab (eds.), *Spectral and High Order Methods for Partial Differential Equations ICOSAHOM 2018*.

135. G. R. Barrenechea, J. Mackenzie (eds.), *Boundary and Interior Layers, Computational and Asymptotic Methods BAIL 2018*.

For further information on these books please have a look at our mathematics catalogue at the following URL: www.springer.com/series/3527

Monographs in Computational Science and Engineering

1. J. Sundnes, G.T. Lines, X. Cai, B.F. Nielsen, K.-A. Mardal, A. Tveito, *Computing the Electrical Activity in the Heart.*

For further information on this book, please have a look at our mathematics catalogue at the following URL: www.springer.com/series/7417

Texts in Computational Science and Engineering

1. H. P. Langtangen, *Computational Partial Differential Equations.* Numerical Methods and Diffpack Programming. 2nd Edition

2. A. Quarteroni, F. Saleri, P. Gervasio, *Scientific Computing with MATLAB and Octave.* 4th Edition

3. H. P. Langtangen, *Python Scripting for Computational Science.* 3rd Edition

4. H. Gardner, G. Manduchi, *Design Patterns for e-Science.*

5. M. Griebel, S. Knapek, G. Zumbusch, *Numerical Simulation in Molecular Dynamics.*

6. H. P. Langtangen, *A Primer on Scientific Programming with Python.* 5th Edition

7. A. Tveito, H. P. Langtangen, B. F. Nielsen, X. Cai, *Elements of Scientific Computing.*

8. B. Gustafsson, *Fundamentals of Scientific Computing.*

9. M. Bader, *Space-Filling Curves.*

10. M. Larson, F. Bengzon, *The Finite Element Method: Theory, Implementation and Applications.*

11. W. Gander, M. Gander, F. Kwok, *Scientific Computing: An Introduction using Maple and MATLAB.*

12. P. Deuflhard, S. Röblitz, *A Guide to Numerical Modelling in Systems Biology.*

13. M. H. Holmes, *Introduction to Scientific Computing and Data Analysis.*

14. S. Linge, H. P. Langtangen, *Programming for Computations* - A Gentle Introduction to Numerical Simulations with MATLAB/Octave.

15. S. Linge, H. P. Langtangen, *Programming for Computations* - A Gentle Introduction to Numerical Simulations with Python.

16. H.P. Langtangen, S. Linge, *Finite Difference Computing with PDEs* - A Modern Software Approach.

17. B. Gustafsson, *Scientific Computing from a Historical Perspective.*

18. J. A. Trangenstein, *Scientific Computing.* Volume I - Linear and Nonlinear Equations.

19. J. A. Trangenstein, *Scientific Computing*. Volume II - Eigenvalues and Optimization.

20. J. A. Trangenstein, *Scientific Computing*. Volume III - Approximation and Integration.

For further information on these books please have a look at our mathematics catalogue at the following URL: www.springer.com/series/5151

Printed in the United States
by Baker & Taylor Publisher Services